SYMPOSIA OF THE ZOOLOGICAL SOCIETY OF LONDON
NUMBER 40

The Zoological Society of London 1826–1976 and Beyond

(The Proceedings of a Symposium held at The Zoological Society of London on 25 and 26 March, 1976)

Edited by

PROFESSOR LORD ZUCKERMAN
The Zoological Society of London

and

Staff of the Publications Department
The Zoological Society of London

Published for
THE ZOOLOGICAL SOCIETY OF LONDON
BY
ACADEMIC PRESS
1976

ACADEMIC PRESS INC. (LONDON) LTD
24/28 Oval Road
London NW1

U.S. Edition published by
ACADEMIC PRESS INC.
111 Fifth Avenue,
New York, New York 10003

Library of Congress Catalog Card Number: 74-5683

ISBN: 0-12-613340-9

PRINTED IN GREAT BRITAIN BY
J. W. ARROWSMITH LTD, BRISTOL

CONTRIBUTORS

BALL, D. J., *Overseer of Reptiles, The Zoological Society of London, Regent's Park, London NW1 4RY* (p. 119)

BELLAIRS, A. d'A., *Department of Anatomy, St Mary's Hospital Medical School, Paddington, London W2; and Honorary Herpetologist, The Zoological Society of London* (p. 119)

BRAMBELL, M. R., *Curator of Mammals, The Zoological Society of London, Regent's Park, London NW1 4RY* (pp. 147, 335)

BULLOUGH, W. S., *Department of Zoology, Birkbeck College, Malet Street, London WC1* (p. 223)

CAVE, A. J. E., *c/o The Zoological Society of London, Regent's Park, London NW1 4RY* (p. 49)

EDWARDS, MARCIA A., *Assistant Editor, The Zoological Society of London, Regent's Park, London NW1 4RY* (p. 253)

FISH, R. A., *Librarian, The Zoological Society of London, Regent's Park, London NW1 4RY* (pp. 17, 233)

GOODWIN, L. G., FRS, *Director, Nuffield Institute of Comparative Medicine, and Director of Science, The Zoological Society of London, Regent's Park, London NW1 4RY* (p. 215)

GREENWOOD, P. H., *Department of Zoology, British Museum (Natural History), Cromwell Road, London SW7 5BD* (p. 85)

HAMILTON, FEONA, *Assistant Librarian, The Zoological Society of London, Regent's Park, London NW1 4RY* (p. 223)

JEWELL, P. A., *Department of Zoology, Royal Holloway College, University of London, Englefield Green, Surrey TW20 9TY* (p. 269)

JONES, D. M., *Senior Veterinary Officer, The Zoological Society of London, Regent's Park, London NW1 4RY* (p. 203)

MANTON, V. J. A., *Curator, Whipsnade Park, Dunstable, Beds* (p. 167)

MATHEWS, SUE J., *The Zoological Society of London, Regent's Park, London NW1 4RY* (p. 147)

MARTIN, R. D., *Senior Research Fellow, The Wellcome Institute of Comparative Physiology, The Zoological Society of London, Regent's Park, London NW1 4RY* (p. 283)

MONTAGU, I., *Old Timbers, Verdure Close, Garston, Watford, Herts* (p. 17)

OLNEY, P. J. S., *Curator of Birds and Editor, International Zoo Yearbook, The Zoological Society of London, Regent's Park, London NW1 4RY* (p. 133)

SMITH, J. E., FRS, *Marine Biological Association of the United Kingdom, The Laboratory, Citadel Hill, Plymouth PL1 2PB* (p. 67)

CONTRIBUTORS

SHORT, R. V., FRS, *Director, MRC Unit of Reproductive Biology, 39 Chalmers Street, Edinburgh EH3 9ER* (p. 321)

TOOVEY, J. W., *Architect, The Zoological Society of London, Regent's Park, London NW1 4RY* (p. 179)

VEVERS, H. G., *Assistant Director of Science and Curator of the Aquarium, The Zoological Society of London, Regent's Park, London NW1 4RY* (p. 105)

ZUCKERMAN, S., OM, KCB, FRS, *Secretary, The Zoological Society of London, Regent's Park, London NW1 4RY* (p. 1)

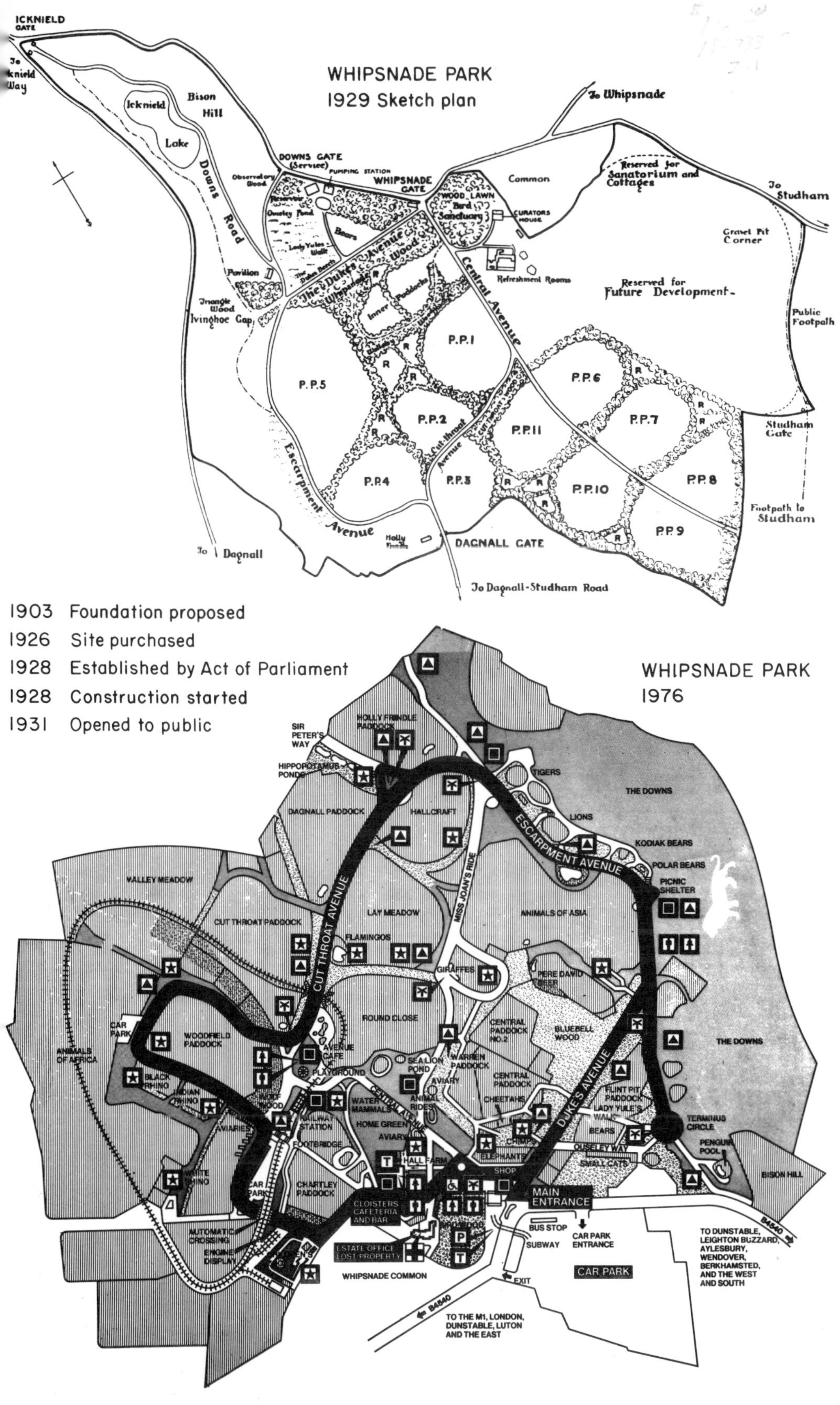

WHIPSNADE PARK
1929 Sketch plan
ICKNIELD GATE
To Icknield Way
Icknield Lake
Bison Hill
Downs Road
DOWNS GATE (Service)
PUMPING STATION
Observatory Wood
WHIPSNADE GATE
Reservoir
Ouseley Pond
Bears
Lady Yules Walk
Pavilion
Triangle Wood
Ivinghoe Gap
The Duke's Avenue
Whipsnade Wood
Inner Paddocks
WOOD LAWN Bird Sanctuary
CURATORS HOUSE
Common
Refreshment Rooms
Central Avenue
To Whipsnade
Reserved for Sanatorium and Cottages
To Studham
Gravel Pit Corner
Reserved for Future Development
Public Footpath
P.P.1
P.P.2
P.P.3
P.P.4
P.P.5
P.P.6
P.P.7
P.P.8
P.P.9
P.P.10
P.P.11
R
Cut-throat Avenue
Escarpment Avenue
Studham Gate
Footpath to Studham
Holly Frindle
DAGNALL GATE
To Dagnall
To Dagnall-Studham Road
1903 Foundation proposed
1926 Site purchased
1928 Established by Act of Parliament
1928 Construction started
1931 Opened to public
WHIPSNADE PARK
1976
HOLLY FRINDLE PADDOCK
SIR PETER'S WAY
HIPPOPOTAMUS PONDS
DAGNALL PADDOCK
HALLCRAFT
TIGERS
THE DOWNS
LIONS
ESCARPMENT AVENUE
KODIAK BEARS
POLAR BEARS
PICNIC SHELTER
VALLEY MEADOW
CUT THROAT PADDOCK
CUT THROAT AVENUE
LAY MEADOW
MISS JOAN'S RIDE
ANIMALS OF ASIA
FLAMINGOS
GIRAFFES
PERE DAVID DEER
CAR PARK
WOODFIELD PADDOCK
ROUND CLOSE
CENTRAL PADDOCK NO.2
BLUEBELL WOOD
THE DOWNS
ANIMALS OF AFRICA
AVENUE CAFE
PLAYGROUND
SEALION POND
WARREN PADDOCK
DUKE'S AVENUE
BLACK RHINO
INDIAN RHINO
AVIARY
CENTRAL PADDOCK
FLINT PIT PADDOCK
LADY YULE'S WALK
WATER MAMMALS
ANIMAL RIDES
CHEETAHS
HOME GREEN
CENTRAL AVENUE
TERMINUS CIRCLE
AVIARIES
RAILWAY STATION
AVIARY
CHIMPS
BEARS
PENGUIN POOL
FOOTBRIDGE
HALL FARM
ELEPHANTS
OUSELEY WAY
WHITE RHINO
SHOP
SMALL CATS
BISON HILL
CAR PARK
CHARTLEY PADDOCK
MAIN ENTRANCE
CLOISTERS CAFETERIA AND BAR
AUTOMATIC CROSSING
BUS STOP
CAR PARK ENTRANCE
B4540
ENGINE DISPLAY
SUBWAY
ESTATE OFFICE LOST PROPERTY
TO DUNSTABLE, LEIGHTON BUZZARD, AYLESBURY, WENDOVER, BERKHAMSTED, AND THE WEST AND SOUTH
WHIPSNADE COMMON
EXIT
CAR PARK
B4540
TO THE M1, LONDON, DUNSTABLE, LUTON AND THE EAST

The Zoological Society
of London
1826–1976 and Beyond

FRONTISPIECE: Guy the gorilla, who arrived at Regent's Park in 1947, is one of the best-known animals in London Zoo.

The Zoological Society of London

ORGANIZER AND CHAIRMEN

ORGANIZER

The Zoological Society of London

CHAIRMEN OF SESSIONS

E. J. W. BARRINGTON, FRS, *Cornerways, 2 St Margaret's Drive, Alderton, Tewkesbury, Glos. GL20 8NY*

C. A. WRIGHT, *Department of Zoology, British Museum (Natural History), Cromwell Road, London SW7 5BD*

S. ZUCKERMAN, OM, KCB, FRS, *Secretary, The Zoological Society of London, Regent's Park, London NW1 4RY*

FOREWORD

by

HRH The Prince Philip
Duke of Edinburgh, KG, KT

President of the Zoological Society of London

At the Annual General Meeting in the 150th anniversary year of the Zoological Society of London I referred to the origin of some of the European menageries. It seemed to me that this might form a suitable introduction to this record of the development of our Society.

Research reveals that the Ménagerie du Jardin des Plantes, Muséum National d'Histoire Naturelle (the Paris Zoo), is recognized to be the senior institution of this kind in post Roman Europe. It was originally founded in the seventeenth century as a "Physic Garden" and developed under the guidance of Buffon as a botanic garden and a centre for natural history studies. A collection of live animals was added in 1794 when the menagerie which Louis XIV had established at Versailles was, for reasons which I think should be fairly obvious, transferred to Paris.

I confess that, when I discovered the seniority of Paris, I found it rather difficult to believe or to accept that the French could ever be more interested in animals than the English, so I caused a little research to be undertaken. It appears that the first known menagerie in this country dates from as long ago as the twelfth century. Henry I assembled it "at his Park at Woodstock wherein he preserved with great pleasure divers sorts of strange beasts which were liberally sent him by other Princes, and it contained lions, leopards and other strange beasts".

This menagerie was still in being a century later when Henry III added to it three leopards presented by the Emperor Frederick II, and soon after that all the animals were moved to the Tower of London. History does not relate exactly how it came about but it seems that the animals then became a charge on the City Fathers because history does relate that the Sheriffs had to foot the bill for their upkeep. Although the value of money was slightly different in those days, a white bear in 1252 was recorded to have cost 4d a day to maintain, and lions and leopards 6d a day.

Three years later, that is in 1255, a house had to be built for an elephant presented by the King of France. The continued existence of the menagerie in the Tower is reflected in the composition of the Royal Household, one of whose officials was Master of the King's Bears and Apes. Considering the number of animals The Queen has been responsible for sending to the Society's collections, it is a title which might suitably be revived for its Secretary.

The Tower menagerie lasted for a total of 600 years, six lions were known to be there in the mid-seventeenth century; it survived right though the Civil Wars, and by the beginning of the eighteenth century there were eleven lions in it, plus two leopards or tigers, three eagles, two owls, two cats of the mountains, and a jackal. The numbers then increased in the late eighteenth century, but fell heavily at the beginning of the nineteenth, being reduced in 1822 to a grizzly bear, an elephant, and a few birds. This rather sorry state of affairs may well be one of the reasons why Sir Stamford Raffles decided to found the Zoological Society, but then a new Keeper, Alfred Cops, imported a wide variety of animals and built up a fine collection, which was eventually transferred to Regent's Park, in 1832.

By this time the menagerie had lost any direct connection with the Sovereign, and the collection of animals which George IV established in Windsor Great Park at Sandpit Gate was a purely personal hobby. One of the animals was the giraffe shown in the picture by Agasse in 1827 which hangs in the Society's rooms at Regent's Park. This animal was captured, like so many others, in infancy after its mother had been shot. Unfortunately, it had lost the use of its legs on the long journey to England and died a year after its arrival. Stangely enough, its brother, captured at the same time, was sent by its captors to the King of France.

It appears that George III was not particularly interested in wild animals, although he was much interested in domestic animals, and imported new breeds to improve the domestic animals in this country. His flock of merino sheep, some of which were smuggled out of Spain, were the foundation stock for the merinos in Australia and New Zealand today.

It is, of course, possible that other European sovereigns made collections of animals at various times but those established by Louis XIV in the seventeenth century and later by the Austrian

Emperor in the eighteenth century, while they predated that of George IV, came into being far later than the first recorded Royal menagerie established in this country.

The tradition of Kings and Heads of State presenting animals to each other and to the Zoo continues to the present day. Various animals were given to the Zoo by The Prince of Wales (later Edward VII), after his tour of India in 1875. There are some splendid drawings published in The Illustrated London News at the time, of sailors struggling to get some ostriches in and out of boats and on board ship for the passage home. Sailors can do almost anything, but trying to get an ostrich into a boat is a bit beyond the normal call of duty. I am glad to say that the most we have ever had to contend with was a sloth in a box and a baby alligator which made the journey home in one of the guest bathrooms in the Royal Yacht Britannia.

A recent gift of animals from one Head of State to another took place in May 1976. The President of Brazil presented to The Queen, three pairs of toucans, one pair of anteaters, one giant armadillo, one two-toed sloth and one pair of blacknecked swans. All, except the swans which went to Slimbridge, are now in the Society's Gardens.

To bring the record right up to date. Just before going to press The Queen presented President Giscard d'Estaing of France with a black labrador gun-dog bred at Sandringham.

Buckingham Palace
July 1976

Philip

ACKNOWLEDGEMENTS

We have made every effort to trace the ownership of copyright material used in this volume, and both we and the publishers believe that all the necessary permissions authorizing publication have been granted by authors, agents and publishers. But in the event of question arising over the use of any material we and the publishers, while regretting any error unconsciously made, will be very pleased to make the necessary corrections in future editions of this book.

CONTENTS

CONTRIBUTORS v
ORGANIZER AND CHAIRMEN OF SESSIONS vii
FOREWORD BY HRH THE PRINCE PHILIP, DUKE OF EDINBURGH . ix
ACKNOWLEDGEMENTS xii

The Zoological Society of London: Evolution of a Constitution

S. ZUCKERMAN

Text 1
References 16

The Zoological Society and the British Overseas

R. FISH and I. MONTAGU

Synopsis 17
Introduction 17
Raffles 23
Cuming 30
Belcher 30
Lowe 34
Blyth, Hodgson, Tickell 34
The Macleay Family 41
Hudson 44
Johnston 44
References 48

The Zoological Society and Nineteenth Century Comparative Anatomy

A. J. E. CAVE

Text 49
References 66

CONTENTS

Early Invertebrate Zoology: Men and their Animals

J. E. SMITH

Synopsis 67
Introduction 67
The London zoologists: scientific meetings 68
Invertebrate zoology in the provinces 71
Research communications: two controversies 75
Invertebrate zoology and zoologists 78
Taxonomy and systematics 78
Natural history 80
Zoology in Australia 80
References 81

The Zoological Society and Ichthyology, 1826–1930

P. H. GREENWOOD

Synopsis 85
Introduction 85
The Museum 86
Fish culture and the aquaria 88
The Society's scientific publications 94
Conclusions 103
Acknowledgements 103
References 103

Management of a Public Aquarium

H. G. VEVERS

Introduction 105
The present aquarium 111
Materials 115
Filtration 116
Maintenance of pH 117
Range of exhibits 117
References 118

Reptiles

A. d'A. BELLAIRS and D. J. BALL

Synopsis 119
Introduction 119

Reptile husbandry today 121
Plan of Reptile House and cages 121
Heat, light and air 125
Feeding 126
Breeding 128
Handling and precautions against bites 128
Ideas for the future 131
References 131

The Policy of Keeping Birds in the Society's Collections 1826–1976

P. J. S. OLNEY

Text 133

Primates and Carnivores at Regent's Park

M. R. BRAMBELL and SUE J. MATHEWS

Introduction 147
Past housing 150
Present and future housing 155
Management 156
Feeding 158
Performance 161
Conclusion 164
References 165

Mammals at Whipsnade

V. J. A. MANTON

Text 167
References 178

150 Years of Building at London Zoo

J. W. TOOVEY

Text 179
References and Sources 197
Notes 198
Appendix A 199
Appendix B 200

CONTENTS

The Maintenance of Animal Health in the Collections

D. M. JONES

Preventive aspects 204
The veterinary services 208
Pathology 210
Clinical medicine 211
Summary 212
Reference 214

Research

L. G. GOODWIN

Introduction 215
Research arising from the maintenance of the collections 216
The research laboratories 218
Sickle cells 219
Botulism 219
Essential fatty acids 220
The future 221
References 222

The Role of Education

W. S. BULLOUGH and FEONA HAMILTON

Introduction 223
The publications 223
The Library 224
The early educational efforts 225
The present school activities 226
Education at the university level 228
Other educational activities 228
The future 231
Reference 231

The Library and Scientific Publications of the Zoological Society of London: Part I

R. FISH

Synopsis 233
Introduction: the early years 1825–30 233
The Library 237
Notes 250

CONTENTS

The Library and Scientific Publications of the Zoological Society of London: Part II

MARCIA A. EDWARDS

Synopsis 253
The scientific publications 253
Proceedings and Transactions 253
Zoological Record 255
Nomenclator Zoologicus 260
Symposia 262
International Zoo Yearbook 263
Conclusions: the future 264
References 266

The Contribution of the Zoological Society of London to Field Studies and Prospects for the Future

P. A. JEWELL

Synopsis 269
Introduction 269
Some examples of past field studies 269
The Society's research institutes 272
Present and future 275
References 280

A Zoologist's View of Research on Reproduction

R. D. MARTIN

Synopsis 283
Introduction 283
Prospects for research on reproduction in zoological collections 285
Monitoring of reproduction in the great apes 292
Comparative aspects of primate placentation 300
Stress and reproduction in captivity 306
Reproduction in an ecological context 311
Conclusions 314
Acknowledgements 315
References 316

CONTENTS

The Introduction of New Species of Animals for the Purpose of Domestication

R. V. SHORT

Synopsis 321
Introduction 321
What determined Man's initial choice of species to domesticate? 322
Do we need new species of domestic animals? 323
Why have we failed to domesticate new species in recent times? 324
Theoretical possibilities for the exploitation of genes from different latitudes and altitudes 324
Improvement of the domestic goose 325
Improvement of the domestic duck 326
Improvement of the domestic sheep 327
Conclusions 332
References 332

An Early Label Found in the Zoological Gardens, Regent's Park, London

M. R. BRAMBELL

Text 335

GENERAL INDEX 337

Symp. zool. Soc. Lond. (1976) No. 40, 1–16.

THE ZOOLOGICAL SOCIETY OF LONDON: EVOLUTION OF A CONSTITUTION

S. ZUCKERMAN

The Zoological Society of London, Regent's Park, London, England

This Symposium needs no new frame within which to set the major events of the Society's history over the past 150 years. The story of our foundation was told in detail by Scherren in the history which he published in 1905, and by Chalmers Mitchell, then the Society's Secretary, in the Centenary volume of 1929. We have also just issued a pamphlet which spells out the broad outlines of the story, while three new books which provide particular slants on various events that stand out during the Society's progress over the past century and a half are also about to appear (Blunt, 1976; Holloway, 1976; Vevers, 1976). What is even more important, the papers which make up this volume have been specially selected to illustrate the development of the Society's educational and scientific activities. What I therefore propose to do in this opening paper is focus on matters and events that relate to our constitutional status, and on a few considerations of which I became aware only after I became the Society's Secretary and which, over the years, have imparted to us our unique character.

For we are unique. We are a national institution. But unlike other national institutions, such as Kew Gardens or the Natural History Museum, we are not in receipt of an annual subvention from the Government. We maintain the national collections of animals, and to that extent we correspond to the essentially governmental bodies which shoulder the same responsibilities in Washington, in Moscow, in Paris, or Peking. But even though the land of which we are the custodians in Regent's Park belongs to the Crown, in law we are a private body. We own one of the most comprehensive zoological libraries in the land, and one which is widely used not only by our Fellows and Associates but also by other scholars. But we receive no support from the Department of Education and Science. We publish more scientific journals than does even the Royal Society, including one, the *Zoological Record*, which is a unique annual bibliography of inestimable value to the worldwide zoological community, and at the same time one of the few bibliographic journals covering a major field of science for which a private body in the United Kingdom still shoulders the international responsibility. But unlike the Royal Society, we receive no grant-in-aid from the Exchequer. We are responsible for an extensive educational scheme, to which the Inner London Education Authority has now linked its own Centre for Life Studies. But this we do as a voluntary service which we have undertaken on our own initiative. We have two world-renowned Research Institutes. Yet we receive no regular subventions from the Government's Research Councils. We are unique because we do all these things in the strict discharge of the terms of our Royal Charter, and we do so as an educational charity that has to find its own means of making ends meet. In the course of our development we became a zoological institution which is pre-eminent in the world because of the breadth and quality of its activities. That is what we propose to remain.

Why then, with all the public responsibilities we now bear, did we get going in the way we have? The short answer is that we came into being to fill a need which at the time was not being satisfied either by the Royal Society, which was founded in 1660 to promote *all* "natural knowledge", or by the Linnean Society which came into existence nearly 50 years before our own foundation, specifically in order to cultivate "the science of natural history in all its branches".

Of the many distinguished names in the list of those who helped Sir Stamford Raffles launch our own Society, the most significant was certainly that of Sir Humphry Davy, President of the Royal Society from 1820 to 1827. Today support from so eminent a source as the Royal would no doubt be regarded as a pre-requisite to the establishment of any scientific institution which aimed to be, for want of a better word, reputable. But in the early 1820s, some might well have questioned this need. Davy had been preceded as President by Sir Joseph Banks, a man who had gained the reputation of being more than a little of a despot (on one occasion "when the Council was taking steps to dismiss him as President, he dismissed the Council." Andrade, 1960). Banks presided over the Royal Society for 41 years, and while it is to his credit that he succeeded in raising the social prestige of the Royal and of bringing respect to the name of British science abroad, at the time it did not seem—any more than it does now— that he was much concerned with the growth of science in general. According to Cardwell (1957) it was because of "dissatisfaction with the lethargic Royal Society and with the tyrannical rule of its President, Sir Joseph Banks" that a number of societies specializing in different branches of science, as well as less formal bodies such as the Lunar Society of

Birmingham, started to spring up at the end of the eighteenth and the beginning of the nineteenth centuries. This movement was, of course, also stimulated by a widening interest in science in general, and in particular by a realization of the significance of scientific knowledge to the growth of the nation's wealth and power. The Linnean Society, as I have said, was founded in 1788. The Geological Society came into being in 1807, the Royal Astronomical in 1820, the Royal Statistical in 1830, and we in 1826. By becoming one of our founders, Davy, as President of the Royal Society, was in effect helping to redress the negative influence of his illustrious and notorious predecessor.

There is a belief that in our foundation we usurped some of the functions of the Linnean Society, which four years before we came into being had allowed a Zoological Club to develop under its wing. This Club was formed "for the study of zoology and comparative anatomy in all their branches, and more especially as they relate to the animals indigenous to Great Britain and Ireland." But presumably because of their dislike of the autocratic way in which they were treated by the then Linnean Council, several of its members joined what would today be called the Promotion Committee of the prospective Zoological Society, with N. A. Vigors, FRS, the Club's last President, becoming the first Secretary of the Society when it was formed. Obviously there would have been no call for a separate Zoological Society if the Linnean had been doing what was expected of it, and if it had been as interested in the promotion of zoology as it was of botany—an interest which reflected that of Joseph Banks, whose concern with this particular field of science played no small part in persuading the Government to establish Kew Gardens as a national institution. In 1887, the year of Queen Victoria's Jubilee, Sir William Flower, then the President of our own Society, wrote that "if the leading Fellows of the Linnean Society had displayed more energy, it might have kept in its hands the principal direction of the biological studies of the country, instead of allowing the Zoological Society, which had since proved so formidable a rival, to spring up, and to absorb so large a portion of its useful function . . ." (Flower, in Scherren, 1905: 2).

What is of greater interest than the question of our parentage is whether we really did set about the promotion of zoology as a science in a way the Royal Society, within whose ambit zoology also certainly fell, did not. The answer is assuredly "yes".

After a brilliant start, and a period in which men like Isaac Newton vitalized the intellectual environment of the country, the

Royal Society gradually became more of what some regarded as a body, membership of which accorded a particular social prestige, than as an institution which should have been urgently concerned with the advancement of science. Indeed, it is not going too far to say that the repercussions of this "backsliding" were felt almost to the dawn of our century—otherwise it is difficult to see why it then became incumbent upon Government to establish the prototype of the Research Council organization whose prime purpose it was to promote the growth and dissemination of scientific knowledge, a function which was implicit in the stated purpose of the Royal, as it is of the Royal's sister institutions in Washington and Moscow.

So far as the biological sciences were concerned, a glance at the contributions to the *Philosophical Transactions of the Royal Society* over the first 25 years of the nineteenth century shows that papers concerned with the physical sciences then outnumbered those which deal with zoological and medical matters by a ratio of somewhere between 5 : 1 and 3 : 1. When, in 1887, the *Transactions* were split into a physical and biological series, it was only possible to publish a modest number of papers on zoological topics. Even the number of zoological papers published in the *Transactions of the Linnean Society*, including those received by the Society's Zoological Club, rarely reached a figure of 20 a year. But the *Proceedings* and *Transactions* of the Zoological Society, launched almost as soon as the Society was formed, prospered from the start, and have continued without interruption since their first appearance nearly 150 years ago. They could hardly have done this if they were competing to fulfil a need which was being catered for elsewhere.

But it was not only through its publications that the Society took the lead in promoting the development of zoology. Our scientific meetings, which, in one form or another, have taken place uninterruptedly since 1830, from the start provided an opportunity for the discussion of zoological matters which otherwise would not have been debated publicly. This symposium is taking place in the seventh hall which the Society has had since the first meeting in its original offices in Bruton Street, and each in turn was in constant use. A post for a professionally qualified official was established in 1829, the year we were granted our first Royal Charter. It was his duty to report on the health and causes of death of animals dying in the Society's collection. Later it also became his formal responsibility to see that no anatomical material was wasted which could otherwise be used by zoologists to advance comparative anatomy and taxonomy, not that there is any reason to suppose that much had been wasted in the preceding years. Early on, Yarrell and Vigors had begun the practice of dissecting creatures which died in the Society's menagerie and in 1830, Richard Owen described the anatomy of a dead orang-utan which he had dissected. The Society also bought a farm at Kingston, Richmond, to experiment on breeding. This was in part a prototype Whipsnade, where creatures which had become pregnant in Regent's Park could be sent to bear their young in quiet. This venture was shut down in 1834 because of cost, only five years after it had begun. A museum was yet another early venture. This was started because the authorities of the British Museum, which had been founded in 1753, were paying little attention at the time to their zoological collection, and because taxonomists in consequence had no reliable home to which to send their type specimens. Darwin sent many of those which he had collected during the course of the voyage of *The Beagle* to our museum, which started in Bruton Street, from where it was moved first to Leicester Square, where John Hunter had had his school of anatomy, and eventually to Regent's Park Gardens, some of the more valuable specimens being exhibited at the Society's rooms in Hanover Square. In its Report for 1855 the Council of the Society noted that

> "the remarkable development of the Natural History Department in the British Museum has for some years past entirely superseded the necessity of maintaining a second Zoological Museum in London".

The Report then went on to say that because of its national importance

> "the Council felt that the first step to be taken was to transfer to the Trustees of the British Museum the whole of the types of species described in the Society's publications, in order that they may be there preserved during the longest possible period for the purpose of reference and identification".

The remainder of the collection was sold to various other museums.

I have referred so far only to the start of the implementation of the Society's scientific interests. But the "useful function" to which Sir William Flower referred in his Jubilee speech of 1887 embodied two major streams of interest, the one professionally scientific, the other amateur and to some extent "social". During most of the 150 years of the Society's history which this Symposium celebrates, these two conjunctional interests—I say conjunctional because it has been ruled legally that the second relates to the first—have complemented each other. But on occasion they have conflicted, and some 20 years ago they collided head-on. The kind of interest in Natural History which Raffles exemplified became enshrined in the second part of the sentence in our 1829 Royal Charter which states as the Society's purpose "the introduction [to England] of new and curious subjects of the Animal Kingdom". The more scientific interest is embodied in the prior phrase of this sentence, "the advancement of zoology and animal physiology". Both objectives were equally represented on the Committee which was set up in 1825 to consider the formation of a Zoological Society, and at this time there was not the slightest hint anywhere that the two concerns of the new Society could possibly conflict. Indeed, in our early days there were no professional zoologists in the sense that we use the term today. Even Darwin, who sprang to prominence decades later, would today be counted as a wealthy amateur scientist. T. H. Huxley, his most celebrated standard-bearer, earned his living as a member of the Geological Survey. University degrees in zoology as such were not given till years after our foundation.

Not only was there no sign that the two interrelated interests set out in our Royal Charter could conflict; at the start the two reinforced each other as the Council strove to fulfil the scientific aspirations which had animated Raffles and Davy. In 1829, the year of our first Royal Charter, the Society's total income was about £12 000, of which some two-thirds came from Members' admission fees and subscriptions, and the remainder from gate charges paid by other visitors. The Council then believed that its income would easily pay for the capital and running costs of the menagerie, Gardens and office. Even when one-fifth of all income was set aside for investment as permanent capital, it was possible to allocate £1000 each year to the upkeep of the museum, and £1000 to the experimental farm at Kingston. Thus at the start 16 per cent of the Society's income was earmarked for scientific activities. Today, when the Society spends more than £2 000 000, some £370 000, or about 18 per cent of the total, is spent on research, education, the library and other scientific work. But now we have to find nearly half of this sum from outside research grants and funds.

Until 1847 the general public enjoyed only a restricted access to the Regent's Park Gardens, which soon became widely regarded as a kind of exclusive preserve for people of fashion, most of whom, it seemed, knew and cared little about animals. Yet even at a time when the Council allowed such non-scientific fund-raising activities as "Promenades", a well attended meeting of Fellows was of one mind "that the Society cannot divest itself of its scientific character, so essential to its dignity and respectability, without violating the Charter of Incorporation". This was in 1841, and more than a century was to pass before the sentiment of the resolution was seriously challenged.

The fact is that there was little chance of the Society's scientific and educational purposes being submerged so long as they remained in the mainstream of zoological interest. And that they did until the early years of our own century, for the good and simple reason that until then the main topics of zoological discussion were comparative anatomy, systematics, taxonomy, and phylogeny, to all of which the Society was in a unique position to contribute. Even though Charles Darwin himself did not like the controversial nature of our scientific meetings, we should never forget that our old meeting rooms in Hanover Square provided a continuing forum for the discussion of his basic ideas about evolution, and that T. H. Huxley played a very prominent part in our affairs.[a]

Obviously the vigour and standing, both scientific and general, of an Institution such as ours depends basically on its governing body, the Council, and on its chief executive, the Secretary, who, in accordance with our Royal Charter, is also a statutory Member of Council. The Charter and Byelaws of 1829 were silent about the question of payments to any Member of Council, but in 1847 the Council departed from the unwritten rule that there should be none because, to quote Scherren (1905), "they could not expect from their Secretary that degree of responsibility to the Society and constant attention to its affairs . . . so long as the appointment was an honorary one". From then and until the end of Julian Huxley's

[a] Unfortunately we no longer possess many letters or other documentary material relating to this period of the Society's development. At the beginning of the Second World War, the Council of the day set up a Committee to consider what to do with the Society's papers. So far as I can make out the Committee never reported back to Council, and most of the documents were dispersed and destroyed. Happily, and because of the vigilance of Mr Fish, our Librarian, we did recover some very interesting letters a few years ago. A lady arrived at our Main Office and asked to see a responsible official about a small collection of letters which she wished to sell. It turned out that they were some that had been preserved by one of the members of this Committee, and who on his death had bequeathed them to her. Some were marked with the stamp of the Society, to which they obviously belonged, and to which the would-be seller then returned them. Among the signatories of these lost letters were the names of Sir Richard Owen, T. H. Huxley, A. R. Wallace and W. H. Hudson.

Secretaryship in 1943, the post remained a salaried one, in spite of the anomaly that the Council, the legal governing body of the Society and employer of its "servants"; the term used in its first Charter and Byelaws, had as one of its Members one of its employees. The point is important not because of the anomaly, which was corrected in the Byelaws made under the Society's Supplemental Charter of 1948, when it was specifically declared that "no Officer or Ordinary Member of the Council shall be appointed to any salaried office of the Society or to any office of the Society paid by fees" (a provision now embodied in the new Charter of 1963), but because as a result the office of Secretary became one of the most important, and indeed sought-after, in the domain of British zoology during the 100 years or so it was salaried. Philip Lutley Sclater, who was Secretary from 1859 to 1902, published hundreds of papers on birds and mammals, and, as Chalmers Mitchell puts it, became "one of the founders of knowledge of the geographical distribution of animals". Under Sclater's direction both the Library and the Society's two scientific publications, the *Proceedings* and the *Transactions*, prospered. Chalmers Mitchell, who followed after a contested election in which the public and the Press took a vigorous interest, and who reigned until 1935, was a distinguished and influential zoologist of the old style, much concerned with structural zoology and with speculations about the phylogeny of animals. But he, like Sclater before him, seemed to be unaware that the experimental approach had already started to revolutionize the biological sciences, and that the transformation was going to be permanent. The Royal Society was more alert, and during the last two decades of the nineteenth century it began to take the centre of the biological stage, largely because of its publication of papers describing advances in knowledge gained from experiments on animals. This, for example, was the period when Sir Charles Sherrington, later to become the Royal Society's President, played a major part in laying the foundations of our understanding of the physiology of the central and peripheral nervous system. Then came the rediscovery of the basic laws of genetics, and so the opening up of a vast new area of discovery. The science of endocrinology emerged; biochemistry started to flower. Anatomy was clearly giving way to physiology. The structural approach to biological knowledge was no longer in a mainstream, even though it still provided a necessary foundation to almost all biological enquiry. But experiment was imparting a new dynamism to the whole subject.

Chalmers Mitchell appeared to be unaffected by the new currents, and the Society's publications during the period of his Secretaryship rarely reflected the new trends of zoological research. Indeed, when I became Anatomist in 1928, no one on the Society's small professional staff was allowed to have a licence to carry out experimental work on animals. This is not to say that the Societies meetings were not "full of meat", or that our publications lacked interest, but younger zoologists were turning elsewhere, and particularly to the new Society for Experimental Biology, whose journal was a much livelier home for their papers. Without anyone realizing what was happening, our own Society was losing the scientific support on which its integrity, and indeed its continuity depended.

Unlike Chalmers Mitchell, Julian Huxley, who became Secretary in 1935, was very much a zoologist of the modern school, and also a man with the widest possible interests. But it soon became apparent that some of the Council of which he was now a Member were anything but sympathetic to his ideas, and equally that he was sometimes impatient in dealing with them as he strove to modernize and revitalize the Society's zoological image. More than that, times were bad, and shortly after taking charge as the Society's main executive officer, it became all too clear that war was on the horizon. Huxley was, however, able to divide the Society's *Proceedings* into two parts, the one dealing with structural, the other with experimental zoology. Unfortunately, by 1942 the Council and Secretary had become openly opposed to each other, and in 1943 Huxley resigned. As a friend, and not knowing much about the rights and wrongs of the case, I joined those Fellows who supported Huxley in the dispute, and then regarded his resignation as a rout of the scientific modernists. Sheffield Neave, who succeeded Huxley, was an entomologist of considerable distinction, and was devoted to the interests of the Society, but there was little he could do in the remaining years of war except keep the Society alive. He was followed in 1953 by Lord Chaplin, an amateur naturalist who, after two years, decided not to seek re-election. Neither he nor Sheffield Neave, nor the Councils of which they were Members, could achieve much that was positive during the difficult post-war years in which they served, and, not surprisingly, the resultant standstill still further weakened the Society's scientific image. In the early 1950s this was also damaged by the very wide publicity accorded to undignified disputes about the management of the Regent's Park menagerie. Matters were coming to a head when I became the Secretary in 1955, and what should have been a healthy scientific and non-scientific symbiosis, both on the Council and in the Fellowship, was threatening to break up the Society.

Our Royal Charter had established us in a framework which today we define as an educational charity. Legally, the Society, as such, has never had the right to do anything other than what the Charter declares our purpose to be—"the advancement of zoology and physiology and the introduction of new and curious subjects of the Animal Kingdom". But in order to raise money, the Council had from time to time, even if unwittingly, permitted certain infractions of the law. The first challenge to our scientific status came as early as 1854. In 1843 the Scientific Societies Act was introduced in order to exempt such bodies from the burden of rates. We enjoyed its provisions for 12 years, but then the Vestry-men of the Parish of St Marylebone, landlords of some of the acres we used, ruled that we were not exempted. The Council decided to challenge their view, but lost in the Courts. No one denied the fact that we were a scientific institution. We lost because members' subscriptions were obviously intended to purchase privileges of one sort or another, and also because we were employing snake charmers to entertain the public, and bands to give colour to the Promenades—activities which could not be considered "exclu-sively" scientific.

The question of our charitable status came up formally in 1930, when George de Arroyave Lopes, a name until then unknown to us, bequeathed to the Society the residue of his estate on the assumption that we were a charitable institution, and on condition that we hung the portrait of his mother in the Council Room and undertook to tend the graves of the family. Were we to fail in these provisions, the money—the sum involved was about £100 000—was to go to St Bartholomew's Hospital. The Will was challenged by a member of the family on the plea that we were not, strictly speaking, a charity. Because we sold refreshments and had employed a band, we did things which were incompatible with the terms of our Royal Charter. The legal questions raised were submitted to Mr Justice Farwell of the Chancery Division of the High Court. Fortunately for the Society, his decision was that we were a valid charitable trust. What his judgment, in effect, said was that our purposes were undeniably educational and therefore charitable; and secondly, that refreshments had to be provided for visitors, and the Society could not be expected to give away food for nothing. So far as giving elephant rides was concerned, Mr Justice

Farwell observed "that it widens a child's mind and in that broad sense is educational, that it should not only see the animal but should ride upon it, having impressed upon it the nature of animal that it is".

This judgment once and for all established the Society as a charity, and this was confirmed under the Charities Act 1961, when for the first time the registration of charities became obligatory.

When I became Secretary in 1955, I was ignorant of all this. Nor was I aware that there was little foundation for the belief, which I innocently assumed to be true, that the Society was still as rich, relatively, as it had been in the earlier part of the century, when the entrance fee at the gates was only 1 shilling. The real situation turned out to be the very reverse. It did not take any lengthy enquiry to make me realize that we were certain to go bankrupt if we did not take speedy remedial action. Moreover, in the forgetfulness of time, certain things were still being done within the Gardens which again contravened the spirit of the laws under which we enjoyed the status of a charity. Above all, I found that on average the Fellows of the Society were costing the Society far more than the value of their subscriptions—which was not the way I conceived a charity should operate. This was because, first, the public was being denied access on Sunday mornings to Crown land which in essence was theirs, a denial which was depriving the Society of revenue which it badly needed to improve the Gardens as well as the conditions of service of its employees—for example, at the time we had no assured pension scheme. Second, the level of a Fellow's subscription at £3 a year was ludicrously low. It had been set at that figure in 1832, and presumably because it could not be changed except by a vote of the Fellows, it was still £3 more than 100 years later!

The basic facts which were presented to the Council, and then to the Fellows, are set out in the Society's Annual Report for 1957. Part of this story is worth repeating here.

> "The site which the Society occupies in Regent's Park is Crown land, and it is held by grace on annual lease at a peppercorn rent. The Society's tenure is a privilege. It is beyond all question that the Society must not exercise its privilege to the prejudice of public rights. But the matter does not end there. The Society is under an obligation to justify and enhance its privilege by adding by every means possible to the educational, scientific and scenic value of the acres which we occupy."

Up to 1847 the greater part of the Society's revenue was entirely derived from Fellows.

> "But by 1850, when the total income of the Society was about £15,000, Fellows' contributions amounted to £6,000, that is to say, 40 per cent. In 1900 the corresponding figures were £29,000 and £11,000—somewhat under 40 per cent. In 1913, the last normal year before the first World War, Fellows' contributions had fallen to about half the sum which the public paid at the gates, and constituted no more than about 25 per cent of the total revenue of the Society. By 1930 the figures had fallen to 19 per cent; by 1938 to 16 per cent".

In 1957 it was 6 per cent and today it is less than 2 per cent.

> "The issue of Sunday opening can be seen in perspective only against the background of the Society's history. It was in 1826 that the Society was first granted a lease of land in Regent's Park, and for 20 years only Fellows, their guests, [or those who were introduced by a voucher] were allowed into the Gardens at any time. This was changed in 1847, when the public was allowed in [merely] on payment at the gates. Although the Council of the time was authorised to express the Crown's approval of the step as a measure which helped to diffuse 'intellectual recreation among the great mass of the people' the reason for the change was essentially financial. The Society had started on a wave of popularity, and almost within ten years of starting it had 3,000 Fellows, income and expenditure being kept more or less in balance until the beginning of the 1840's. The number of Fellows had then started to decline sharply, and as a result, the Society's revenues had become insufficient to cater for even its limited activities. The Society's finances were also little helped by such innovations as Promenade Days and other privileges, for which Fellows paid extra."
>
> "In 1847 there was never any question of the public being admitted for payment on Sunday, for the simple reason that, until 1932, when the Sunday Entertainments Act was passed, such a step would have been illegal. In 1940, the Council realised that the Society's revenues were likely to be very adversely affected by the war, and in consequence decided to open the Gardens to the public on Sundays after 1 p.m., and to Service personnel and their friends throughout the day. When hostilities ended in 1945, the ruling was altered, and the time of opening delayed until 2.30 p.m. This later opening did not prevent Sundays from being the day on which the public made proportionally their greatest contributions to the Society's revenues."
>
> "In 1956, receipts from the public on Sunday afternoons amounted to more than £40,000, i.e. to 19 per cent of all gate receipts, as compared with a total annual gross revenue from

Fellows (including entrance fees, subscriptions and purchase of tickets) of £29,000. Against this gross revenue from Fellows has to be set an amount equal to the counterpart value of Fellows' visits to the Gardens, leaving a net revenue from Fellows of only about £15,000.

"In a six months experimental period in 1957, in which the hour of admission of the public was advanced from 2.30 to 1 p.m., members of the public paying at the turnstiles on Sundays formed a significantly greater proportion of the total of paying visitors than they did in any one of the corresponding periods of the preceding ten years. The average increase in the proportion of Sunday visitors was approximately 15 per cent." (Ann. Rep. 1957)

On top of all this the question of the rating of the land occupied by the Society, both in Regent's Park and Whipsnade, had again become an urgent issue. Under the new Rating and Valuation Act of 1955 we became threatened with the imposition of heavy new charges. Council obviously had to consider how this could be lessened and, once again, the question of applying for exemption from rates as granted to other scientific societies, such as the Linnean, was considered. The legal advice we received indicated, however, that we did not fulfil two essential requirements of the relevant act (Scientific Societies Act, 1843).

"First, subscriptions from Fellows do not form a sufficiently high proportion of the Society's total income; second, the amount of the subscriptions do not in themselves sufficiently exceed the value of the privileges derived by the Fellows personally from the Society. Only a substantial increase in the amount of voluntary subscriptions, and a decline in the proportion of those subscriptions from which the Fellows derive a personal benefit would, so Council is advised, make it possible for the Society to plead that it was eligible for relief under the Act."

"The stage had been reached in the history of the Society in which the Fellowship was threatening to be a financial impediment to the discharge of the purpose of the Royal Charter, and was therefore threatening the very existence of the Society itself." (Ann. Rep. 1957)

In the light of all the facts it was clearly the responsibility of the Council, as it was within its powers, to declare the Gardens open to the public throughout Sundays. And this is what it did. Not unexpectedly the decision was resented by a group of Fellows who, sensing further changes to come, and not without support from a small number of members of Council, soon organized themselves into a so-called "FZS Committee", under the Chairmanship of Mr R. J. Knowles, to fight the Council, which in compliance with our

Byelaws had to put the next changes it had in mind and which involved substantial amendments of the Byelaws, to a vote of the Fellowship as a whole. The first change that was proposed was to raise the annual subscription from £3 to £8. The second and more important one was to divide the Fellowship into a scientific and non-scientific class, election to the former necessitating certain prescribed professional qualifications. A non-voting class of subscribers called Associates was also proposed. The whole purpose of these changes was to strengthen the Fellowship in such a way that it would be better able to discharge the purposes of the Society's Charter, which essentially meant the advancement of zoology.

The reaction was immediate. The Press was inundated with reports and letters claiming that everything was wrong with the way the Council and its staff ran the Gardens, Whipsnade and the menageries. Council was told that while "Fellows did not mind having a few professional zoologists on the Council", in fact there were too many. Every charge that was made in what became a public propaganda battle could be easily shown to be false, as was also the statement, made to the Press, that

> "the Society's prestige had fallen very low in the eyes of the Directors of continental Zoological Gardens."

This particular charge was later referred by us to the International Union of Directors of Zoological Gardens, by which it was vigorously denied.

The General Meetings of the Society at which Council's proposals were discussed were stormy and the amendments to the Byelaws were finally put to the vote at the Ordinary General Meeting on 16th April 1958; 1788 Fellows voted in favour of the Council's proposals and 1227 against, voting either in person or by proxy.

The "FZS Committee" promptly challenged the decision, claiming that the construction of the Byelaw defining a majority—which read that any proposal by the Council to alter or repeal a Byelaw

> "shall be deemed to have been confirmed if the majority of Fellows entitled to vote shall vote in its favour and for this purpose voting may be in person or by proxy"

—implied that a majority had to be more than half of the total number of Fellows in the Society, whether or not they all voted.

The Council was taken to the High Court, where the interpretation which Mr Knowles was maintaining was upheld. Council then decided to appeal, and the Court of Appeal ruled that the words "majority of Fellows entitled to vote" meant

> "if the majority of Fellows who attend this meeting in person or by proxy and are not disqualified from voting shall vote".

In the course of the judgement it was pointed out that the construction advanced by Mr Knowles not only resulted in absurdities but was also inconsistent with and repugnant to the Society's Charter. During the discussion on the award of costs Lord Justice Romer observed that it could hardly be said that Mr Knowles had sued the Society "in order to benefit it".

Following this decision, Mr Knowles, together with a large number of Fellows, most of whom had supported him, resigned from the Fellowship. Three years later the Society was granted a new Charter which corrected the ambiguity in the definition of a majority, and to which was appended a new set of Byelaws which defined the way in which Ordinary and Scientific Fellows should be balanced on the Council, and which also ordained that while the President and Treasurer could belong to either category of Fellow, the Secretary had to be a Scientific Fellow "or an Ordinary Fellow having the qualifications for election as a Scientific Fellow".

In my view the consequences for the Society if its non-scientific element in the Knowles campaign had won would have been its death. From then on the practical difficulties of changing any Byelaw, or even the Charter, would have been insuperable. What is more, abuses of the Society's charitable purposes would soon have seen to it that we would have been denied any benefits in taxes and rates which we enjoyed as a charitable scientific Society which also managed a national institution. Regent's Park could never have been "commercialized", as was at the back of the minds of some of those who opposed the Council, for a variety of reasons into which there is no need to enter here.

Even before our legal status had been upheld and then redefined, the Council had embarked on a major programme of reconstruction, details of which can be read in our Annual Reports from the end of the fifties onwards. Conditions of service were revised and a non-contributory pension scheme introduced. Over half of the Regent's Park Gardens has since been rebuilt, and major developments undertaken at Whipsnade. New journals have been launched, and two research laboratories, as well as an animal hospital, established. A formal educational programme was organized. Our professional staff has multiplied. We can now confidently claim that we are once again in the mainstream of biological science.

The Society's resolution of 1841, which I have already quoted, is no less valid today than it was more than a century ago. The Society cannot divest itself of its scientific character without violating its Charter of Incorporation. For science is at the very heart of our constitution.

References

Andrade, E. N. da C. (1960). *A brief history of the Royal Society.* London: The Royal Society.
Blunt, W. (1976). *The ark in the park.* London: Hamish Hamilton.
Cardwell, D. S. L. (1957). *The organisation of science in England.* London: Heinemann.
Holloway, Sally (1976). *The London Zoo.* London: Michael Joseph.
Mitchell, P. Chalmers (1929). *Centenary history of the Zoological Society of London.* London: Zoological Society of London.
Scherren, H. (1905). *The Zoological Society of London.* London: Cassell & Co.
Vevers, Gwynne (1976). *London's Zoo.* London: Bodley Head.

Symp. zool. Soc. Lond. (1976) No. 40, 17–48.

THE ZOOLOGICAL SOCIETY AND THE BRITISH OVERSEAS

R. FISH and I. MONTAGU

The Zoological Society of London, Regent's Park, London and *Old Timbers, Garston, Herts, England*

SYNOPSIS

The Zoological Society owes its foundation to one of the British whose main life-work was performed overseas, and the British overseas played a large part in the Society's development. To illustrate this theme brief accounts are given of the careers of ten such men, from the founder, Sir Thomas Stamford Raffles, to Sir Harry Johnston, who died in 1927. These show the variety of their origins and professions, and the range of their contributions to the Society.

INTRODUCTION

The theme is an appropriate one for an opening paper to this symposium, for it was one of the British overseas whose conception and energy inspired the foundation of the Society and planned its direction; and it was the response of British overseas that was to prove so essential a factor in its development.

The turn of the eighteenth/nineteenth and the early nineteenth century, when the situation was growing ripe for such a society, was a period of European expansion. Particularly from Britain, trade connections were building with "far-flung" parts of the world. On the profits of such trade the growth of industry was speeding up at home. The needs of trade and industry sent British administrators abroad, to India and South East Asia, to North and South America, to Australia and Africa: aim—the promotion of stable conditions protecting trade. Land enclosures and poverty led to emigration. Travel, especially by sea—though still precarious—became comparatively faster and safer. Thus opportunity occurred for Europeans, and especially British, to encounter animals strange to them.

In a sense this is a mere statement of the obvious. An interest in strange animals and unfamiliar plants has constantly been associated with outward-looking trading groups and empires. Without the expansion and encounters such development facilitated such an interest would never have been aroused. It was a feature of ancient Egypt (domestication and deification), of Greece and the Middle East, of Rome and the China of the Mings.

Menageries were a result of such contacts, the strange animals being a convenient and appreciated gift or tribute from subordinate king or satrap to emperor, or a proper exchange between equal monarchs. And traces of this hallowed custom still spin off a benefit to the Society's menageries to this day. Trading companies and friendly rulers still present animals to Her Majesty, which by her grace often find their way to the Society's gardens. Animal gifts exchanged with the People's Republic of China climaxed in the generous presentation, through Mr Heath, of the noble pair of giant pandas. And not many years ago the President of this Society, on his Pacific travels, received from the proprietor of an island several splendid pairs of birds and, enquiring what would be appreciated in exchange, complied with the request to send "one of Mr Churchill's hats".

The new features in the old situation that were present in the early nineteenth century were twofold. First, the scientific attitude, pioneered by the Royal Society, and associated with the growth in many fields of the same spirit. To define it? Study, accurate description, classification, experiment, collegiate exchange and discussion of evidence and experience. Second, the presence abroad of administrators, soldiers, settlers, venturers, many of them of the highest calibre, often interested in the world around them and now activated by the age's scientific curiosity.

To those at home, an exotic animal imported aroused wonderment and little more. To those who met the animal in *its* home and context, it could arouse a deeper interest, the drive to comprehend and share new knowledge with others who could further understanding.

It was from those with new opportunity that ultimately were drawn the Corresponding Members of the Zoological Society (CMZS). Almost immediately following its foundation the Society dispatched a circular by hundreds to all corners of the earth. It appealed for animals, exaggerating, no doubt, the ease of their collection and despatch—a barrel of brine was suggested as the adequate container for bodies (Fig. 1). And in the next years thousands of specimens came back, a few live for the menagerie, more dead for the museum, plus notes of observation of the fauna of the countries where the CMZS lived and worked.

TABLE I

The "British overseas" of whom accounts are given in this paper

Name	Dates of birth and death	Age at death	Area of activity	Connection with the Society
Raffles	1781–1826	44	S.E. Asia	Founder
Cuming	1791–1865	74	S. America *et al.*	CMZS
Belcher	1799–1877	78	Africa, Pacific, Atlantic, S.E. Asia	FZS
Lowe	1802–1874	71	Madeira	CMZS
Blyth	1810–1873	63	India	CMZS
Hodgson	1800–1894	94	India, Nepal	CMZS Silver Medal
Tickell	?–1875	?	India, Nepal, Burma	CMZS
Macleay family	1767–1891	81,73, 82,81,71	Australia	5 × CMZS
Hudson	1841–1922	71	S. America	CMZS (FZS)
Johnston	1858–1927	69	Africa	FZS CMZS Gold Medal

For this paper we have made a selection of the distinguished names from among the many British abroad helping the Society (Table I).

The second column shows that the order is roughly chronological and covers the first 100 years of the Society's existence.

From the third column it might be a fair inference that zoology is a healthy occupation. Recall that, in the age when most of those in the Table flourished, the expectation of life was much less than it is today, yet nearly all approached or even exceeded the biblical allotment—the one exception being the founder whose life ended suddenly and abruptly with a brain tumour, conjecturally a possible consequence of the tropical rigours that carried off one wife and three elder children and made his second wife and himself desperately ill.

Why this longevity? Perhaps it has something to do with the actuarial observation of US insurance companies that many business men do not long survive their retirement; the habit of a lifetime gone, they simply fade away. Zoology, by contrast, is an

Zoological Society.
33, Bruton Street
1827.

Sir,

I take the liberty with the sanction of
of sending to you the last Report of the Zoological Society.

It is possible that in the course of your residence at opportunities of promoting our views and objects may occur to you, and that you may be able to send to us occasionally and at a very inconsiderable expense, specimens of subjects in Zoology of much curiosity and interest.

Living specimens of all rare animals, and particularly of such as may possibly be domesticated and become useful here, will be much valued by us: and above all, varieties of the Deer kind, and of gallinaceous Birds. but beyond this preserved insects, reptiles, birds, mammalia, fishes, eggs and shells, will be gratefully received. And I may mention that where a more scientific method does not occur, the promiscuous immersion of any number of subjects in a tub of strong brine (feathers, bodies, & all) will be sufficient for preservation, not quite effectual perhaps for the skins in all instances, but perfectly so for purposes of dissection and comparative anatomy.

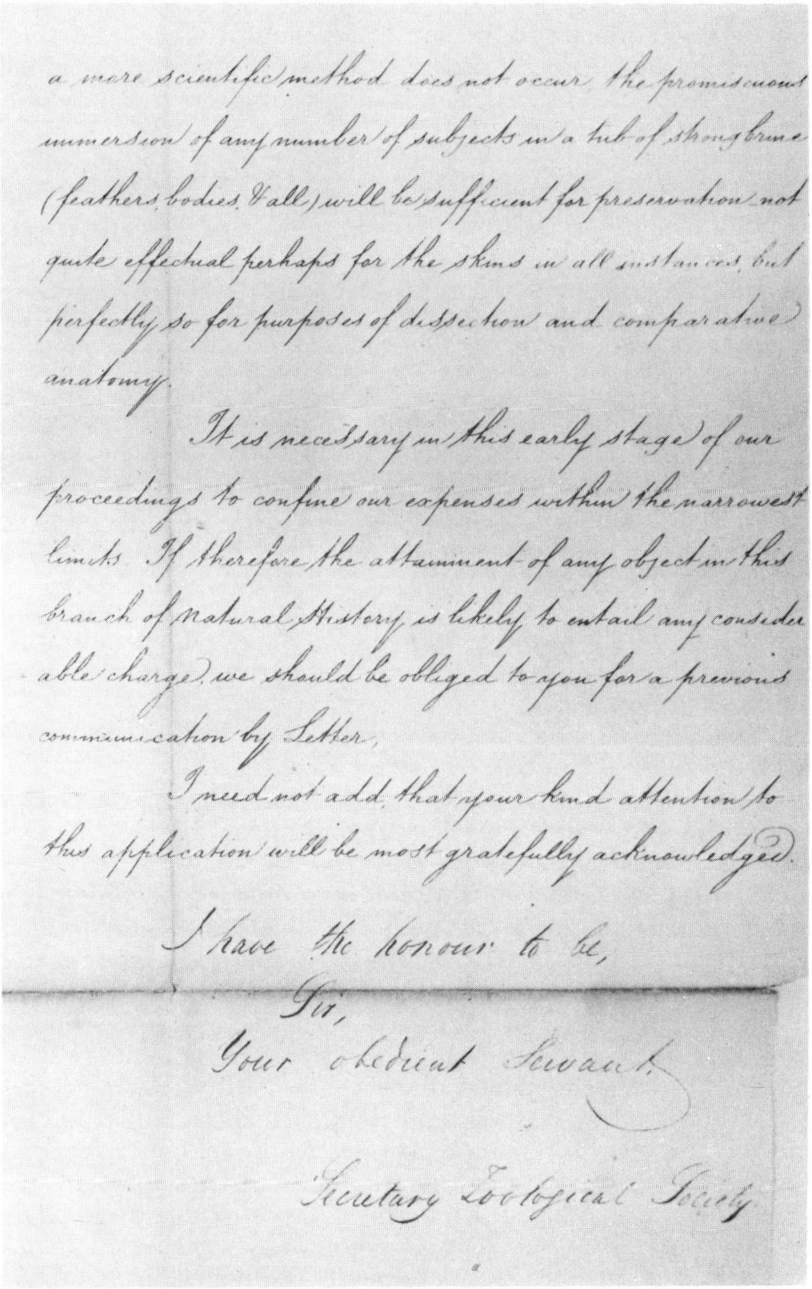

It is necessary in this early stage of our proceedings to confine our expenses within the narrowest limits. If therefore the attainment of any object in this branch of Natural History is likely to entail any considerable charge, we should be obliged to you for a previous communication by Letter.

I need not add, that your kind attention to this application will be most gratefully acknowledged.

I have the honour to be,
Sir,
Your obedient Servant,

Secretary Zoological Society

FIG. 1. Circular sent to Corresponding Members of the Zoological Society in 1827.

interest that animates to the last. Incidentally, four at least of our list are like some other famous zoologists in that they originally took up "Natural History" as weakling youngsters recommended activity in the open air.

The fourth column shows that, between them, the activities of our choices covered a great part of the globe.

What is highly significant is the variety of origins and professions represented. While the British social structure of the period was certainly not what has come to be implied by the current use of the word "democratic", it was at least more flexible and mobile than that of many of its contemporaries, and this perhaps was a feature of its strength. Some four of our choices were what is called "well born" and "well educated", products of public school and university, born from families of administrators, lairds, ranchers. Some were self-taught and "self-made". Three were orphans, and yet others were compelled by circumstance to start earning early.

No fewer than half were born outside the home country—even so early, British abroad of a second generation.

Four became Colonial Governors or Resident Commissioners; one was an admiral, one a colonel, one a clergyman; one was a sailmaker's apprentice, one opened a chemist's shop and two started as junior clerks.

Three, at least, as a sideline, were explorers, and filled in their blank spaces on the map.

Five were good artists, who illustrated their works, all competently and some beautifully; a demonstration of the value of this so precious talent to the working zoologist.

What is clear is that in those days an interest in zoology cost money rather than made it. Where the "silver spoon" was absent at birth, they had to make a living by some other career and make of zoology an extra labour. There was only one professional zoologist among the lot of them—and he had a hard time. All—including the pro—were, in the deepest and most literal sense, amateurs from youth up, and that sustained them.

Partly by chance, it has come about that these names fall into three groups. First four very early ones, whose zoological work was connected with their travels by sea. Then three who spent their working lives in India, and then three more that carry us to this century.

Except for the founder—of course—(and Belcher, who, as a naval officer, was nominally based in London) all were CMZS. We examine them seriatim.

RAFFLES

Sir Thomas Stamford Raffles, b.1781, d.1826

His biography is well known, so here the brilliant and all-too-brief career may be abbreviated. Born on his father's ship (his father a sea captain) off Jamaica. The father died when he was 10. Came to England as a child with his mother and siblings. Temporary clerk in the employ of the East India Company at 14. At 24 sailed to Penang as an assistant secretary, at 30 Lieutenant-Governor of Java.

At the age of 35, in 1816, he came back briefly on a visit to home and Europe (after visiting St. Helena to call on Napoleon on the way).

One year later he was back in Sumatra and Governor of Bencoolen there. One year more and he had founded Singapore as a British trading entrepôt. Two years further on and disastrous illness and death struck at his whole family. Another three years and he was back in England at 43 with a great reputation, influential friends, jealousies within the company and financial troubles. He had only two years left. But these are the two that are important here. In this short time he founded the Society and set it on course.

We are going to cover two points. His mystery portrait and a little-noticed aspect of the foundation.

The portrait—what was Raffles like? In what we may presumably call the Holy of Holies of the Society—The Council Chamber—hangs his picture. A young man of byronic beauty (Fig. 2). But is it, can it be, a portrait from the life? When was it painted? The National Portrait Gallery has a copy, but it does not know.

Raffles was in England up to 24 years old—when he was too young and unimportant to have his portrait painted—then briefly at 35 to 36—and again from 43 to 44 when he already had painful headaches and must at least have been showing signs of wear, quite unlike the youth in the portrait. We know that when he came in 1816 (at 35) he was somewhat lionized and several portraits were painted, some equally chocolate-boxy. We know that then, as since and even to the present era, it has been the custom to flatter men of worth by smoothing lines out and beautifying them (indeed several artists have made short reputations and large fortunes out of this speciality). Nor does the costume help much; at first glance it looks like something out of amateur theatricals, for Hamlet maybe,

FIG. 2. Sir Thomas Stamford Raffles. Portrait in the Society's Council Room attributed to Lonsdale. Presented to the Society in 1946 by Mr A. Spencer Moore.

FIG. 3. Sir Thomas Stamford Raffles. Engraving from a miniature owned by Mrs Raffles.

reminding one that surely the togas on the contemporary statutes of statesmen in the House of Commons were never worn in life by those exalted pseudo-senators. On careful inspection it does indeed look like a copy of the famous (and duly toga-ed) Chantrey bust, the face smooth and lineless after the marble original. Could the open-neck collar have been official dress of a Lieutenant-Governor in the hot South East?

There is another far more convincing portrait, dark, intense, a man credibly of will-power and demonic energy, also purporting to be Raffles (Fig. 3), which is *known* to have been made in 1817. It is not possible that the man who so looked at 36 could have looked like the Council picture, after grievous misadventures, at 43. We must conclude therefore that portrait was either a compliment in lifetime or an idealization afterward.

Now for the foundation. Raffles in boyhood, including his period as a temporary clerk attracting the attention of his superiors by assiduous application to his duties, had ever revered facts and accumulated as wide a knowledge by reading as he could. As a colonial administrator he at various times employed two French zoologists and two English botanists to record the natural history of his realms of sway, and with one of the latter aides he himself made an expedition into the interior of Sumatra that disclosed that peculiar object—sometimes called the "biggest flower in the world"—*Rafflesia arnoldi*, a stinking monster the coloured plaster image of which for so long decorated the main hall of the Natural History Museum at South Kensington.

During his brief European trip in 1817 he had visited the Jardin des Plantes in Paris, and it is undoubtedly then that he conceived the notion of something like it, but better, for Britain, as Britain deserved.

Now in 1824, returned once more and with no presentiment of how little time was left, he immediately set about making the dream real. As Fellow of the Royal Society, he approached its President, Sir Humphry Davy, for co-operation. Davy responded favourably.

A VIP's committee was set up with Raffles as chairman and, as members, influential noblemen and politicians; a subscription list was opened, the Society came into being and Raffles was appointed its first President.

But certainly not all was plain sailing. Firstly, the enthusiasts had to overcome the frosty welcome so often in store for every good new idea. For example, cold water on the project from the *Literary Gazette*:

> "Those friendly to it . . . declare it to be worthy of an enlightened country—and anticipate wonderful improvements from its being carried into effect; while those of an opposite way of thinking laugh at it as a wild speculation . . . . Like too many of our modern associations and companies, this is extremely sonorous on paper; but, alas for the execution of the design—is it not altogether visionary?".[a]

And they jeered at it as an attempt to make "an Ark in London". But behind the scenes there was another issue, concealing a rift between Raffles and Davy. A famous phrase in the ZSL Charter (1829), calls the project:

> ". . . a society for the advancement of Zoology and Animal Physiology and the introduction of new and curious subjects of the Animal Kingdom".

Most people, with hindsight, think this last point refers to importing animals for zoo purposes. It does not. It is the old story of "applied" versus "pure" science. A false antithesis of course. Because as often as not the first, attempted in a hurry with too few data, comes to nothing, while the second may have unexpected, far-reaching results.

Raffles wrote to a cousin:

> "Sir Humphrey Davy and myself are the projectors; and while he looks more to the practical and immediate utility to the country gentlemen, my attention is more directed to the scientific department."

A neat contrast, recalling the no less categorical distinction formulated for a later generation by the Douanier Rousseau to separate the only two painters he thought worth mentioning: "I represent the modern, Picasso the classical".

Clearly the Prospectus, their joint work, is a compromise. Raffles thought the best line beating the big drum.

After complaints that here in Britain "we possess no great scientific establishments for teaching or elucidating zoology, and no public menageries," the opening explains how the nation's possessions, colonies, fleets and varied intercourse with every quarter of the globe give it greater facilities for collection and study than any other nation, but, alas—

> "the student of Natural History, or the philosopher who wishes to examine animated nature, has no other resource but that of visiting and profiting by . . . "

[a] Quoted in: Mitchell, P.C. (1929). *Centenary history of the Zoological Society of London*: 13–14. London: ZSL.

Here Raffles in his original drafting wrote: "... the magnificent institutions of a neighbouring and a rival country."

Very understandable no doubt, when we remember that Napoleon was still only three years dead and ten defeated, and that Raffles' feats in South East Asia had all along included the responsibility of coping with the fringe effects of global war. His visit to Paris and the Jardin des Plantes that he had been so stirred to emulate had occurred only two years after Waterloo. But, after all, the reinstated Capet kingdom was now a British ally, sharing peace-keeping, so this formula was too much for the committee and it was replaced by the anodyne: "... of neighbouring countries" with insertion of a matching claim that "in every other part of Europe ... something of this kind exists" (which just didn't happen to be true) while in Britain there were no "public menageries or collections of living animals where their nature, properties and habits may be studied". Finally the ringing declaration: "This opprobrium to our age and nation" must be removed by foundation of the Zoological Society. As Churchill once put it: "Patriotism by the imperial pint".

Davy's contribution, on the other hand, preferred an appeal to practicality. Number one of the "great objects" that followed in the prospectus is clearly his. This priority "should be": "The introduction of new varieties"—note the exact phrase familiar from the charter—"breeds and races of animals for the purpose of domestication or for stocking our farmyards, woods, pleasure-grounds, and wastes".

How far did Davy really disagree with Raffles? The latter certainly thought he did and wrote to intimates about it several times, e.g. "... at this very time I am a little at issue with Sir Humphrey Davy as to the share which science is to have in the project", and, again, "I look more to the scientific part." He wanted to revise the prospectus but did not get his way.

On the other hand, Humphry Davy, whom Raffles always misspelt if not deliberately, at least consistently, and who was nothing if not a scientist-politician, accustomed to dealing with panjandrums and the influential—only too ready (in those days as in some others) to confuse the "scientific" with the airy-fairy or the mere gratification of curiosity—may have been wise in seeking to stress the immediately useful as more likely to appeal to the purse of the contemporary VIP.

Malthus, then a lecturer at Haileybury, was all the rage. In Britain the first three decades of the nineteenth century saw population increases of 14.6 per cent, 17.6 per cent, and 15.8 per cent—enough to frighten anyone. The prospect of a novel (and profitable) means of increasing food output could have been the lever that removed obstacles.

Anyway, Davy got Raffles his land in Regent's Park out of the Crown—at a rental of £6 6s 0d per acre per year, and this joint prospectus enlisted the necessary support of nobility and gentry.

So Raffles realized his dream, the scientific society, and its course was charted:

(i) a museum, removed from site to site at first with the society's offices, later hither in Regent's Park, until its accumulation burst all bounds and, somewhat reluctantly, in 1855 its contents were ceded, some to better those of the British Museum and some to spread the gospel in the provinces;
(ii) a base, centre for researches and publications to share zoological knowledge throughout the world; and
(iii) the menagerie he never lived to see.[b]

Within a dozen years of Raffles' death, the *Penny Magazine* (1837) was declaring: "The Zoological Gardens in the Regents Park, for picturesque beauty, far surpass the Jardin des Plantes of Paris".[c]

How this would have pleased Raffles! And it went on to quote the *Quarterly Review*;

> "As we walk along the terrace, commanding one of the finest suburban views to be anywhere seen, let us pause for a moment while 'the sweet south' is wafted over the flowery bank musical with bees, whose hum is mixed with the distant roar of the great city."

Today the roar is louder, the bees out-hummed, but the influence of the first of our overseas Britons still prevails.

[b] As can be seen from the circular (Fig. 1), Davy's pet projects were not neglected either. The Society started an acclimatization farm and a pond, but both were given up in a few years. Too little was known then of the costs and complexities that make for failure, the snags that face success (as, later, rabbits showed in Australia and grey squirrels here). But it is not a theme the Society has by any means abandoned. Only a few years back its research laboratories were studying the palatability and nourishing qualities of wild animal proteins, and a paper on the possibility of new domestications is included in this symposium.

[c] *Penny Magazine*, 16th December 1837.

CUMING

Hugh Cuming, b.1791, d.1865

The dates show that Hugh Cuming must have been one of the first British overseas to respond to the brine tub circular.

He was born in Devon, in a sea-coast village where men of his name and ancestry still live. He began as a sailmaker's apprentice and it was not until the age of 26 that a venturesome disposition and an interest in sights and places sent him forth. He first set up in Valparaiso at his old trade and indulged a passion for collecting seashells. Encouraged by visiting naval captains, including Fitzroy of the *Beagle*, and having sufficiently prospered after seven years, he built a yacht and cruised among the Pacific archipelagos, corresponding with the Society about his collections.

In 1836 he decided to come home. First he sent off a batch of rare Chilean animals to the menagerie. Also he sent what the Society's records describe somewhat oddly as "a complete series of Crustaceans" for the museum. Then he negotiated about and deposited with the Society an immense conchological collection containing "fully 6,000 species, and generally from three to ten specimens of each". This was estimated to be worth over £3000—a huge sum for those days. In return he was to become Curator at a salary of £200 a year (Fig. 4). To gauge such figures we should note that the Society's report for that year recorded that in the nine months it had been open in new premises the museum had been visited by 3660 persons paying a total of £38 17s 0d.

Then he set off, sailing across the Pacific and back round the world to London. After arrival he further presented the museum with mammal skins from Malacca and birds from the Philippines, besides a number of living birds and mammals also collected on the journey. Later he topped this with a collection of Philippine bats.

The Cumingian collection is now incorporated in that at South Kensington, whither it was finally transferred. Cuming himself contributed prolifically to the PZS (*Proceedings of the Zoological Society of London*) but he never took up the proposed curatorship. Perhaps he did not wish to settle within four walls.

BELCHER

Admiral Sir Edward Belcher CB, b.1799, d.1877

Belcher was born in very different circumstances—in Nova Scotia, a second-generation Briton-abroad, son of its Governor.

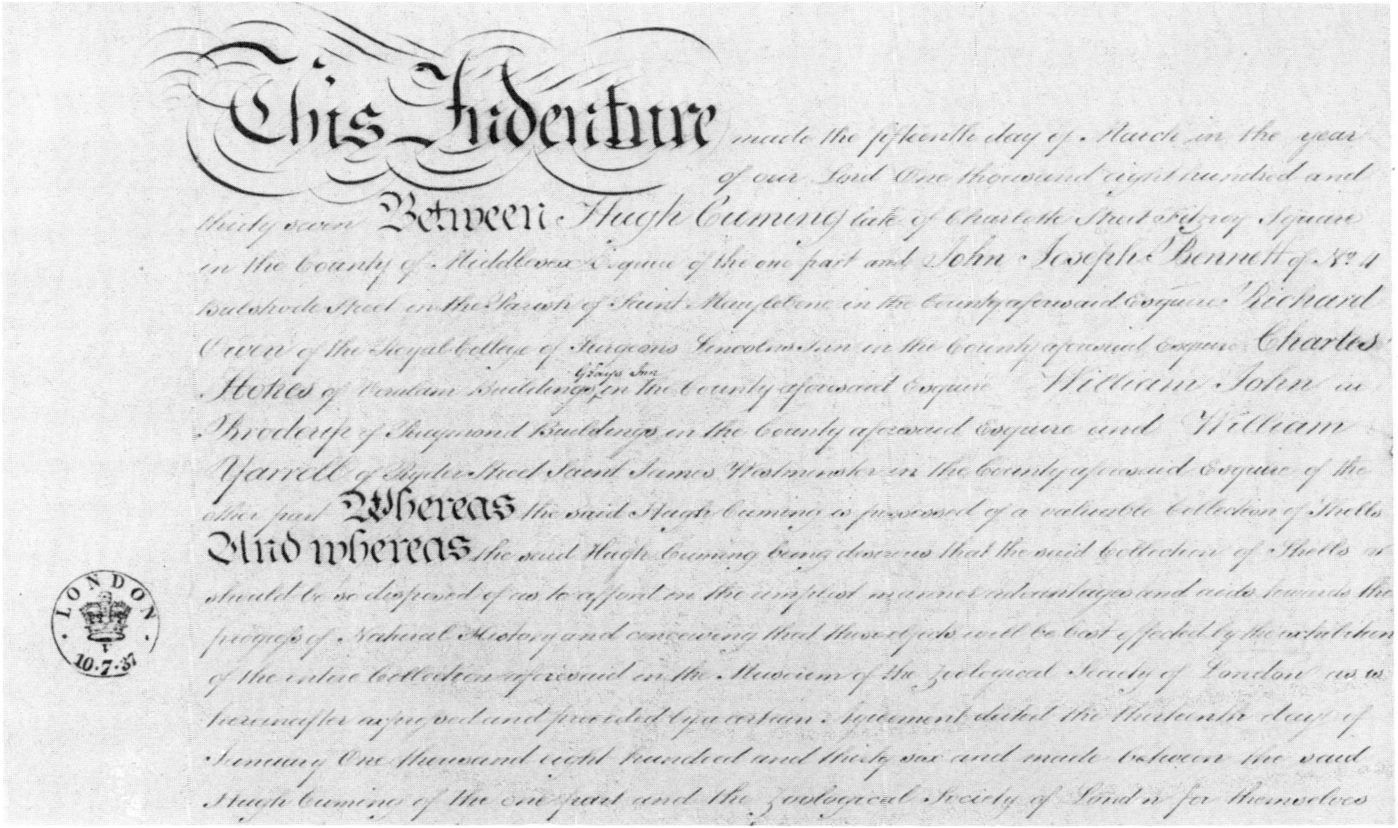

This Indenture made the fifteenth day of March in the year of our Lord One thousand eight hundred and thirty seven Between Hugh Cuming late of Charlotte Street Fitzroy Square in the County of Middlesex Esquire of the one part and John Joseph Bennett of No 4 Bulstrode Street in the Parish of Saint Marylebone in the County aforesaid Esquire Richard Owen of the Royal College of Surgeons Lincolns Inn in the County aforesaid Esquire Charles Stokes of Verulam Buildings Grays Inn in the County aforesaid Esquire William John Broderip of Raymond Buildings in the County aforesaid Esquire and William Yarrell of Ryder Street Saint James Westminster in the County aforesaid Esquire of the other part Whereas the said Hugh Cuming is possessed of a valuable Collection of Shells And whereas the said Hugh Cuming being desirous that the said Collection of Shells should be so disposed of as to afford in the amplest manner advantages and aids towards the progress of Natural History and conceiving that those objects will be best effected by the retention of the entire Collection aforesaid in the Museum of the Zoological Society of London as is hereinafter agreed and provided by a certain Agreement dated the thirteenth day of January One thousand eight hundred and thirty six and made between the said Hugh Cuming of the one part and the Zoological Society of London for themselves

FIG. 4. Part of the agreement between Hugh Cuming and the Zoological Society regarding the collection of shells.

He entered the Navy at 13, became a midshipman at 17, and steadily ascended the naval ladder, specializing in the work of surveying, himself commanding and carrying out survey voyages in all the Seven Seas. He sailed in the *Blossom*, the *Aetna*, the *Sulphur* and the *Samarang*.

He charted coasts in the Pacific, the Behring Straits, on the west and north coasts of Africa, in the Irish Sea, on the west coast of North and South America. Then his ship was ordered home by the same route that Cuming had taken only a few years earlier. From Singapore *en route* he was ordered back to China and took part in the not very glorious actions in the Canton river. Beginning the homeward journey, he surveyed on his way coasts of China, Borneo, the Philippines and Taiwan, arriving at last in England in 1847.

He might almost have had a subsidiary profession as animal collector. From everywhere that Belcher went, specimens were sure to go. Skins to the museum, live animals to the menagerie. Remember the conditions on board in that era, the tiny working space available, the problems of live transport. Yet he was one of the most prolific donors the Society was ever privileged to have. He contributed four papers to the PZS and wrote a book about every voyage.

So far very good.

At last, alas, in 1852 he was appointed to command the expedition in search of Sir John Franklin, vanished with crews and vessels while searching for the North West Passage. Not an unsuitable-seeming choice for the relief perhaps, but it did not work out:

> "The appointment was unfortunate, for though an able and experienced surveyor he had neither the temper nor tact for a commanding officer under circumstances of peculiar difficulty. Perhaps no officer of equal ability has ever succeeded in inspiring so much personal dislike, and the customary exercise of his authority did not make Arctic service less trying. Nor did any happy success make amends for much discomfort and annoyance, and his expedition is distinguished from all other Arctic expeditions as the one in which the commanding officer showed an undue haste to abandon his ships when in difficulties, and in which one of the ships so abandoned rescued herself from the ice and was picked up floating freely in the open Atlantic".[d]

[d] Quoted from the article in the *Dictionary of National Biography* by John Knox Laughton (1830–1915), the naval historian.

FIG. 5. Captain Belcher's own sketch of one of the disasters suffered by his arctic expedition. HMS *Assistance* is blown out of winter quarters. From Belcher (1885).

Sad. But perhaps not so bad. Since he was already CB and Sir Edward, and later became Vice Admiral and then Admiral. Or did this mean that the Admiralty itself was worse? Probably only that the promotion system was based on seniority—and, so long as you lived, you went up. He wrote out of the disaster his usual book. His own drawing of it is reproduced as Fig. 5.

Anyway, he was a wonderful friend to the Zoo, from all corners of the world.

LOWE

Rev. Richard Thomas Lowe, b.1802, d.1874

Lowe is the fourth of the early contributors whose zoological propensities could be indulged through connections with seas and islands.

Educated at Cambridge University, he took holy orders in 1825. He was one of those many—including Darwin— for whom irregular health had provided an excuse and an opportunity to indulge a passion for travel and nature. At the same age as Cuming he left England, to seek—as any modern tourist does—the sunny climate of Madeira. In two years he compiled and published a paper on the primal fauna and flora of Madeira. Two years later he became English chaplain on the island and so remained for 22 years, sending back numerous specimens for the museum, including a collection of land and freshwater shells and several collections of fishes, and contributing to the PZS and the TZS (*Transactions of the Zoological Society of London*) (Plate 1 shows one of his fishes as illustrated in the TZS). Then he came home.

But if Madeira and its climate seems brought so close to us by modern travel and package tours as to be hardly worth including among the list of far-flung spots frequented by other of our paladins, in the last century journeying was different. Raffles himself was wrecked and faced dangers for himself and family several times. In 1874, now in his seventies, Lowe felt a yen for the sunshine and the natural discoveries of his youth.

The ship that bore him foundered off the Scilly Isles and went down with all aboard.

BLYTH, HODGSON, TICKELL

Edward Blyth, b.1810, d.1873; Brian Houghton Hodgson, b.1800, d.1894; Colonel Samuel Richard Tickell, ?, d.1875

We now come to the next batch: perhaps the three most distin-

Plates

Plate 1. *Chaunax pictus*. Drawing by the Rev. E. H. Woodall of a specimen of a new genus of the Lophidae caught by the Rev. R. T. Lowe off the coast of Madeira, described by him in the *Transactions of the Zoological Society*, vol. 3 pt. 4, and presented to the Society's Museum.

Plate 2. A. Hodgson's drawing of *Martes flavigula* of Nepal. B. Hodgson's drawing of *Aquila bifasciata.*

Plate 3. A. Tickell's drawing of the Indian kingfisher *Alcedo bengalensis*. B. The iguanid *Calotes mystaceus* by Tickell.

Plate 4. A. Johnston's watercolour of the Okapi, the first illustration of this animal seen in Europe. Exhibited at the Scientific Meeting of the Zoological Society on 7th May 1901. B. Mrs Georgina Lopes: the picture which hangs in the Society's Council Room.

Plate 1

PLATE 2A

PLATE 2B

Plate 3A

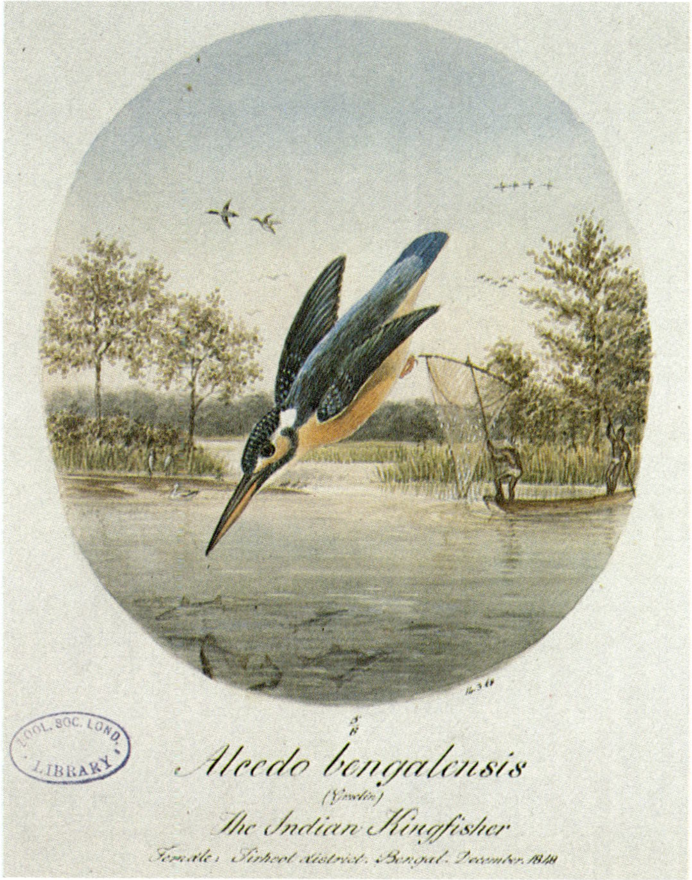

Alcedo bengalensis
(Gmelin)
The Indian Kingfisher
Female. Pirhool district. Bengal. December 1840
ZOOL. SOC. LOND. LIBRARY

Plate 3B

Sauria
Iguanidæ
Calotes mystaceus
Definition
Family.

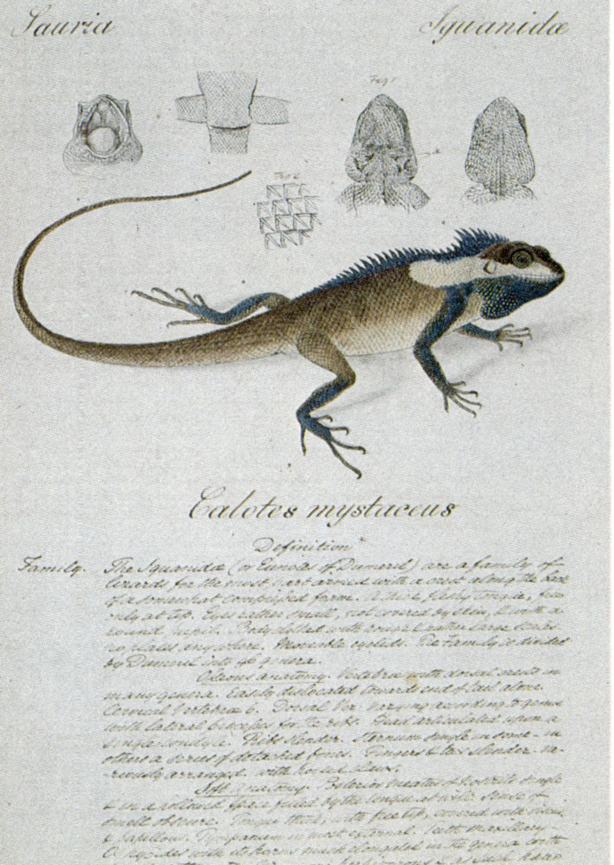

Plate 4A

P.Z.S. 1901, vol. II. Pl. I.
H. H. Johnston pinx
P. J. Smit lith.
THE OKAPI.
(OKAPIA JOHNSTONI)
(reduced one half from Sir Harry Johnston's original painting)
Mintern Bros. imp.

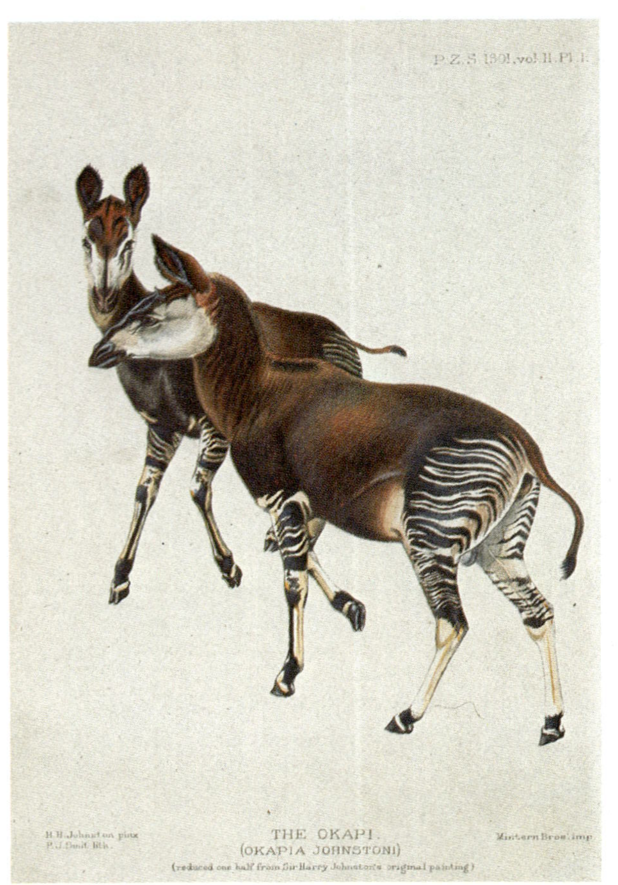

Plate 4B

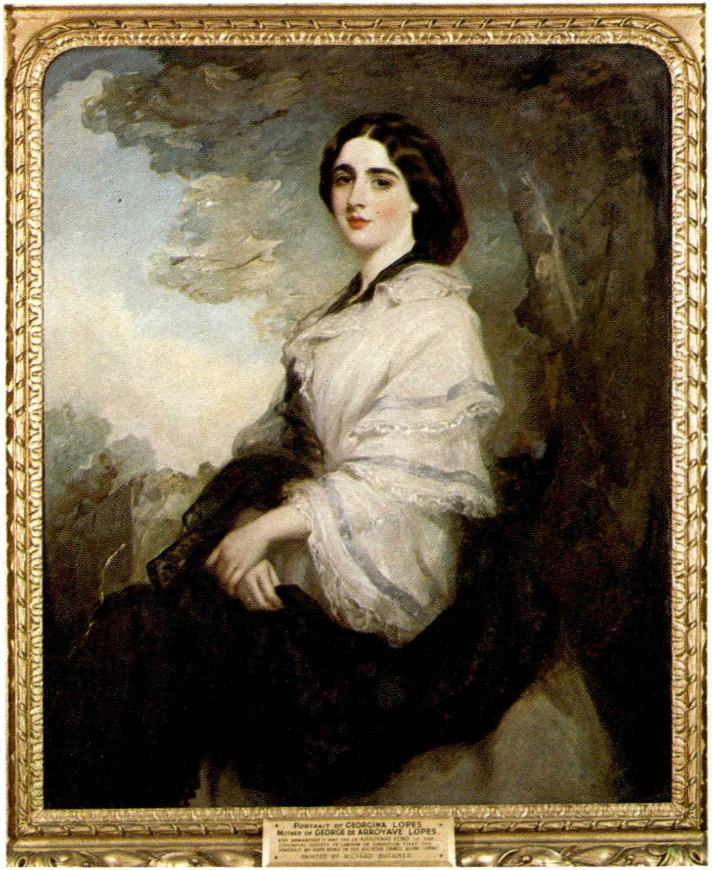
Portrait of GEORGINA LOPES
Mother of GEORGE DE ARROYAVE LOPES
PAINTED BY RICHARD BUCKNER

guished of the Society's numerous CMZS in India. Like Raffles himself, all three were connected with the East India Company.

FIG. 6. Edward Blyth. From Blyth (1875). Reproduced with permission from the Asiatic Society of Bengal.

Unlike Raffles however, all three began their Eastern activity in Bengal. Hodgson went to Nepal later, Tickell to Arakan. Blyth was the only one to see his stay out in Calcutta, on the sweltering plains. All three were remarkable zoologists.

Blyth

First Blyth, our lone professional (Fig. 6 shows him, quite the hairiest of our heroes). It was said of him:[e]

> "Ill paid, and subjected as he was to ceaseless humiliations, he felt that the position he held[f] gave him opportunities for that work which was his mission, such as no other then could, and he clung to it with a single hearted and unselfish constancy nothing short of heroic".

The ruling passion started early and so did his tribulations.

Blyth was born in London, the son of a Norfolk farmer who died when Blyth was ten, leaving the widow with four children to bring up. He did not like his schooling, he liked only natural history, and neglected the former for the latter. At 21 he invested his small patrimony in a druggist's business in Tooting. His sister relates:[g]

> "Never was any youth more industrious: up at three or four in the morning, reading, making notes, sketching bones, colouring maps, stuffing birds by the hundred, collecting butterflies and beetles—teaching himself German sufficiently to translate it readily, singing always merrily at intervals."

He even took rooms to be near the libraries and the British Museum. Naturally the shop went bust.

For years he scraped along, on the fringes of the Zoo: contributing to natural history journals and the PZS, translating Cuvier with added notes, presenting specimens to the Society's museum.

In 1841, at 31, he jumped at the chance to go to Calcutta as Curator of the Museum of the Asiatic Society of Bengal. Here—cataloguing, writing, examining, for a stint of over a score years of sultry climate—by 1863 he had worked himself out of health and had to come home. But all that time he initiated, and largely filled the pages of local zoological journals. He wrote and published books still standard on the local fauna. Home once more was no different—largely a new period of contriving, largely in poverty and the pages of the PZS.

Darwin respected him and often consulted him. Gould described him as "one of the first zoologists of the time and the founder of that science in India".

[e] Hume, Allan (1874). In memoriam in *Stray feathers* 2.

[f] In Calcutta.

[g] Blyth, E. (1875). *Catalogue of mammals and birds of Burma, with a memoir and portrait of the author.* Introduction: iv. (*J. Asiat. Soc. Bengal* Pt 2, extra no.). Hertford.

Professionals of today will be interested in this tribute to Blyth, written by one of his partisans after his retirement:[h]

> "Talking of Mr. Blyth's great services in the cause of science, during his twenty years *slavery* (for it was little else) in Calcutta, one cannot help feeling that there is a sort of retributive justice, even in this world.
>
> The pompous Jacks in office, who alternately neglected and attempted to patronise Mr Blyth (whose invaluable services they pretended to remunerate by a pittance less than they would have presumed to offer to their French cooks) are now either dead or dragging out their crotchety existence in some old-Indian-peopled watering-place; in either case unknown and unhonoured; while the name of Blyth, the naturalist, is known, and respected, by men of science throughout the civilised globe".

Retributive justice after retirement may sound all right, but a trade union in his active life-time would possibly have been better.

Zoologically speaking, the other two may have accomplished less but what they did achieve, as only a spare-time activity, was, if uncompleted, yet enormous. (And that it was left unfinished was not their doing.)

The two men present a striking contrast with one another, yet each in his own style had similar gifts.

Hodgson

Hodgson started his education at Haileybury, then directly a training-college for the East India Company, and at 18 went out to Calcutta to start at first as a clerk. Ill health obliged his transfer to the hills after one year. At the age of 20 he became Assistant to the British Resident in Nepal. At 33 he himself became the Resident (Fig. 7 shows him in his prime).

All but two of the 38 years 1820 to 1858 Hodgson spent in the Himalayan region making copious notes on the fauna, sending back data for the PZS and specimens for the collection; incidentally, a first endeavour to send back pheasants failed, when all died, some as soon as they left the cool of the hills, some waiting shipment from Calcutta, some on shipboard. A second attempt backed by Prince Albert, as the then ZSL President, with the help of the Governor General and a keeper (Mr Thomson, head keeper) sent

[h] Hume, Allan (1869). *My scrap book or rough notes on Indian oology and ornithology.* Footnote on p.182.

FIG. 7. Brian Houghton Hodgson at the age of 72. From Hunter (1896).

out from London, ended successfully and brought Hodgson and eight others the Society's Silver Medal.

At last Hodgson returned home to reach the ripe age of 94.

Tickell

Of Tickell (Fig. 8), on the other hand, no one knows when and where he was born, or anything about him until he joined the

FIG. 8. Samuel Richard Tickell. From Hume, A. O. (1889). *The nests and eggs of Indian birds,* 2nd edn. Oates, E. W. (Ed.). London: Porter.

Bengal Native Infantry in 1829. After active service, and promotion to Colonel, he took civilian employment with the East India Company in 1843 and four years later was transferred to Arakan.

Here, clearly, in the intervals of administrative duty and from his own account, he was a mighty hunter, sending notes and skins to Calcutta and publishing reports. He was friendly with Blyth, and planned collaboration with him on a faunistic work, but later the two quarrelled and Tickell accused Blyth of pinching bird skins he had sent him and naming and describing them himself. He retired in 1865 after 18 years in Arakan.

Was Tickell a peppery colonel in the mould of those castigated by Allan Hume (see above p. 37)? Certainly he was very much more and his fate grievous.

Back home Tickell settled first in France and then the Channel Islands, contributing copiously articles to the *Field* on game birds and wild fowl under one pseudonym, and on sport and other natural history matters under another. In 1870 exposure while fishing off Brittany affected his eyesight and gradually he became blind, five years later dying in Cheltenham after great suffering.

The extraordinary link between these two men, however, resides in the major written scientific work that each produced, and both of which remained in manuscript. Each of them deposited the manuscript in the Zoo library and each was awarded Honorary Fellowship in recognition of doing so. For almost a century these two vast works have lain disregarded in our library.

The work of Hodgson consists of seven enormous volumes of notes on the fauna of the Himalayas. Each volume measures 23 × 16in. and is 3in. thick. Five volumes are on birds, two on mammals. Every huge page is covered with minute, meticulous handwriting, difficult to read, interspersed—every square inch of space occupied at all angles—with accurate and beautifully composed and coloured paintings of the animals concerned and detailed comparative drawings of their anatomical parts (Plate 2).

Tickell's magnum opus is different but no less voluminous—14 manuscript volumes only less in size than Hodgson's (12 are 18 × 11·5 × 1·5 in. and two are quarto). But whereas Hodgson's are plain untouched notes, Tickell's are in the form of a written-up memoir he had hoped to see published, and its neglect must have been an added disappointment to his final years. Tickell, too, has copiously illustrated his own work, but whereas some drawings and paintings (e.g. Plate 3) show him no less capable as an anatomist and colourist than Hodgson, others reveal the difference in charac-

FIG. 9. Vignette at the end of a chapter in Tickell's *Ornithology.*

ter of the two men. He was ambitious also, to illustrate the animal in its setting (Plate 3) or to illuminate the menace (Fig. 9) or comedy (Fig. 10) of his narrative by little graphic comments (these latter in the style our forebears called "good enough for *Punch*").

At any rate here are rich quarries—comparable in their obscure lot, and even to a certain degree, in extent of national loss, to Turner's neglected bequest of water-colours—which no Indian natural history should ever ignore and which are only now beginning to be explored and indexed—still waiting to be fully and adequately mined.

Our last three instances carry us to the present century.

THE MACLEAY FAMILY

Alexander, b. 1767, d. 1848; sons William Sharp 1792–1865, Sir George 1809–1891 and James Robert 1811–1892; nephew Sir William 1820–1891.

The first, the remarkable Macleay family, provides, in only two generations, a bridge that spans almost the entire period.

FIG. 10. Another vignette in Tickell's *Ornithology*. After 15 days' march into the jungle the Shikaree breaks the news that the gunpowder has been left behind.

Alexander the patriarch (Fig. 11), head of the Ross-shire family, was born 14 years before Raffles.

FRS, secretary of the Linnean Society and an original FZS (the year the Zoological Society of London was founded) at the age of 58, he left for Australia to become Colonial Secretary of New South Wales, served in that post 12 years, stayed on, and six years later became Speaker of the New South Wales legislature. He had already served in youth as Provost of Wick and Deputy-Lieutenant of Caithness. Three sons and a nephew followed. The average of the ages they attained works out at over 77 and all five became CMZS.

Alexander published a letter in the PZS on the habits of the kiwi and sent to the Society's museum one of the earliest skins of the bird to be received in England. He was said to have had the finest collection of insects then in private hands, took it to Australia with

FIG. 11. Alexander McLeay. From a painting by Sir Thomas Lawrence in the possession of the Linnean Society. Reproduced by kind permission of the Linnean Society.

him, became first President of the Australian Museum and established a fine botanical garden.

The activities of the next generation included sheep farming, exploring, authoring scientific papers, a friendship with Cuvier and St. Hilaire and a role in the suppression of the slave trade in Africa and the West Indies. The eldest son enlarged his father's insect collection and cared for his father's garden. The nephew was also an entomologist and became first President of the Entomological Society of NSW (now the Linnean Society of NSW).

Is it fanciful to see in this one family the embodiment of the Society's role, from its beginning, in promoting and encouraging the spread of like activities among aficionados of zoology however far afield?

HUDSON

William Henry Hudson, b.1841, d.1922

To many, Hudson (Fig. 12) is known only as an ornament of English literature, who, born in Buenos Aires, younger son of an expatriate rancher, came to England at 33 and spent the latter part of his life, in some discomfort and poverty in Notting Hill, writing such classics as "*A naturalist in La Plata*", "*Birds and men*", "*Far away and long ago*" and eventually had a controversial sculpture by Epstein called "Rima" erected in Hyde Park to his memory.[i]

What many do not know is that he was, in fact and on record, a zoologist, indeed a CMZS, having at the age of 15 suffered a heart attack and become yet one more of those favoured by a physically precarious youth that gave him licence to indulge his love of nature, in this case running almost wild about the pampas, sending contributions to the PZS and skins to California and the Society. A friend of Sclater, by 1898 the success his hard-written recollections of scenes of childhood and South American nature had won, enabled him to rejoin the Society as FZS.

JOHNSTON

Sir Harry Johnston, b.1858, d.1927

Johnston (Fig. 13) brings the wheel full circle. Like Raffles he had an early interest in science. Like Raffles, he was an Empire trouble-shooter—though in Africa, not in South East Asia (perhaps, indeed, in date he was one of the last of the great empire-builders). Like Raffles, on retirement he played a big part in the history of the Zoological Society, for it was almost exactly halfway in its 150 years' existence—after 75 years, at the turn of the century—that he became Secretary of its Reorganization Committee, and as such ran the Society during the interregnum after Sclater finished and before Chalmers Mitchell was appointed.

[i] The name is that of a nature spirit mentioned in his writings. It was repeatedly daubed with paint by vandals during the outcry raised in the early twentieth century by Epstein's work.

FIG. 12. W. H. Hudson. From Roberts, M. (1924). *W. H. Hudson—a portrait.* London: Eveleigh Nash & Grayson.

FIG. 13. Sir Harry Johnston at St. John's Priory. From Johnston (1929).

Unlike Raffles, he began, as a boy, with the Zoo, whereas Raffles ended with its beginning.

At 14 he painted animals in the Zoo and dissected dead specimens there. From 17 to 19 he studied zoology at Regent's Park and anatomy at the Royal College of Surgeons. Followed travel and sketching in Majorca, Spain and France and—once again it occurs in the story—a sojourn abroad for health reasons, in this case in Tunis.

At 24 the ZSL recommended him to an expedition to south Angola, from where he went on alone into the Congo Basin. Surviving this, he undertook a scientific mission to Kilimanjaro for the Royal Society.

Apprenticeship over, so to speak, he felt strong enough for battle and joined the Foreign Office. From now on Africa claimed him thoroughly: Vice-Consul in Cameroon and the Niger Delta, Consul in Mozambique, Commissioner for south Central Africa, Consul-General in Tunisia, Special Commissioner in Uganda. Seventeen busy years suppressing slavery, establishing protectorates (e.g. in Nyasaland and what is now Zambia), reconciling conflicts. He left with a high reputation, not least, and maybe surprisingly—because to do so in Africa has always been difficult—in Africa itself. Meanwhile, a loss to zoology? But no. As with Raffles he always seemed able to maintain his interest; perhaps also his artistic skills. He is said to have saved from extinction the water-antelope known as Speke's tragelaph, he discovered the five-horned giraffe and he made the expedition on which the skin of the okapi was discovered and painted it alive (Plate 4a). For this and his "efforts to promote the advance of zoological discovery in the several posts he has occupied in various parts of Africa"—the Society's Gold Medal.

Perhaps, before this paper terminates, we should pause to wonder why no woman figures in our list of the contributory British abroad. The indubitable answer is: lack, not of capacity, but of opportunity. Any history of the second half of the Society's 150 years would be impossible without many exemplary women zoologists, jostling for inclusion. But in the social climate of the Society's formative years it took money, or authority, inherited or acquired, to indulge a passion as a scientist or traveller, and women seldom enjoyed either. If anyone points to exotic exceptions, such as Hester Stanhope or Mary Kingsley, it must be answered in rebuttal that the eccentric romanticism of the one, and the real botanical accomplishments of the other, were both more acceptable socially for ladies than zoology. The reproductive habits of animals are more unspeakably suggestive, the dissection or stuffing of animals more messy, than the same study of or close contact with plants, and so the former could hardly at that time have ranked as an approved occupation for females. The brine-tub circular of the Zoological Society's founding fathers was addressed exclusively to men.

Not to be exclusive now, therefore, we must slightly extend our mandate from application to the British abroad, only, so as to include a personality among the British *from* abroad. The Secretary in his introduction has already spoken of the second mystery portrait in the Council Chamber, that of Mrs Georgina Lopes, mother of the Society's unknown benefactor in 1930. We include it here (Plate 4b) as a reminder that no science, even zoology, can live from scientists alone, that contributions are welcome also, indeed essential, in another form, and that if past generosity has made them never lacking in a crisis, yet we still have many vacant walls in our large buildings; and that there can be no doubt the Treasurer will welcome similar commemorations on the same conditions, especially if they are as beautiful as the lady in question, now exchanging glances with our handsome Raffles.

References

Belcher, Sir E. (1855). *The last of the Arctic voyages*: being a narrative of the last expedition on HMS *Assistance*, under the command of Captain Sir Edward Belcher, CB, in search of Sir John Franklin, during the years 1852–53–54, with notes on the natural history by Sir John Richardson, Professor Owen, Thomas Bell, J. W. Salter and Lovell Reeve. London: Reeve.

Blyth, E. (1875). *Catalogue of the mammals and birds of Burma* . . . with a memoir [by A. Grote] and portrait of the author. *J. Asiat. Soc. Bengal*, pt 2, extra no., August.

Dictionary of National Biography [compact edition] 1975. London: Oxford University Press.

Hudson, W. H. (1918). *Far away and long ago: a history of my early life*. London: Dent.

Hunter, Sir W. W. (1896). *Life of Brian Houghton Hodgson, British Resident at the Court of Nepal.* London· Murray.

Johnston, Alex. (1929). *The life and letters of Sir Harry Johnston.* London: Cape.

Wurtzburg, C. E. (1954). *Raffles of the Eastern Isles.* London: Hodder and Stoughton.

Zoological Society of London. Annual reports, 1829 onwards.

Sundry contemporary journals.

Symp. zool. Soc. Lond. (1976) No. 40, 49–66.

THE ZOOLOGICAL SOCIETY AND NINETEENTH CENTURY COMPARATIVE ANATOMY

A. J. E. CAVE

The Zoological Society of London, Regent's Park, London, England

The modern zoologist, in an age of nuclear physics, molecular biology, and incredibly sophisticated exploratory apparatus, may well fail to appreciate the wholly different world of his nineteenth century predecessors, the then prevalent philosophical outlook and the factors influencing contemporary zoological research. Broadly speaking, whereas twentieth century biological research is fundamentally experimental in approach, its nineteenth century counterpart was essentially morphological, a difference which merits explanation and understanding.

The outlook of eighteenth century zoology had been revolutionized in differing fashion by both Linnaeus (1707–1778) and John Hunter (1728–1793), with permanent effect upon subsequent zoological enquiry. Linnaeus had introduced the new, now universal, system of animal classification which brought logical order into a previously unregulated subject. Henceforth animals were to be classified by assignment to precisely defined categories (species, genus, family, etc.), each determined by the possession of certain diagnostic anatomical characters. External characters alone soon proved too fallible a guide to taxonomic status and increasing attention had perforce to be paid to an animal's internal structure, in brief, to its total morphological constitution. Only in the light of such information could superficial resemblances of external form, occasioned by convergence or other phylogenetic factors, be detected and their irrelevance made manifest. Hence a great impetus to the anatomization of every available animal form and the rise of comparative anatomy as a principal zoological discipline.

John Hunter's zoological interest, epitomized by his unique Museum, was functional (physiological), not taxonomic. From his incomparable knowledge of animal morphology he was led to proclaim the doctrine that animal structure was but the tangible expression of animal function and the key therefore to its understanding. Hunter was primarily concerned to understand the mode of operation of the vital processes—locomotion, respiration, temperature regulation, reproduction and the rest—throughout the entire animal kingdom; he saw in the morphological machinery subserving these processes the surest guide to their manner of action and to appreciation of how some common functional end might be attained, in different animal forms, by different anatomical arrangements. Hunter therefore diligently pursued comparative anatomy to make manifest the operation of the various body systems and to demonstrate the complementary association of form and function. At the hands of his disciples, Hunter's doctrine was carried into nineteenth century zoology, the course and scope of which it virtually determined. Its influence is apparent in the enormous volume of comparative research undertaken during that century and in the resultant unprecedented augmentation of knowledge of animal organization.

Philosophical considerations apart, subsidiary factors influenced nineteenth century concentration upon the morphological approach in zoological research.

Firstly, the infant basic sciences were unable to provide the technology essential to the opening of alternative research fields, thus leaving investigation to be conducted by a relatively simple (even primitive) armamentarium, of which the light microscope was the apotheosis. Secondly the advent of Darwinism inevitably stimulated morphological enquiry if only in the search for data capable of interpretation as evidence supportive of the new evolutionary hypotheses. Thirdly, throughout the century rare and new forms of animal life, both vertebrate and invertebrate, reached this country in great profusion, thanks to Empire expansion and the overseas exploration of new territories. Such new forms required urgent identification and classification and on what other basis than that of morphology.

Eighteenth century British zoological science had remained largely the province of the amateur-patron, independent of its pursuit for livelihood. The most notable of these (John Hunter, Joseph Banks, James Edward Smith) had assembled or acquired invaluable research collections, had subsidized zoological explorations, had trained a rising generation of professional scientists and had stimulated public interest in, and respect for, biological science. They thus influenced profoundly the development of British zoology and bequeathed to their scientific heirs a seminal tradition.

By the 1820s, however, their day had passed and their collections had been transferred elsewhere—the Hunterian to the new Royal College of Surgeons, the Banksian to the British Museum and the Linnean to the Linnean Society of London. Source material for zoological research was thus largely restricted to these institutions, wherein a somewhat proprietorial custodianship did not always facilitate access thereto.

The increasingly numerous body of professional zoologists lacked any unifying organization, any common forum for their meeting and discussion and any dependable source of material assistance in their individual researches. To such men, and to those influential in public life, this disquieting situation demanded speedy resolution. Such was duly achieved by the establishment (1826) of the Zoological Society of London.

This new Society, at its first General Meeting (29th April 1826) under Raffles' chairmanship, announced its intention of providing for contemporary zoological needs by establishing "a collection of living animals, a museum of preserved animals with a collection of comparative anatomy, and a library devoted to the subject". And by its subsequent Charter (1829) the Zoological Society was formally committed to "the advancement of Zoology and Animal Physiology and the introduction of new and curious subjects of the Animal Kingdom".

Comparative anatomy (morphology) was thus to constitute the Society's fundamental scientific concern. New or rare animal forms were to be exhibited in its Menagerie and anatomized upon their death. All this is obtrusively apparent in the proposed establishment of "a collection of comparative anatomy" and in the Charter term "Animal Physiology". At that date, the word "physiology" had not acquired its modern experimental connotation but still retained its eighteenth century meaning of the morphological apparatus essential to the various body functions. In a non-experimental era, knowledge of such morphological apparatus was the principal key to understanding of animal function, and the body machinery itself was considered to be, if not explicative, at least indicative, of the operation of such function. Hence it was the prime business of the morphologist to "describe with accuracy and delineate with fidelity" the several parts, organs and body systems which collectively made up the entire animal. What human anatomy was to medicine morphology was to zoology—the inescapable basis of all understanding of the subject.

The goal of comparative anatomy was not the production of morphological memoirs as ends in themselves but the establishment of such memoirs to illustrate the manner in which the various vital functions discharged their respective office in different animal forms. (The memoirs themselves maintained the highest degree of

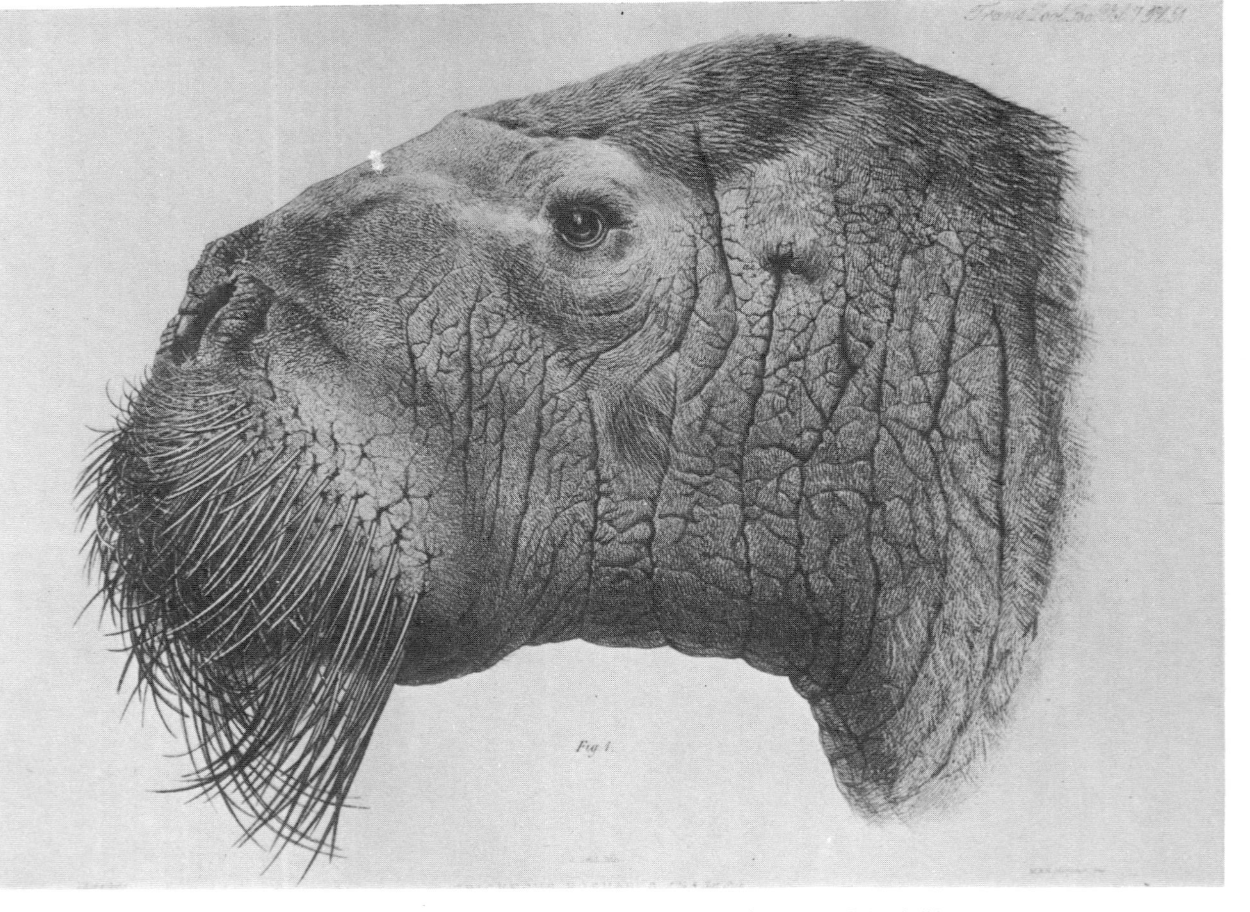

FIG. 1. Walrus, external characters. (From *Trans. zool. Soc. Lond.* **8**: Pl. 51.)

FIG. 2. Aye aye, external characters. (From *Trans. zool. Soc. Lond.* **5**: Pl. 14.)

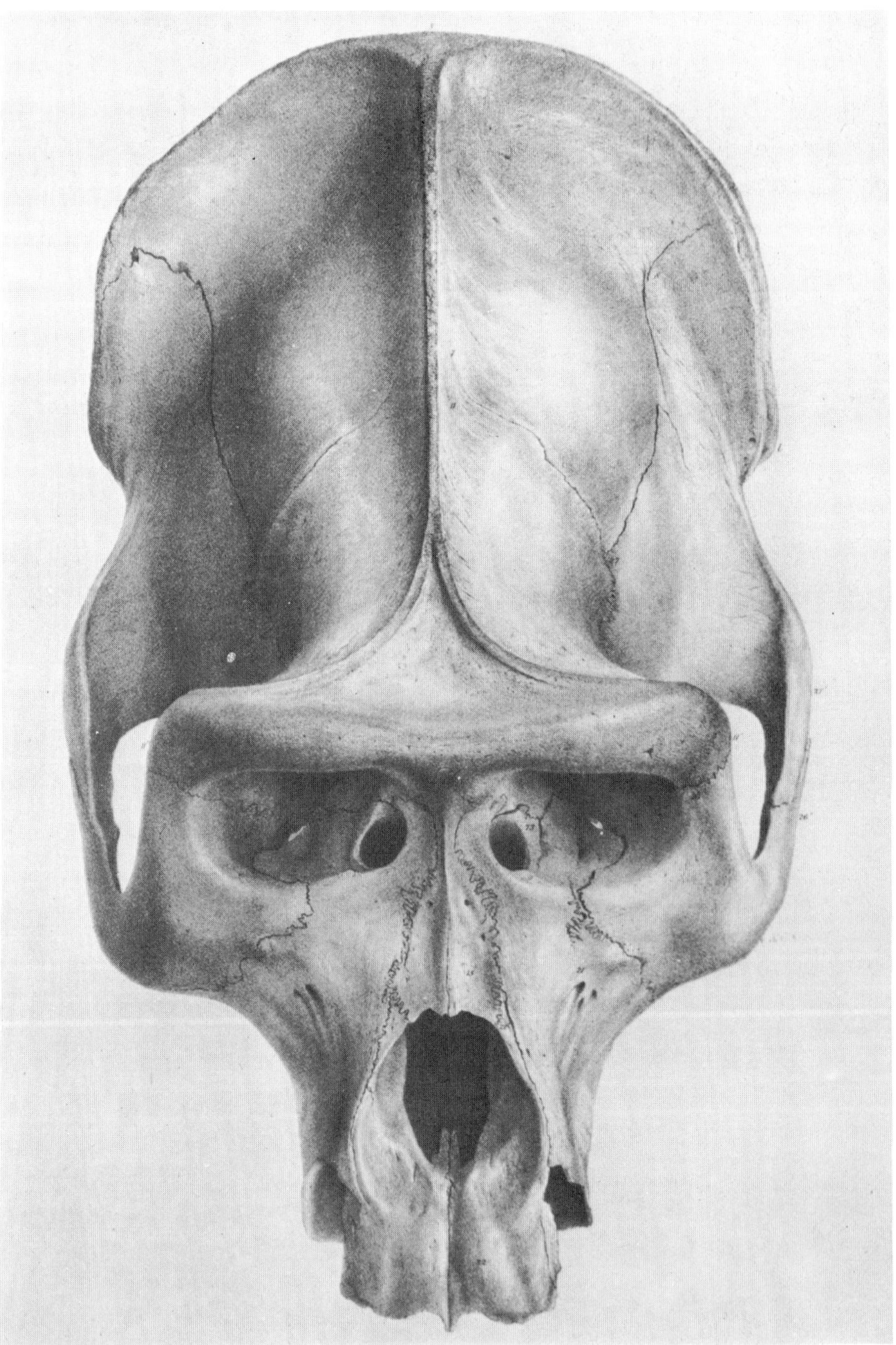

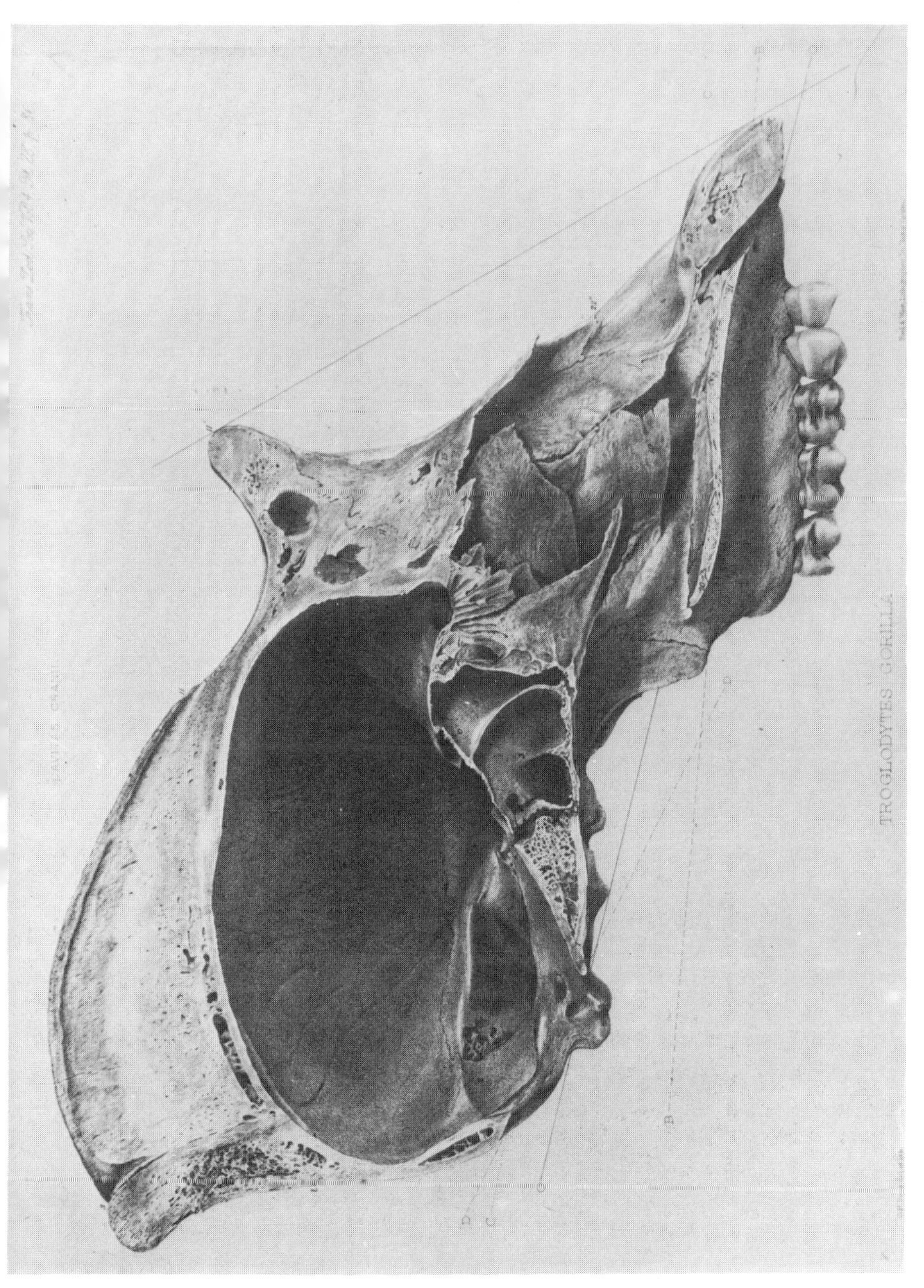

FIG. 3. (and opposite) Gorilla skull. (From *Trans. zool. Soc. Lond.* **4**: Pls 27, 28.)

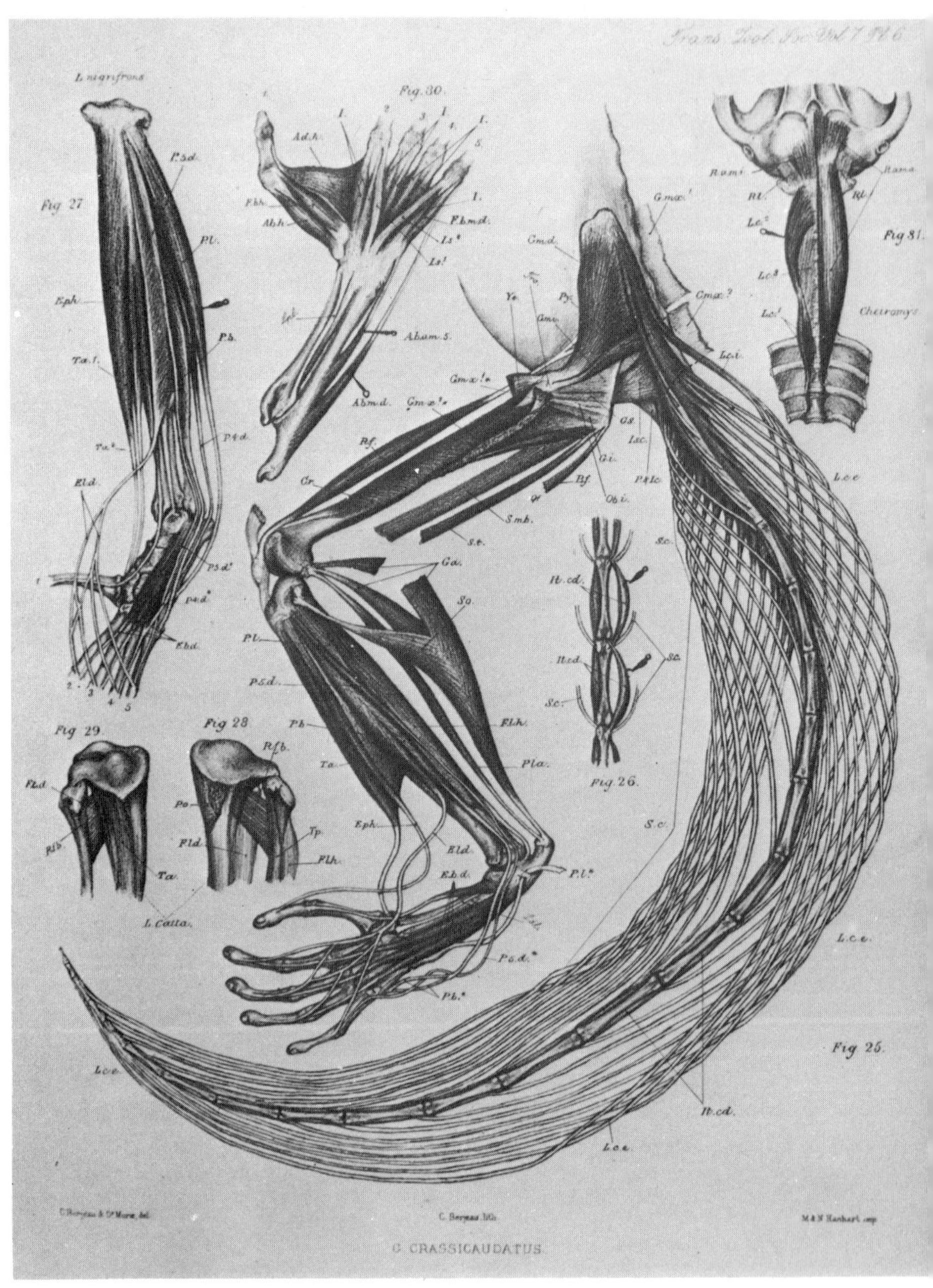

FIG. 4. Muscles of *Galago*. (From *Trans. zool. Soc. Lond.* **7**: Pl. 6.)

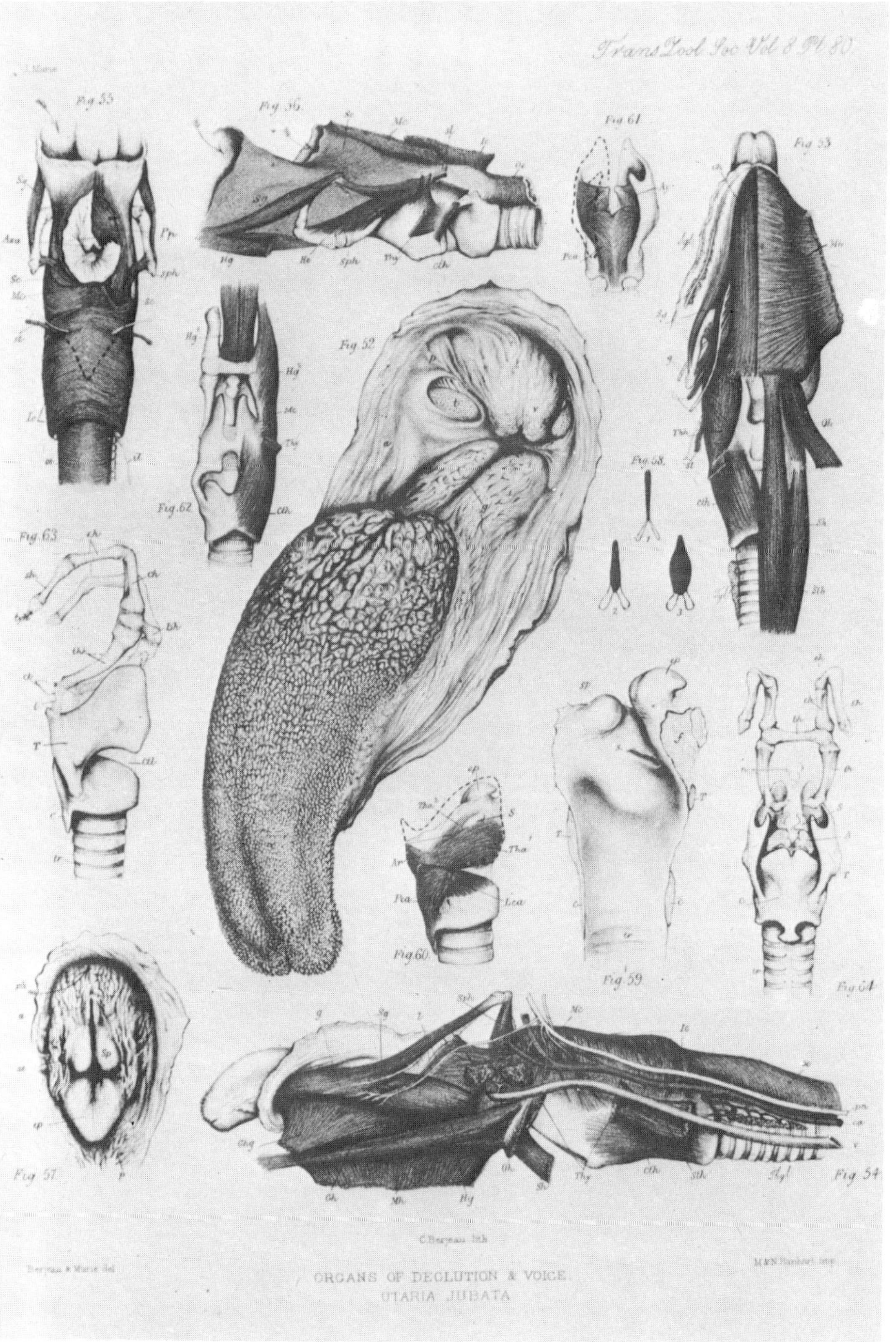

FIG. 5. Viscera of *Otaria*. (From *Trans. zool. Soc. Lond.* **8**: Pl. 80.)

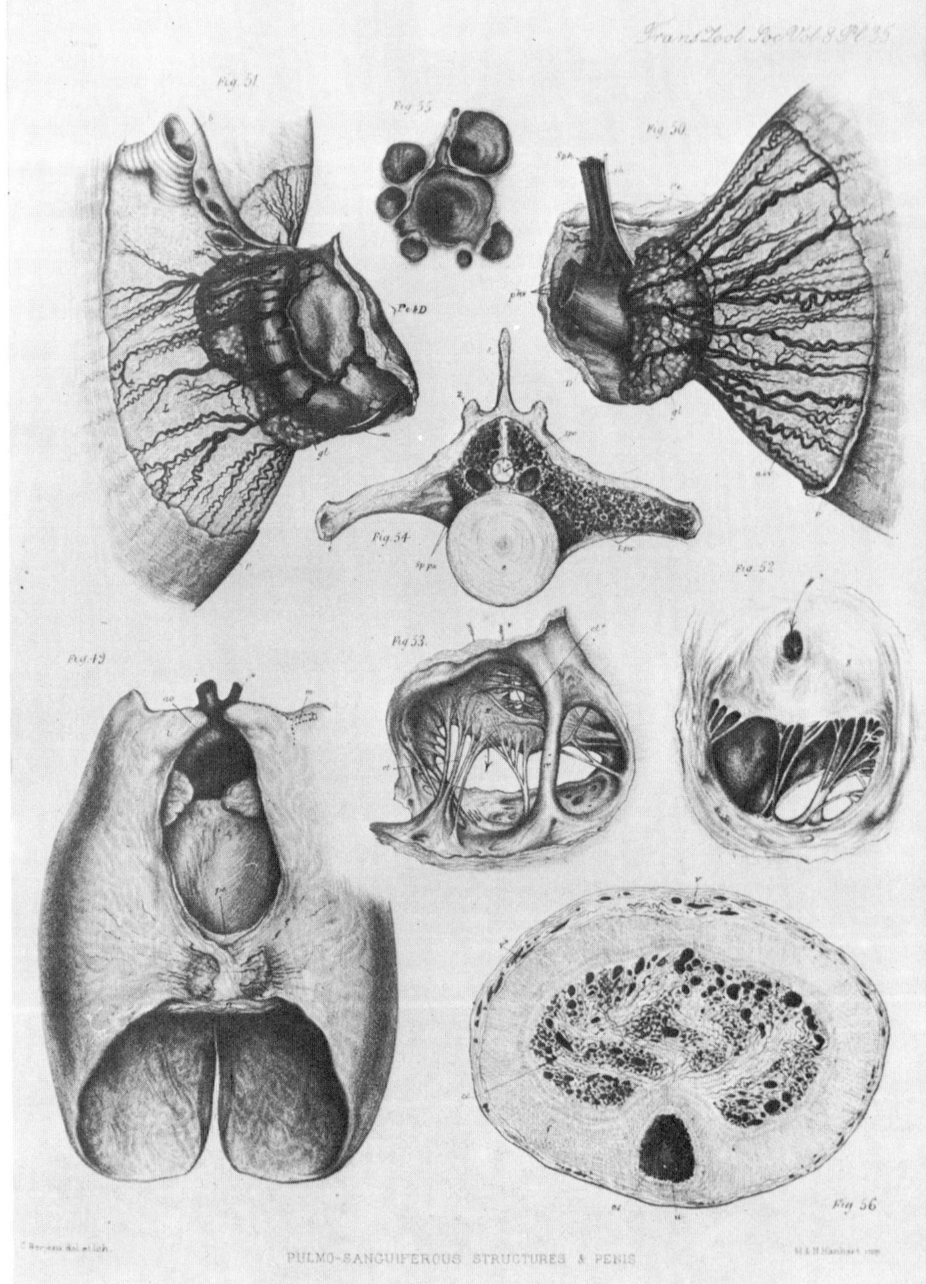

FIG. 6. Viscera of *Globiocephalus*. (From *Trans. zool. Soc. Lond.* **8**: Pl. 35.)

anatomical accuracy, matched by the quality of the illustrations, examples of which are indicated in Figs 1–6.)

The Zoological Society met its commitments by establishing (a) a menagerie (1828) of non-domesticated animals, (b) a museum (1826) of preserved specimens, (c) regular scientific meetings (1830), and also a library and prosectorial facilities. It duly erected the first-ever reptile house (1849), aquarium (1850) and living insect house (1881). Its menagerie duly exhibited animals new to science [e.g. Arabian oryx (1856), hairy fronted muntjac (1884), Shaw's monkey (1892)] or animals unseen alive in Europe either previously or since the Roman circus days [e.g. kiang (1831), echidna (1845), tree kangaroo (1848), hippopotamus (1850), shoebill (1860), aye aye (1862), black rhinoceros (1868), Javan rhinoceros (1874), hammerhead (1880), siamang (1898), gaur (1899)]. The menagerie supplied the Prosectorium and indirectly the Museum. This last, wherein John Gould (1804–1881) had worked, was at one time superior to the British Museum, which received it on disbandment (1852–1856) (Thomas, 1906:8), as its most historically important accession to date. The anatomical utilization of menagerie material produced an increasing output of memoirs requiring some appropriate medium of publication. Accordingly the Society instituted its own *Proceedings* (1830) and *Transactions* (1833) which achieved an immediate, world-wide prestige. In 1886 the Society resumed responsibility for the indispensable *Zoological Record.*

From the outset the Society's principal contribution to zoological science was effected by the individual labours of its Fellows, both amateurs and professionals. The Society never contemplated any coverage of the entire field of morphological investigation by its own officers. Yet the need for the appointment of some permanent official capable of prosecuting comparative anatomical research became gradually but persistently apparent following an initial period during which Richard Owen (1804–1892) had given honorary service in this capacity. In the Society's earlier years comparative material from its menagerie lay at Council's direct disposition, which, in effect, meant its reservation to Owen, the personification of comparative anatomy throughout the first half of the nineteenth century. Owen, whose name is writ so large in British zoology, was then Hunterian Professor and Assistant Conservator of Museum in the Royal College of Surgeons, outstandingly able and industrious and *persona grata* with Council as with the Court and the eminent in public life. Already in 1826 he was anatomizing Society material and thereafter attended the Gardens assiduously, wherein by his personal prosectorial labours he acquired that impressive range and depth of morphological knowledge which is exemplified by his classic *Comparative anatomy and physiology of vertebrates* (1866–1868). So greatly was Owen's utilization of Prosectorium material appreciated that Council ruled (3rd June 1840) that Professor Owen "be allowed to dissect whenever and whatever he liked, when deaths occurred in the Gardens, and to have precedence over every one else".

In 1856, however, Owen transferred from the College of Surgeons to the British Museum (as Superintendent of the Natural History departments) whereafter official duties and an increasing devotion to palaeontological study saw the lessening and ultimate cessation of his personal activity in the Society's Prosectorium. Followed a considerable but unspecified period during which Prosectorium research material lay neglected, with loss to the Society of desirable anatomical memoirs and thereby some implicit threat to its prestige.

Thomas Henry Huxley (1825–1875) notified his fellow Council members of this deplorable wastage of material and Council promptly commissioned Huxley, W. H. Flower, J. E. Gray, George Busk and the Secretary (P. L. Sclater), to advise collectively concerning "the mode by which animals dying in the Gardens might best be disposed of with the most benefit to Zoological Science". Council accepted the advice tendered and promptly instituted (1865) the proposed Zootomical Committee, comprising the Secretary and three members of Council. This permanent committee was to regulate the disposal of Prosectorium material and to supervize the work in the Prosectorium of an officer new to the Society's establishment, the Prosector, who should be responsible both for pathological routine and anatomical research. Thus the Society proposed to make its own direct, or domestic, contribution to comparative anatomy, in complementation and encouragement of that made by the generality of Fellows, who might henceforth rely upon the Prosectorium for much of their essential research material. The Prosector was a salaried, full-time officer of the Society available for consultation and advice no less than for material aid, and the Prosectorship was to play a long and honourable role in the advancement of nineteenth century comparative anatomy.

The Zootomical Committee continued from 1865 until 1903 when, among other administrative changes, it metamorphosed into the Prosectorial Committee (1903–1960) with a membership no longer restricted to Council members but including also other Fellows distinguished in anatomy, physiology or pathology.

The Minutes of the Zootomical Committee reveal recurrently the names of W. H. Flower, Alfred Newton, George Busk, St. G. Mivart, Osbert Salvin and Albert Gunther (some of whom appear in Fig. 7): those of the Prosectorial Committee (1903–1920) the successive names of Oldfield Thomas, C. S. Tomes, W. R. Ogilvie-Grant, R. H. Burne, Havelock Charles, John Bland Sutton, S. F. Harmer, E. W. MacBride, J. P. Hill, G. Elliot Smith, F. Wood Jones and R. I. Pocock. To anybody at all familiar with the recent history of British zoology such names are sufficient evidence of the Society's sustained concern for the advancement of morphological study in all branches of zoological science.

The first official Prosector appointed (1865) was the able, widely travelled and experienced pathologist, James Murie, MD (1832–1925), who held office until 1870. This lustrum was deprived of its full potentiality by an unfortunate antagonism between Secretary and Prosector. The former insisted upon the production of anatomical memoirs for publication and was unsympathetic towards the latter's pathological burden. Menagerie mortality was unduly high (over 4000 post-mortems in five years) and Murie attributed this—doubtless justly, but undiplomatically—to official ignorance and bad management. An invidious situation was resolved by Murie's resignation (1870) on health grounds, by which date he had accomplished, and meticulously documented, an incredible amount of pathological routine and had enriched the *Transactions* with classic anatomical memoirs (Murie, 1871a,b, 1874a,b,c) on the manatee, pilot whale, pinnipedes, kagu and other animals.

The second Prosector appointed (1871) was Alfred Henry Garrod, BA, LSA (1846–1879), a brilliant Cambridge graduate who rapidly achieved recognition and distinction. Garrod, who had qualified medically, brought with him a personally trained pathology attendant (William Ockenden) from King's College Hospital to whom he delegated much of the pathological routine, thus freeing himself for undivided attention to comparative anatomy and to the production of memoirs satisfying Secretarial demands. His untimely death was a grievous loss to the Society and to British zoology itself, and when it befell over 70 memoirs stood to his credit. The interim had seen his election to the Professorship of Comparative Anatomy in King's College London (1874), to the

FIG. 7. A sketch which appeared in *Punch* in 1885 showing a meeting of the Zoological Society at Hanover Square. After a drawing by Harry Furniss.

Fullerian Professorship of Physiology in the Royal Institution (1875) and to the Fellowship of the Royal Society (1876).

A third Prosector was appointed in 1879. This was William Alexander Forbes, BA (1855–1882), another convert from medicine to zoology, Fellow of St John's College, Cambridge, and a noted scientific ornithologist. He contributed to avian and mammalian anatomy, but spent much time in collecting trips abroad—to the Americas (1881) and to West Africa (1882)—and during this last he died on the Upper Niger River.

In 1884 the fourth Prosector was appointed—Frank Evers Beddard, MA (1858–1925), an Harrovian and Oxford honours graduate in Natural Science, already distinguished for his work upon the isopod crustaceans from the *Challenger* Expedition. Beddard held office for 31 years (1884–1915) and in character, erudition and catholicity of zoological interest personified the ideal of the late Victorian zoologist. His researches embraced earthworms (for work on which he was elected FRS), parasites, crustaceans, avian myology, reptilian morphology, mammalian splanchnology and the cerebral vasculature of vertebrates. In 1902 appeared his popular monograph *Mammalia* in the authoritative *Cambridge Natural History* by which date he could claim authorship of nine books and over 230 memoirs.

At his retirement (1915) the prevailing conditions of World War I precluded immediate appointment of any successor: Prosectorium business lay first (1915–1919) in the honorary charge of Professor F. Wood Jones (1879–1954), later (1919–1921) under the Directorship of Professor R. T. Leiper (1881–1960).

Meantime Charles F. Sonntag, MD (1888–1925), of the Anatomy Department of University College London, was appointed (1919) as "Anatomist to the Society" and thus served as Prosector until his early death in 1925. The benefits of the liaison thus established between the Society and a progressive University department so impressed Council that the Prosectorship was, technically, abolished and replaced by a Research Fellowship. This Fellowship, tenable for three to five years, was intended to be held concurrently with a University teaching post and was designed to encourage young men of promise to embark upon a career in some department of zoology. This Fellowship was held successively by John Beattie, MD (1926–1927), the present Lord Zuckerman (1928–1932) and Dr Burgess Barnett (1933–1937). Thereafter it lapsed, as the Society faced the unpredictable hazards of World War II. After that War, the last Prosector was appointed (1950) in the person of William Charles Osman Hill, MD (1901–1975), who served until 1962.

Under whatever title, the twentieth century Prosectors maintained the tradition of their predecessors, pursuing personal research in comparative anatomy, assisting and advising visitor-workers in the Prosectorium and ensuring the provision of requested material to accredited institutions and individuals.

The post-war outlook in zoology differed radically from what for so long had been regarded as the traditional. Revolutionary developments in the basic physical sciences had produced an exploratory armamentarium of enormous (and previously unimaginable) sophistication, making possible access to zoological fields hitherto unexplored because technically unexplorable. There followed, understandably, a general desertion of the study of gross structure for that of the cellular and the intracellular and an eagerness to explore the operation of vital processes at the molecular level. In so changed a climate of interest the demand for Prosectorium research material virtually ceased and in 1960 the Prosectorial Committee came to an end.

So long as morphological investigation had remained the primary approach in zoological enquiry, so long had the Zoological Society supported and encouraged such investigation with all the means at its command—had been culpable indeed not to have done so.

From its beginning it had introduced "new and curious subjects of the Animal Kingdom" into its Menagerie, had provided for their efficient anatomization after death, had employed an official Prosector to conduct significant anatomical investigations, had supplied regularly otherwise unobtainable research material to a host of individuals and institutions and had provided for the publication of anatomical memoirs, whether by Fellows or others, in its *Proceedings* and *Transactions*. The pages of these publications are eloquent testimony to the Society's achievements, direct or indirect, in the nineteenth century development of comparative anatomy. Those pages, incidentally, contain certain memoirs representing all that is presently known concerning the organization of numerous animal forms, information which, but for the Society's multifarious sponsorship, might well be wanting today. With the currently continuous and increasingly rapid disappearance of so many animal forms this information increases in value and importance.

From the Society's inception also, its Scientific Meetings were the acknowledged forum wherein zoologists could meet for the presentation and discussion of topics of mutual interest: they functioned, in effect, as the Parliament of British zoology. Its extramural relationships, largely unofficial and based upon its Fellowship alone, proved of immense value to those working in museums, university departments, medical schools and other scientific institutions, the nineteenth century history of which is inseparable from that of the Society itself.

It is no exaggeration to say that, through the medium of its Fellowship and by its consistent scientific policy, the Zoological Society of London played a unique and historically important role in the fosterage of nineteenth century zoology. Fidelity to its Charter obligation to "the advancement of Zoology and Animal Physiology" found expression in an overall enlargement of knowledge of vertebrate and invertebrate animal structure unmatched qualitatively and quantitatively during any corresponding era in the history of zoological science. In the absence of the Society's inspiring influence and without its continuous encouragement and assistance of innumerable workers, nineteenth century zoology might well have followed some other, but less distinguished, course.

In this present, late twentieth century, day of concentration of zoological interest upon the operation of biological phenomena at the nucleo-molecular level, the grosser morphological studies of the nineteenth century may, to some, appear to lack real significance and the Society's preoccupation therewith to have been a somewhat unfortunate irrelevance. But any such view is naïvely superficial, at once a misunderstanding of the true goal of zoology and a mis-reading of its history.

For zoological science is committed to securing the maximal knowledge of every animal form as a living entity and essential to such business is knowledge of the total morphological constitution of each. Assessment of this total morphological constitution is achieved by anatomical investigation, so that comparative anatomy remains fundamental in every branch of zoology, and the anatomical approach as essential as any other. Upon the sound foundation of morphology depends all accurate taxonomy, all interpretation of the fossil record, all clue to animal ecology and behaviour and all reliable guidance in the solution of problems arising in the consideration of phylogenetic and evolutionary history.

Whereas in the past anatomical study was the principal, or only, means of exploration available to the zoologist, today it is but one of several approaches to understanding of the animal world, all mutually complementary. It preserves its utility—together with its venerable tradition—and integrates harmoniously with the newer, more sophisticated, methods of zoological investigation.

References

Beddard, F. E. (1902). *The Cambridge Natural History*. **10**. *Mammals*. London: Macmillan.

Murie, J. (1871a). Researches upon the anatomy of the Pinnipedia. Part 1. On the Walrus. (*Trichechus rosmarus* Linn.). *Trans. zool. Soc. Lond.* **7**: 411–464.

Murie, J. (1871b). On the dermal and visceral structures of the kagu, sun-bittern and boatbill. *Trans. zool. Soc. Lond.* **7**: 465–492.

Murie, J. (1874a). On the form and structure of the Manatee (*Manatus americanus*). *Trans. zool. Soc. Lond.* **8**: 127–203.

Murie, J. (1874b). On the organisation of the Caaing whale, *Globiocephalus melas*. *Trans. zool. Soc. Lond.* **8**: 235–299.

Murie, J. (1874c). Researches upon the anatomy of the Pinnipedia. Part 3. Descriptive anatomy of the sea-lion *Otaria jubata*. *Trans. zool. Soc. Lond.* **8**: 501–582.

Owen, R. (1866–1868). *Comparative anatomy and physiology of vertebrates*. 3 vols. London: Longmans Green.

Thomas, O. (1906). *History of the collections contained in the Natural History Department of the British Museum* **2** (Mammals). London: Trustees British Museum.

Symp. zool. Soc. Lond. (1976) No. 40, 67–83.

EARLY INVERTEBRATE ZOOLOGY: MEN AND THEIR ANIMALS

J. E. SMITH

Marine Biological Association of the United Kingdom, Plymouth, England

SYNOPSIS

The foundation of the Zoological Society in 1826 and the decision four years later to hold fortnightly meetings to receive and prepare reports on all matters connected with zoology proved to have a powerful and beneficial influence on the progress of invertebrate zoology both in Britain and overseas.

The zoologists who regularly supported the scientific meetings over a number of years were almost all, whether amateurs or professionals, good field naturalists. Many were actively engaged in research, and some in the preparation of monographs on the taxonomy and natural history of animals, including a number of the invertebrate groups. Many were influential in public affairs, while their scientific interests seem to have brought them in frequent communication with the leading amateur zoologists in various parts of Britain, with whom they exchanged information and specimens in the creation over the country of a web of active and close working relationships.

This essay reviews the contributions of the London and provincial zoologists to invertebrate zoology over the period from about 1830 to 1860, discusses some of the major scientific communications and discussions, and examines through selected examples the role of the Society in the early development of invertebrate zoology in Britain and overseas.

INTRODUCTION

The historian G. M. Young, in the preface to his essay *Victorian England: Portrait of an age, 1831 to 1865* (Young, 1964), describes how, in thinking what he should say about the procession of events between the passing of the Reform Bill and the death of Palmerston, and how he should say it, found himself asking the question: what is history about? And the conclusion he reached was that the central theme of history is not so much what happened, but what people felt about events when they were happening. What was the conversation of people who counted? Who were they? What were the assumptions behind their talk? And what became of it all?

The period of which G. M. Young speaks coincides almost exactly with the first 40 years of the history of the Zoological Society and, for our present purposes, may be taken to begin when, as recorded in the Minutes of Council of July 21st 1830,

> "On consideration of the advantages likely to accrue to the Society by cultivating an extensive correspondence on subjects of Natural History, it was resolved that a Committee be appointed, to be entitled 'The Committee of Science and Correspondence', for the purpose of

> suggesting and discussing questions and experiments in animal physiology, of exchanging communications with the Corresponding Members of the Society, of promoting the importation of rare and useful animals, and of receiving and preparing reports on matters connected with Zoology".

Whilst it is evident that not all these purposes were framed with invertebrate animals primarily in mind, the opportunities the meetings offered for the reporting of new observations, for the exchange of ideas, the exhibiting of specimens and, perhaps most important of all, of getting in touch with other people of like interests, evoked a strong response from invertebrate zoologists. Who then were the people who came together to talk about invertebrate animals or wrote to the Society about them; what did they talk and write about; and what was the nature and range of their conversation?

THE LONDON ZOOLOGISTS: SCIENTIFIC MEETINGS

The printed summaries of the proceedings of the scientific meetings of the Committee of Science and Correspondence, and from 1833 of the general membership of the Society, record the names of the chairmen at the fortnightly meetings and give some hints of the more regular attenders. Amongst those most frequently in the chair at the early meetings were William Yarrell, John Edward Gray and Richard Owen; and Thomas Bell seems to have been one of the more active founder members (Figs 1–4). Each was to remain influential within and outside the Society for many years and there is clear evidence that these distinguished members of the newly-founded Society were able, as indeed were many of their colleagues, powerfully to influence and to benefit the progress of invertebrate zoology through their high scientific standing and their involvement in public affairs.

This is well exemplified, and in a somewhat unexpected way, in the case of Yarrell who was in his mid-forties when he joined the Society as a founder member. A well-to-do business man who owned a newspaper agency, he was born in London and lived throughout his life within the parish of St. James's. Yarrell published numerous short papers on the anatomy, systematics and natural history of fishes and birds before completing the notable and well known *A history of British fishes* (1833) and a little later, *A history of British birds* (1843). Through his business connections, Yarrell had become friendly with the publisher John van Voorst whom he persuaded in 1833 to accept the *British fishes* as the first of the magnificent series of *Histories* familiar to this day to all zoologists. Nearly all of the subsequent *Histories* deal with invertebrate animals. They begin with George Johnston's *British zoophytes*, published in 1838, and later volumes include Forbes & Hanley, *British Mollusca* (1848–1853); Forbes, *British starfishes* (1841); Bell, *British stalk-eyed Crustacea* (1853); Gosse, *British sea-anemones* (1860); Jeffreys, *British conchology* (1863–1869); Bate & Westwood, *British sessile-eyed Crustacea* (1868); and Hincks, *British hydroid zoophytes* (1868) and *British marine Polyzoa* (1880). In an age when

FIG. 1. William Yarrell.

FIG. 2. John Edward Gray. Reproduced by kind permission of the Linnean Society.

men had come increasingly to appreciate the nature, meaning and implications of scientific enquiry, and to take pleasure in the search for a greater knowledge of the natural world, the immense surge of interest in natural history which generated the scholarly and attractively presented descriptions of animals in the *Histories* was reflected in their immediate appeal to specialists and well informed general readers alike. A clear indication of the strength of this response is to be found in the records of the membership of the Ray Society, founded in 1844 for the purpose of publishing books on natural history for circulation to its members. The membership in

FIG. 3. Richard Owen as a young man. From: Owen, R. (1894). *The life of Richard Owen*. London: John Murray.

1845 was 650; by 1847 it had grown to 868. Thereafter the numbers declined slowly but progressively to some 500 in the 1860s and, though there was a later recovery, the numbers never again attained the strength of the early years (Curle, 1954).

INVERTEBRATE ZOOLOGY IN THE PROVINCES

The authors of the van Voorst *Histories*, many of whom were closely associated with the Zoological Society and its early scientific meetings, generously acknowledged the help they received from

FIG. 4. Thomas Bell. Reproduced by kind permission of the Linnean Society from Gage, A.T. (1938). *A history of the Linnean Society of London*. London: Linnean Society.

their fellow naturalists throughout Britain, and the copious biographical notes which accompany the acknowledgments present a vivid picture of the intricate web of their communication. It is of interest to glimpse this first through Yarrell's *British fishes* and to explore some parts of the web through other of the early *Histories*.

Yarrell acknowledges the gift of specimens and notes on the natural history of some marine fishes from Dr Jonathan Couch, the formidable but kindly physician of Polperro, himself a distinguished ichthyologist; from Dr George Johnston of Berwick, also in medical practice, three times mayor of Berwick, the mainstay of the Berwick Natural History Society and founder of the Ray Society who, as the minutes of that Society record, "... though living at a distance from London, the seat of the meetings of the Society, to the last took an active interest in all the publications and proceedings of the Society and acted in the capacity of one of its Secretaries" (Curle, 1954); and from Mr W. Thompson, Vice-President of the Belfast Natural History Society. Dr Johnston, two years later, in the preface to his *British zoophytes* in turn thanks both Couch and Thompson for specimens; warmly acknowledges the help given by Dr J. E. Gray of the British Museum and, among other names, refers to Mr J. V. Thompson, Inspector of Hospitals and resident in Cork, Mr Bean of Scarborough and Mr C. W. Peach in the coastguard service at Fowey. And finally, in continuing the sequence of the *Histories* but one stage further, Thomas Bell, now Professor of Zoology in King's College, London, Secretary of the Royal Society and President of the Linnean Society, dedicates his *Stalk-eyed Crustacea* to Professor Richard Owen "the faithful and unchanged friend of many years" and, casting back over the years, recalls the help given to him by Dr W. E. Leach, a Plymothian by birth, the most eminent carcinologist of his day and former Keeper of Zoology at the British Museum; by Mr Spence Bate (of Bate & Westwood's *History*), a dentist in practice in Plymouth; by Mr J. V. Thompson (who had also aided both Yarrell and Johnston); by Mr Richard Quiller-Couch, son of Jonathan Couch and a surgeon in Penzance; and makes reference to the "Ill-fated Mr Cranch" who had died when on an expedition in the Congo and who was born near Kingsbridge, Devon. Cranch, who was an avid marine zoologist, like Hugh Cuming the great collector and creator of the famous cabinet so often referred to in the meetings of the Zoological Society, had his love for natural history kindled by the deservedly respected Colonel George Montague, an ornithologist and the pioneer of marine biology in Britain, who lived at Knowle House, Kingsbridge and who took an interest in the two young men (Dance, 1966; Monod, 1970).

The naturalists in all parts of the country interested in invertebrate animals, discovering new species, writing of their observations and telling their friends about them were—with the exception of the few who, like Owen, Forbes, Bell and Gray, held full-time posts in universities or in institutions such as the British Museum—professional or business men in practice or retirement. There were also among them many men of modest means and of more limited formal education who were fine field naturalists and on whom those better qualified to record and to communicate new discoveries were often dependent for their material and first-hand field observations. Jonathan Couch speaks of the skill with which his friend William Loughran, employed in the coastguard service at Polperro, dissected fish and prepared their skeletons (Bertha Couch, 1891), and it is clear from their published acknowledgments that Loughran assisted numerous naturalists, for example Spence Bate and J. O. Westwood when they were preparing their monograph on the sessile-eyed Crustacea, by loans from his collections and by providing notes on the habits of the living animals. Among the better known amateur zoologists of his day was Charles William Peach who, like Loughran, was employed in the coastguard service. Dr George Johnston, who knew him well, spoke of Peach's interest in natural history having been awakened by the sight of a fine specimen of the hydroid *Antennularia antennina* on the chimney-piece of the inn at Weybourn, in Norfolk, where Peach had his first job (Johnston, 1838). He was subsequently moved to Cornwall, first to Gorran Haven and then to Fowey where he became associated with the Couches, father and son. Dr Richard Q. Couch was particularly warm in his commendation of Peach for the help he gave him during the writing of *A Cornish fauna. Part III The zoophytes and calcareous corallines* published in 1844.

There can be no doubt that the naturalists who were living in London and meeting regularly at the scientific meetings of the Society were as well aware, through the visits and correspondence of their colleagues in the country, of what men like Loughran and Peach were doing as they were of their colleagues' often more sophisticated research. The wealth of their recorded acknowledgments is in itself evidence of this; and, in the world of everyday affairs, it is notable that when Loughran was transferred from the coast-guard at Polperro to Portloe, Jonathan Couch, who clearly missed his help, arranged, through the representation of Dr Gray in London, to have him moved back to Polperro. And, when Jonathan Couch notes in his journal that on June 20th 1848,

> "I received a visit from Mr Alfred Tennyson, the Poet, it is not without interest that he came to Polperro in a Boat, with Mr Peach and others; and after viewing our Scenery in all directions and taking Tea at our house, they all rowed back to Fowey in the evening" (Bertha Couch, 1891).

Zoology is not only about animals. It has and always will depend as much upon the ideas which zoologists have about them, and on their exchanges and discussions one with another. The Zoological

Society, through its membership and particularly through the participants in its scientific meetings, clearly played an important part in the development of these relationships and not least in its early years when the interest in and knowledge of invertebrate animals was rapidly expanding.

RESEARCH COMMUNICATIONS: TWO CONTROVERSIES

The first occasion on which substantial reference was made to invertebrate animals at the scientific meetings of the Society was the meeting of the Committee of Science and Correspondence on December 28th 1830; and it would have been difficult to have found a more exciting and potentially controversial communication than was read to the meeting. The reverberatory discussions which lasted many years involved contributors almost all of whom have already been mentioned by name. The record of the meeting runs:

> "W. Yarrell, Esq., in the Chair . . . A letter was read, addressed to the Secretary of the Society by J. V. Thompson, Esq., dated 'Cork, Dec. 16, 1830'. In it Mr Thompson urges, in support of the universality of a metamorphosis among the *Crustacea*, that he has ascertained the newly hatched animal to be a Zoea in eight genera of the *Brachyura*, viz. *Cancer*, *Carcinus*, *Portunus*, *Eriphia*, *Gecarcinus*, *Thelphusa*?, *Pinnotheres*, and *Inachus*; and in seven Macrourous genera, viz. *Pagurus*, *Porcellana*, *Galathea*, *Crangon*, *Palaemon*, *Homarus*, and *Astacus*. 'These embrace all our most familiar native genera of the Decapoda'."

The information reported by Thompson was founded mainly, but not entirely, on observations he himself had made following his notice of a brief report by the Dutch naturalist Slabber, made as early as 1778, in which he described and figured small creatures subsequently named *Zooea taurus*. Slabber claimed to have seen these creatures metamorphose into more elongate individuals of which he gave drawings but which he did not name. In the spring of 1822 Thompson found in Cove (Cork) harbour large numbers of larvae of a zooea type which he succeeded in rearing to small crab-like creatures with an extended abdomen and which resembled creatures which the distinguished carcinologist Dr Leach had described as of the genus *Megalopa*. In the following year (1823) Thompson succeeded in hatching the eggs of the edible crab, *Cancer pagurus*, into zooeas and he later reared the eggs of the common shore crab, *Carcinus maenas*, and of crabs of other genera

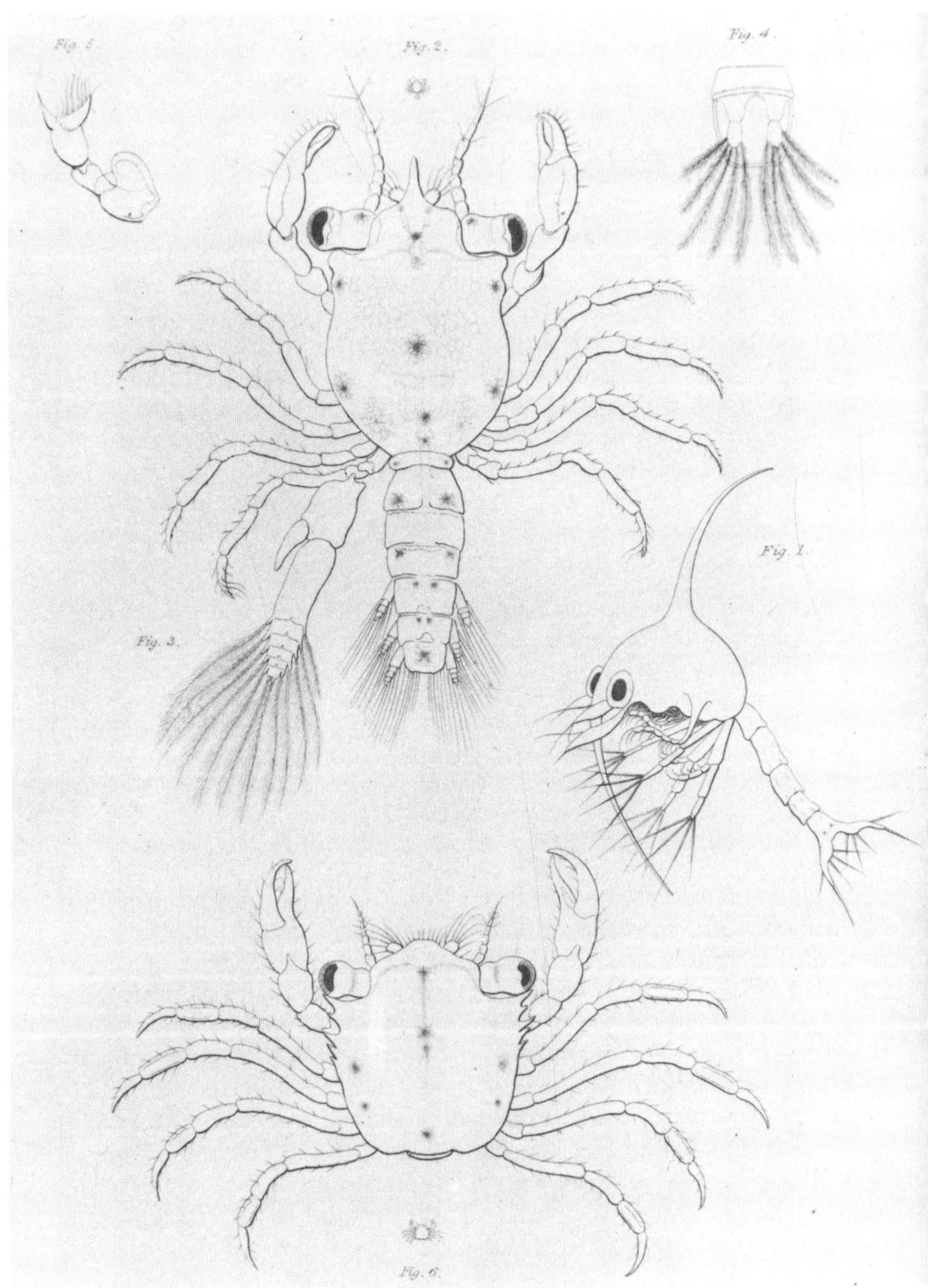

FIG. 5. J. V. Thompson's developmental series of crustacean larvae. Reproduced with permission from Thompson (1835).

through zooea and megalopa stages to juvenile crabs (Fig. 5). Thus the creatures previously described as animals of different genera were shown to be in reality different stages in the life history of the crab (Bell, 1853).

Thompson's communication to the Zoological Society was not received without some disbelief. Dr J. E. Gray, then Assistant Keeper at the British Museum, later (1833) described to the Society how, when on holiday in Devon, he had reared the eggs of the barnacle *Balanus* and found the emergent larvae to resemble miniature adults; and two years later Mr J. O. Westwood (of Bate & Westwood), later to become Hope Professor of Entomology in Oxford, read a paper before the Royal Society in which he stated categorically that

> "no exception occurs to the general law of development in the Crustacea—namely, that they undergo no change of form sufficiently marked to warrant the application to them of the term *metamorphosis*" (Westwood, 1835).

Although Thompson was wrong in assuming that all Crustacea undergo a metamorphosis during their development, for indeed a few that are in his list do not, the occurrence of larval stages was confirmed in an increasing number of species by subsequent workers and by none more convincingly than by Richard Q. Couch. Thomas Bell (1853) wrote of him in his *History*

> "... my friend Mr Richard Q. Couch of Penzance ... dissatisfied with the uncertainty and contradiction of former testimony, resolved to investigate the matter for himself; and this he effected with a degree of acumen and perseverance which characterises all his researches and by which the truth of the doctrine is fully established as regards the genera *Cancer, Pilumnus, Carcinus, Portunus, Polybius, Maia, Galathea, Homarus* and *Palinurus*— a goodly number to have been investigated by one observer—and of some of these he watched every change".

Gray, who had originally mistaken the sequences in the barnacle development, generously withdrew his earlier reservations about Thompson's observations; Westwood did not openly do so.

Another, and a somewhat bizarre, controversy generated in the early discussions of the Society concerned the origin of the crinkled parchment-like shell of the female paper nautilus *Argonauta argo* L. and the much smaller *A. hians* found in most warm seas including the Mediterranean and parts of the Atlantic. J. E. Gray who was an exceptionally good zoologist, highly productive during his lifetime of a great number of publications covering many subjects and fields of zoology, was again unfortunate in being found on the wrong side of an argument. In 1831 and 1834 he presented two papers which maintained the view that the animal which inhabits the argonaut shell is a parasite without any means of depositing or forming a shell of its own but which enters the shell either by expelling the original inhabitant or taking vacant possession of it. The view was put forward that the original inhabitant was a heteropod mollusc related to *Carinaria*. Gray was strongly supported in this view by the distinguished French zoologist M. de Blainville. Professor Richard Owen, and others besides, would however have none of it, and Owen in a short note in the *Proceedings* in 1836 firmly proposed that the shell belonged to the female nautilus (in this instance *A. hians* Lam.) and not to any other creature. The matter was finally settled when in 1839 Owen made known to the Society the admirable observations of Madame Jeannette Power who made a continuous study of a number of specimens of *Argonauta* in her vivarium at Messina. Her observations showed that the young *Argonauta* when first emergent are naked but that, at the age of 10–12 days, the shell begins to form as a secretion of the web-like extensions of the two dorsal arms (Cooke, 1895).

INVERTEBRATE ZOOLOGY AND ZOOLOGISTS

The large number of papers, short communications and exhibits dealing with invertebrate animals brought to the notice of the Society through the early scientific meetings must necessarily be referred to selectively in a somewhat summary fashion.

Taxonomy and systematics

By and large the main need of the time was to describe, name and classify the great numbers and variety of animals that were daily being discovered through the vigorous exploration of the world's oceans and continents and which were being assembled in private and institutional collections in Britain. Two sources of collected material appear mainly to have been used in preparing the reports communicated to the Society, namely the museum collections of the Society and the Cuming collection.

The first notice in the Society's records of the collections of Hugh Cuming, Devon-born naturalist, sail-maker, successful man of business in South America for several years and world-wide traveller, in the records of the Society is contained in the *Proceedings* of the meeting of February 28th 1832, when

> "Specimens were exhibited of numerous Mollusca and Conchifera hitherto undescribed, and which form part of the collection made by Mr H. Cuming during a voyage undertaken in 1827, 1828, 1829 and 1830, for the purpose of obtaining subjects in natural history on the western coast of South America, its adjacent islands and many of those which form the principal Archipelago of the South Pacific Ocean. The specimens exhibited on the present occasion constitute the first portion of the collection which extends in these classes to upwards of four hundred new species, the whole of which Mr Cuming proposes to bring before the Committee from time to time, as the descriptions of them are completed".

The material was worked on by a variety of specialists to whom Cuming generously gave free access. They included, among British malacologists who presented their findings to the scientific meetings, Adams, Gray, Hanley, and the Sowerbys, as well as a number of French, German and other European scientists. There are references in the *Proceedings* extending over 30 years (from 1832 to 1862) to papers deriving from the Cuming collections and, at a rough cast, some 30 in all.

The extent to which the invertebrate collections that rapidly accumulated within the Society's museum were used for taxonomic research is less readily determined. Many of the specimens were sent by Corresponding Members who had been invited to send material, some, without doubt, were specimens exhibited at meetings, and many were from private collections donated to the Society. The task of their curation must have been considerable and finally overwhelming, for in the 1850s the Society decided to disperse its collections, and to transfer the more scientifically valuable parts of it to the British Museum. It is recorded in the *History of the Collections* (1906) that

> "The collection thus received, numbering nearly 1500 examples, forms with that of the India Museum the most important addition from a historical point of view that the Museum has ever received. It contained all, or nearly all of the specimens described in the early days of the Society in its *Proceedings*, and the whole collections of many of the founders of Zoology in Great Britain".

It seems probable that some of the more sophisticated early essays in revisionary taxonomy reported in the *Proceedings*, such as J. E. Gray's *On the Genera distinguishable* in Echinus *Lam.* (1835), were founded on these collections.

Natural history

Under this heading are included field observations of the biology and behaviour of invertebrate animals and, for convenience but less conventionally, sundry other kinds of observations on living animals, including laboratory studies. One of the earliest studies of this latter kind notified to the Society (in 1833) were the observations of Dr Robert Grant, Professor of Zoology in the newly founded University College, London, on the metachronal beat of the comb-plate cilia of the comb-jelly *Pleurobrachia pileus* (O. F. Müller) which is often very common in British coastal waters and on the distribution of the fine fibres within the jelly which he described as nerves. Charles Darwin who had known him when they were at Edinburgh University says of Grant

> "... my senior by several years ... he published some first rate zoological papers, but after coming to London as Professor in University College he did nothing new in science, a fact which has always been inexplicable to me".

Grant, who had earlier discovered the direction and course of water flow through sponges and after whom the genus *Grantia* was named, communicated two further papers on the anatomy of cephalopod molluscs and on the stranding of pelagic gastropods and siphonophores in the same year, but does not otherwise seem to have taken an active part in the affairs of the Society.

Almost all the numerous notes reporting on the ecology and habits of invertebrates concern marine animals. There are in fact rather few references in the early years to terrestrial groups and these are almost invariably arthropods or molluscs. They include, for example, notes on the Lepidoptera in the collections (Doubleday, 1847), on new genera of myriapods (Newport, 1842) and on new species of land snails from Jamaica (Pfeiffer, 1845). Terrestrial invertebrates appear to have received considerably more attention from about the 1860s onwards, a period which however is not within the time span of this review.

Zoology in Australia

It would not be appropriate to conclude this brief survey without some mention of the way in which the interest of the Society in invertebrate zoology was in its early years represented overseas through its Corresponding Members, and of the knowledge of and interest in the Society displayed by its overseas members. The strength of these relationships is perhaps nowhere better shown than in the Society's connections with the growing band of capable amateur naturalists who were among the steady stream of settlers leaving Britain to make a new life in Australia a century and a half ago. These connections are of course best illustrated and were most quickly forged through the passage of information and the dispatch of specimens of the unique mammals, birds, reptiles and other vertebrate animals of Australia. But although the rich variety and abundance of invertebrate life on the continent and within its surrounding seas was as yet virtually unknown there are indications in the records of the early scientific meetings of the Society that there were naturalists in Britain, shortly to emigrate to Australia, whose major interests had been in invertebrate animals. One such was a Mr J. B. Harvey who lived in the Teignmouth area. In the period 1834–1836 he made various communications to the Society on the marine animals of South Devon, and often exhibited specimens. In 1838 he left for Australia to live in or near Adelaide carrying with him a letter and vote of thanks from the Society for his scientific contributions to its meetings, and he later corresponded with the Society, as is evidenced by a letter published in its *Proceedings* in 1843. Another Corresponding Member in Australia who, in this instance, reported at two meetings in 1837 observations on the phosphorescence of the oceans and the siphonophores seen during a journey from England to Sydney, was Mr George Bennett, the first Superintendent of the Australian Museum in Sydney. Whether or not he was in Sydney when Thomas Henry Huxley arrived there as Assistant-Surgeon in the *Rattlesnake* I have not as yet been able to ascertain. It is well known that Huxley had more interesting things to do when in Sydney than visiting museums. There is little doubt however that he expressed to Miss Henrietta Heathorne (as he is known to have done to his mother) his concern that papers he had sent Professor Forbes with a request that he communicate them for publication had not been acknowledged.

> "So (Huxley says) I shall certainly send him nothing more, especially as Mr Macleay, (of this place, and a great man in the naturalist world) has offered to get anything of mine sent to the Zoological Society" (Leonard Huxley, 1903).

References

Bate, C. Spence & Westwood, J. O. (1868). *A history of the British sessile-eyed Crustacea.* London: John van Voorst.

Bell, T. (1853). *A history of the British stalk-eyed Crustacea*. London: John van Voorst.

British Museum (Nat. Hist.) (1906). *History of the Collections*. II. London: British Museum.

Cooke, A. H. (1895). *Cambridge Natural History: Molluscs*. London: Macmillan.

Couch, Bertha (1891). *The life of Jonathan Couch, FLS of Polperro*. Liskeard.

Couch, Richard Q. (1844). *A Cornish fauna. Part III The zoophytes and calcareous corallines*. Truro.

Curle, Richard (1954). *A bibliographic history of the Ray Society*. London: Ray Society.

Dance, S. Peter (1966). *Shell collecting: an illustrated history*. London: Faber and Faber.

Doubleday, E. (1847). On some undescribed species of Lepidoptera in the Society's collection. *Proc. zool. Soc. Lond.* **1847**: 58–61.

Forbes, Edward (1841). *A history of British starfishes*. London: John van Voorst.

Forbes, E. & Hanley, S. (1848–53). *A history of the British Mollusca and their shells*. 4 volumes. London: John van Voorst.

Gosse, P. H. (1860). *Actinologia Britannica. A history of British sea-anemones and corals*. London: John van Voorst.

Grant, R. (1833). On the nervous system of *Beroë pileus*, Lam., and on the structure of its cilia. *Proc. zool. Soc. Lond.* **1833**: 8–9.

Gray, J. E. (1831). Observations on the animal (*Ocythoë*) found in the shells of the genus *Argonauta*. *Proc. zool. Soc. Lond.* **1830–1831**: 107–108.

Gray, J. E. (1833). On the reproduction of Cirripeda. *Proc. zool. Soc. Lond.* **1833**: 115–116.

Gray, J. E. (1834).). Arguments in favour of the parasitic nature of the animals found in the shells of the genus *Argonauta*, Linn. *Proc. zool. Soc. Lond.* **1834**: 120.

Hincks, T. (1868). *History of British hydroid zoophytes*. I. London: John van Voorst.

Hincks, T. (1880). *A history of the British marine Polyzoa*. I. London: John van Voorst.

Huxley, Leonard (1903). *Life and letters of Thomas Henry Huxley*. I. London: Macmillan.

Jeffreys, J. G. (1863–69). *British conchology*. **2–5**. London: John van Voorst.

Johnston, G. (1838). *A history of the British zoophytes*. I and II. London: John van Voorst.

Monod, Th. (1970). John Cranch. Zoologiste de l'expedition du Congo (1816). *Bull. Br. Mus. (Nat. hist.)* Hist. Series **4** (1): 1–75.

Newport, G. (1842). On some new genera of the Class Myriapoda. *Proc. zool. Soc. Lond.* **1842**: 177–181.

Owen, R. (1836). On the shell and animal of *Argonauta hians*, Lam. *Proc. zool. Soc. Lond.* **1836**: 22–24.

Owen, R. (1839). On the paper nautilus (*Argonauta argo*). *Proc. zool. Soc. Lond.* **1839**: 35–48.

Pfeiffer, L. (1845). Descriptions of new species of land-shells from Jamaica, collected by Mr Gosse. *Proc. zool. Soc. Lond.* **1845**: 137–138.

Slabber, M. (1778). *Natuurkundige verlustigingen behelzende microscopise waarneemingen van in- en uitlandse Water- en Land-dieren*. Haarlem: Bosch.

Thompson, J. V. (1835). On the double metamorphosis in the decapodous Crustacea, exemplified in *Carcinus maenas*, Linn. *Phil. Trans. R. Soc.* **1835**: 359–362.

Westwood, J. O. (1835). On the supposed existence of metamorphosis in the Crustacea. *Proc. R. Soc. Lond.* **3**: 341–342.
Yarrell, W. (1836). *A history of British fishes.* London: John van Voorst.
Yarrell, W. (1843). *A history of British birds.* London: John van Voorst.
Young, G. M. (1964). *Victorian England: Portrait of an age, 1831 to 1865.* Oxford: University Press.

Symp. zool. Soc. Lond. (1976) No. 40, 85–104.

THE ZOOLOGICAL SOCIETY AND ICHTHYOLOGY, 1826–1930

P. H. GREENWOOD

British Museum (Natural History), London, England

SYNOPSIS

The Society has contributed to ichthyology in three major ways: through its Museum (1826–1855), its activities in fish culture and aquarium studies, and perhaps most significantly, through its scientific publications (including, of course, the *Zoological Record*). These various activities are reviewed and evaluated.

INTRODUCTION

My brief in preparing this paper for the Society's sesquicentennial meeting was to review early work done on fishes. Two questions immediately arose: where does the period "early" end, and how is one to interpret the word "work" in this particular context? The second question is, perhaps, better answered first.

Fishes do not seem to have attracted quite the same attention from prosectors and other staff members, or from research-minded Fellows, as did many other animals maintained in (or lost from) the Society's collections. There were, of course, exceptions. E. T. Bennett (Secretary, 1833–1836) and the Rev. R. T. Lowe (a Corresponding Member, 1833–1874) described many new species of fish which were deposited in the Society's Museum. Bennett, for example, published 26 papers between 1827 and 1835, mostly in the Society's journals, and all concerned with fishes in the Museum. Then there was the Society's interest in fish culture (particularly salmoniculture), work in which J. Murie (Prosector, 1865–1870) and Frank Buckland (a Fellow, 1862–1880) were much involved.

In all honesty, however, this direct involvement of the Society in work on fishes must be reckoned as rather slight. But, if one considers the contributions made to ichthyology through the Society's publications, a very different picture emerges, and the contents of the *Proceedings* and the *Transactions* must accordingly loom large in any review of work done.

That 1930 was the centennial year of the *Proceedings* is, in fact, the chief reason for my choice of 1930 as the year to end an "early" period in the history of the Society, now well into its second century of life. Another, and more personal, reason for choosing 1930 is that I wanted to include in this review those years when foundations were being laid for my own particular fields of research (and personal interests too must be held responsible for what some may consider undue eclecticism in the choice of papers and research singled out for detailed comment).

In short, the Society's contributions to ichthyology should be considered from the viewpoints of its Museum (1826–1855), its fish culture and aquarium activities (1830–1832 and 1853–1930), and finally, its scientific publications and meetings (1830–1930).

THE MUSEUM

There is ample evidence from early records of the great importance attached to the creation of a Museum as an integral part of the proposed Zoological Society (see Mitchell, 1929).

At first sight it seems paradoxical that so much emphasis was put on the building of a second zoological museum in the same city as the country's national zoological collections, then housed in the British Museum at Bloomsbury. However, in the early part of the nineteenth century curatorial standards in the Natural History section at Bloomsbury were very unsatisfactory, material was not always readily available for study and the identification of new collections was a very slow process (see Gunther, 1975, for a wide-ranging and sympathetic account of the British Museum's problems). Quite rightly, the sponsors of the Zoological Society felt that better services and curation should be provided, especially for the identification and subsequent care of collections made in distant countries, the sort of services many of its hoped-for members would require.

Once the Zoological Society was founded, its Museum soon began to grow in importance and size, and attracted to it, rather than the British Museum, a large number of fish collections. Some of the more important were those of R. T. Lowe (Madeira), Sir Stamford Raffles (Malaysia; see catalogue prepared by E. T. Bennett, in Raffles, 1830), W. H. Sykes (India), Dr Sibbald (Ceylon), C. Telfair and J. Desjardins (Mauritius), H. Cuming (Chile) and Felipe Poey (Cuba). Most of these collections were described, and specimens often illustrated, in one or other of the Society's publications, and all were substantial contributions to a developing knowledge of fishes and of taxonomic ichthyology.

With its growth (there were 600 specimens of fishes by 1836), the Museum began to encounter some of the difficulties that had beset the Natural History section of the British Museum, but still it did not attract quite the same intensity of criticism as did that body. In 1836 Charles Darwin, pondering on the subject of collections made during the voyage of the *Beagle*, wrote

> "... The Zoological Museum is nearly full, and upwards of a thousand specimens remain unmounted. I daresay the British Museum would receive them, but I cannot feel, from all I hear, any great respect even for the present state of that establishment".

[Unfortunately, he did not give them to the Society either; see Darwin (1887: 213).]

By 1850, affairs in the British Museum had improved greatly and the Society's Council were able to record the satisfaction they felt at this progress, the state of the collections then

> "... presenting so striking a contrast with their condition at the time when the Zoological Society was founded, renders the maintenance of our own Museum as a separate collection no longer an object of the importance it formerly possessed".

(See Mitchell, 1929.) Accordingly, arrangements were made to sell or give away the various parts of the collection, mainly to the British Museum but also to smaller institutions.

Here an element of misfortune (and mystery) enters the story. The British Museum's registers show that in late 1855 and early 1856, 349 fishes were either purchased from (285 specimens) or presented by (64 specimens) the Society. Apart from the generic name of a specimen, a note as to whether it was purchased or donated, and sometimes a very general locality reference, no other information is recorded in these registers. It is particularly unfortunate that the Society's Museum numbers did not remain with the fish specimens, or if they did, were not noted in the registers; without these numbers it is now impossible to correlate individual specimens with data available in the Society's catalogue. From a taxonomist's viewpoint this omission is a tragedy since one can neither tell, with any certainty, from whose collections the specimens came, nor can type specimens be distinguished from non-typical material, and the value of this potentially very important collection is thus greatly reduced. The only guide to a specimen's taxonomic status now remaining is whether or not it was donated by the Society. Typical specimens were given to the Museum whereas the others were sold. (See relevant section of the Report for 1855, quoted in Mitchell, 1929: 102–103.)

Quite how this unfortunate curatorial bungle came about is still a mystery, and the more so since it did not happen with specimens of other groups transferred to the Museum at the same time; evidently in some sections of the Museum the Council's optimism was somewhat unfounded.

FISH CULTURE AND THE AQUARIA

Although a Fish House (or more formally, an Aquavivarium) was only opened in 1853, the Society since its inception had been concerned, at least on paper, with the practical aspects of fish keeping. Doubtless the deep interest in salmonid fishes and their culture shown by a founder member and Councillor, Sir Humphry Davy, was an important influence in this respect (see his *Salmonia*, 1820).

A prospectus outlining the objectives of the proposed Zoological Society, dated 1st March 1825, notes that it was to be concerned with "... introducing and domesticating new breeds and Varieties of Animals such as Quadrupeds, Birds or Fishes likely to be useful in Common Life, ..." and that members "... might, at a reasonable price be furnished with living specimens, or the ova of fishes and birds". (There is no evidence, according to Scherren (1905), that any member ever did receive fish eggs.)

In pursuance of these objectives the Society tried to lease land at Carshalton in Surrey, which a subcommittee, comprising Sir Stamford Raffles and Sir Humphry Davy, had found "... adapted for the purpose" of the "... preservation of Fish and Wild Fowl". (Record of the meeting held on 26th April, 1826.) Negotiations for the lease fell through, and I can find no further records of activity in this field until the Society opened its Richmond Park farm in 1829. Here, according to a report of Council read to a meeting in 1829, there were "... very abundant springs" and "... some excellent ponds".

Attempts to raise a number of coarse fish species (carp, goldfish, flounders and eels) at the Richmond Park farm, however, were unsuccessful and the scheme was abandoned in 1832. According to Scherren (1905: 43) the ponds and the water were unsuitable for this type of fish farming.

The Society's early interest in practical pisciculture is also reflected in a paper by J. B. Arnold published in the first part of the *Proceedings of the Committee of Science and Correspondence* (1830–1831). Arnold described in some detail the various experiments he carried out in Guernsey on what he called the "Naturalization of sea fishes in a lake chiefly supplied with fresh water". The experiments were successful and culminated in a number of species, including turbot, brill, plaice, sole, bass and smelt, being transferred to freshwater ponds.

Between the closure of the Richmond Farm ponds and the opening of the Fish House in 1853, I can find no published records for any research on fish or fish culture being carried out under the Society's aegis. Even after the Fish House became functional, several years passed before the results of certain experiments on the artificial fertilization and raising of salmon and trout were published in the *Proceedings*. However, in 1863 Frank Buckland (PZS, 1863)[a] did read three papers concerned with fish culture and the transport of fishes, for which at least part of the work was carried out in the Gardens.

Buckland was also instrumental in supplying the Fish House with fertilized salmonid ova on several occasions, and one such gift gave rise to research and thoughts which were published at some length by James Murie, then Prosector for the Society.

In 1868 Murie exhibited specimens of salmon (*Salmo salar*) which he had raised from ova presented by Buckland in 1863. These fishes, according to Murie (PZS, 1868), showed distinct signs of arrested development, a result he thought of their being kept in freshwater beyond the stage at which seaward migration would have occurred in nature. Albert Günther, curator of fishes at the British Museum and the unchallengeable authority on British ichthyology, did not believe the eggs to be those of a salmon (despite Buckland's assurances to the contrary), and maintained (PZS, 1868) that they were probably trout eggs, or even from some interspecific hybridization. In Günther's opinion, Murie's data on the supposed effects of arrested growth were inadmissible. Murie, in turn, could not accept Günther's identifications, and replied later that year (PZS, March 26th, 1868) with a more detailed paper describing the characteristics of arrested growth. In this paper he also began to develop ideas on the taxonomic implications of arrested growth, ideas which were fully developed in a much longer and more philosophical work published in the *Proceedings* for 1870. Here Murie definitely challenges the validity of many so-called species of salmonid fishes, and suggests that these were

[a] Papers appearing in the Society's *Proceedings* or *Transactions* are not listed in the References on p. 103, since their subject matter and authorship are obvious from the context in which they are mentioned. Instead, they are cited in the text as PZS (*Proceedings of the Zoological Society of London*) or TZS (*Transactions of the Zoological Society of London*), followed by the year of publication or, where necessary, by the volume and the year in which that part of the volume appeared.

merely what would now be called ecophenotypes. Very reasonably he argues that controlled experiments (i.e. tank—compared with wild-raised specimens) could settle many such debatable points.

Murie's 1870 paper also contains some interesting thoughts and speculations on problems of speciation and extinction, concepts which did not seem, at least publicly, to greatly exercise the minds of contemporary ichthyologists. In some respects Murie's ideas were ahead of their time, and this may account for their being overlooked by his contemporaries working in the field of salmonid systematics. That Günther questioned the identity of Murie's material may also have contributed to this neglect. Most surprisingly Francis Day (PZS, 1884 and 1885) writing on the races and hybrids of Salmonidae (a series of papers that in many respects also savours of the latter-day "new systematics") does not cite Murie's papers and experiments, nor even the phenomenon of arrested growth. Day's approach to the problem was rather like that of Murie, and he was certainly not likely to be overawed by Günther's established position (see below, p. 95).

From 1870 onwards little research on fishes appears to have been carried out in the Gardens. In the *Proceedings* for 1906 and again in 1912 the American ichthyologist Bashford Dean published papers describing in detail the behaviour of Australian lungfishes (*Neoceratodus forsteri*) which he had studied in the Society's fish tanks. These papers are still the most detailed ones available on the feeding, breathing and locomotion of *Neoceratodus* in captivity (and may perhaps be responsible for the still repeated misconception that lungfishes breathe air through their nostrils).

Although not directly connected with the Fish House, mention should be made of a long and detailed paper written by a distinguished Fellow of the Society, J. E. Gray (PZS, 1856), on the West African lungfish *Protopterus annectens* (Fig. 1).

The animals Gray described were kept in tanks at Crystal Palace, and had been sent from the Gambia in their encysted, aestivating condition. Like Bashford Dean's paper on the Australian species, Gray's account gives much information on the feeding habits, locomotion and aerial respiration of these fishes. In addition to the purely descriptive account, Gray also speculates on the phyletic relationships of the species (which one senses he would rather not include amongst the Pisces) and gives a synopsis of the air-breathing fishes then known.

The paper is illustrated with a figure, drawn from life by G. H. Ford, which in my opinion is one of the best that has ever been

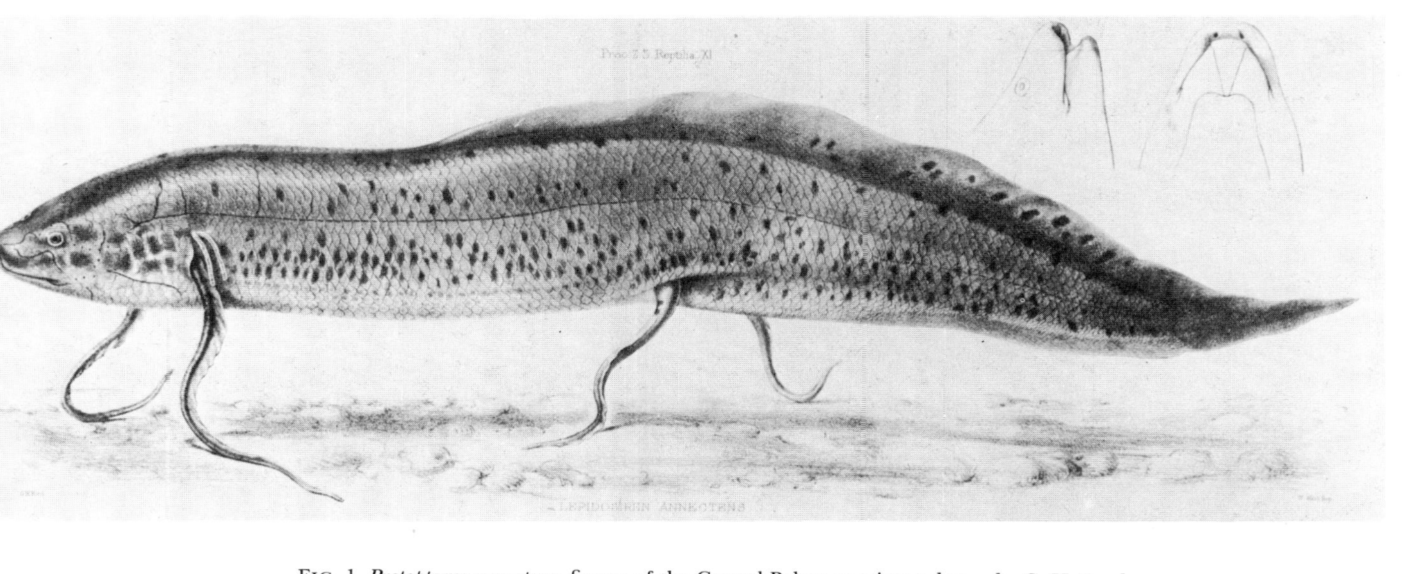

FIG. 1. *Protopterus annectens*; figure of the Crystal Palace specimen drawn by G. H. Ford and published in PZS (1856).

FIG. 2. G. H. Ford. Reproduced by permission of Mr A. E. Gunther.

produced for this species, incorporating as it does morphometric accuracy with lively impressionism. Ford (1809–1876), a South African, ranks among the outstanding ichthyological artists of any period (Fig. 2); fortunately much of his work, both in black and white and in colour, is preserved in the *Proceedings* and *Transactions* of the Society as illustrations to papers by Günther and Gray. (For further details of Ford, see Gunther, 1975, especially p. 268.)

As an appendix to Gray's paper there is a letter from A. D. Bartlett (the Superintendent of the Gardens) describing the release of the fishes from their cocoons, their subsequent behaviour and their extremely fast growth (about nine inches in three months).

Gray's paper, incidentally, belies a later claim by E. G. Boulenger (PZS, 1913) that the first live *Protopterus* to be exhibited in Europe was displayed by the Society in 1913 (a slip that may have arisen from Gray's use of the then correct generic name *Lepidosiren* for the *Protopterus* at Crystal Palace).

When the Fish House opened in 1853 it was the first public aquarium in Europe, and at least one authority on the history of aquaria (Atz, 1971) credits it with initiating the "... first home aquarium craze in England and Scotland".

A guide book to the Fish House exhibits (or more correctly, possible exhibits) was published in 1860. Its author, E. W. H. Holdsworth, a Fellow of the Society, was a strong champion of the "balanced aquarium" concept, and he expounded in detail and with great care on this subject in his Fish House booklet. Theoretically, a "balanced aquarium" is one in which, provided there is sufficient light, the carbon dioxide given off by the animals would be utilized photosynthetically by the plants, and the oxygen produced in this process would be utilized in turn by the animals.

Unfortunately the "balanced aquarium" concept is a great oversimplification, as the newly inspired aquarium-keeping public soon discovered. There was, as a result, a marked decline in the hobby after about 1862, but some amateur aquarists persisted, keeping the hobby alive and even spreading it abroad, especially to Germany. Certainly the Society can claim, through the Fish House and Holdsworth's booklet, to have fathered a hobby that is now world-wide in its practice, is of considerable economic importance, and has, in many ways, contributed to the advance of professional ichthyology.

The first recorded photograph of a living fish was taken in the Fish House in 1854 (Mitchell, 1929), the Council noting the great advantage "... to the study of Ichthyology deducible from this application of the art". Regrettably, their comments were overlooked at the time, or if they were taken up there is no tangible record of the results; equally regrettable is the fact that no copies or later reproductions of the original photograph are now in existence.

The present aquarium opened to the public in 1924. Rather surprisingly it has been virtually unused for any ichthyological research, at least in the sense of researches on the fishes themselves. (However, see Vevers, this volume, for an account of the several other research activities conducted in the aquarium.) Perhaps its most significant contribution, apart from providing valuable anatomical material to various institutions, has been in the meticulously kept records of fish longevity. Many of the data used by S. S.

Flower (PZS, 1925 and 1935) in two most useful papers on that subject were obtained from the Aquarium.

THE SOCIETY'S SCIENTIFIC PUBLICATIONS

One of the Society's most important contributions to the study of fishes was, and still is, through the medium of its journals (and, by implication, through the high standards set by its Publications Committee and Editors).

Between 1833 and 1850 a number of major taxonomic works were published, both in the *Transactions* and in the *Proceedings* (or, from 1830 to 1832, the latter's predecessor, *The Proceedings of the Committee of Science and Correspondence*).

Of special note are Lowe's monograph on the fishes of Madeira (a synopsis published in PZS, 1836, the main paper in TZS **2**, 1839, and a supplement in TZS **3**, 1842), his four shorter papers describing new species and new records from Madeira (PZS, 1840, 1843, 1846 and 1851), Richardson's two major papers on the fishes of Australia (TZS **3**, 1848) and another lengthy paper by Richardson in the *Proceedings* for 1850.

Sykes' two papers (PZS, 1838 and TZS **2**, 1841) on the fishes of Deccan were the first major papers on freshwater fishes to be published by the Society. The present-day ichthyologist will feel great sympathy with Sykes when he comments (PZS, 1838) on the difficulties he experienced in classifying Indian Cyprinidae, a commentary that indicates what little progress we have made in that particular sphere. He will also wonder why some recent workers did not heed Sykes' warnings on the unreliability of using barbels as characters for defining the genera of cyprinid fishes (see Schultz, 1957).

The next 20 years (1851–1871) show a marked increase in the number of taxonomic papers published in the *Proceedings*; only one major paper (Günther, TZS, 1869 on fishes collected by Dow, Godman and Salvin in Central America), however, was printed in the *Transactions.* The majority of these papers were by Francis Day (principally on Indian freshwater species) and by Albert Günther [on fishes from a number of countries outside Europe, including one paper (PZS, 1861) with some interesting observations on the zoogeography of Panamanian coastal fishes]. There was also an important paper by Sir John Richardson (PZS, 1856) on fishes from Asia Minor and Palestine, and another by the Dutch ichthyologist Pieter Bleeker (PZS, 1861) in which the labroid fishes are reviewed and classified to genera, with the type species of all the author's genera clearly designated.

One of Albert Günther's papers during this period (PZS, 1862) deals with the taxonomy of British charrs (*Salvelinus* spp.). It contains two sentences which are of particular interest since they clearly show the attitude of a museum taxonomist of the period, and also that some of his worries are still with us:

> "... We will not enter into a fruitless investigation as to the possibility of the differences we observe in those fishes being induced by those physical peculiarities of the localities indicated by Mr Thompson. We will take and examine them as they are, and as they will be, as long as zoologists of the present species of man exist, provided that human interference does not put a premature termination to the whole tribe".

Günther's long-standing disagreements with Francis Day (see Gunther, 1975: 409–414) erupted publicly in the pages of the *Proceedings* for 1871.

A paper by Day, innocuously entitled "Remarks on Indian fishes" (PZS, 1871:634–638) is, in fact, a reply, always polite but very vinegary in tone, to criticisms Günther had made (in the *Zoological Record* for 1869) of an earlier paper published in the *Proceedings* for 1869. Günther not only criticized the taxonomic procedure but also implied that Day was a slipshod and incompetent ichthyologist. Day defends himself well in the 1871 paper, and concludes his defence by writing "... There are several omissions in the 'Record' but on them I do not propose offering any remarks...".

Günther was quick to respond, with a paper (received and read on the same day, December 5th, 1871) whose acidity would be difficult to match; his closing remarks were crushing: "... I feel that for the future it will be undesirable to employ my time taking notice of similar communications to the Society" (PZS, 1871). And neither did he, except merely to note the publication of Day's papers in the "Record". The two men (Figs 3 and 4) were only reconciled, through Günther's initiative, a few months before Day's death 28 years later (see Gunther, 1975).

Trivial though this squabble may seem in isolated retrospect it did have one unfortunate consequence for taxonomic ichthyology; after 1870, no more of Day's specimens were lodged in the British Museum until after the reconciliation with Günther. Instead, his material went to museums throughout the world, and many

FIG. 3. Albert Günther. Reproduced by permission of the Trustees of the British Museum (Natural History).

important type specimens cannot now be identified as such (see Whitehead & Talwar, 1976).

Both Günther and Day contributed to the Society's *Proceedings* for many years although their major contributions, as was usual at that period, were contained in monographs and books published elsewhere.

From 1870 until the early 1900s the Society's journals carried many papers on taxonomic ichthyology, together with a leavening of others that could be classified as natural history notes (but none the less valuable since they brought field observations to the notice of bench-workers at a time when few museum men took to the field). Most significantly there was, during this period, a marked increase in the number of anatomical and comparative morpholog-

FIG. 4. Francis Day. Reproduced by permission of the Trustees of the British Museum (Natural History).

ical papers. Previously these had been surprisingly few and far between, some notable exceptions being T. Spencer Cobbold's paper on the cranial osteology of *Protopterus annectens* (PZS, 1862), A. Günther's description of a skeleton of *Luvarus imperialis* (PZS, 1866), and Francis Day's account of swim-bladder morphology and anatomy in Indian catfishes (PZS, 1871), one of the few instances of an author of that period using deeper-lying characters in alpha-level taxonomy.

After about 1870 the Society published a number of important (and now classical) works on fish anatomy.

T. H. Huxley contributed two such papers. One (PZS, 1876), on the anatomy of *Neoceratodus forsteri,* the Australian lungfish, also included an outline classification of fishes, which animals Huxley divided into four "equivalent and distinct natural assemblages", the Dipnoi, Ganoidei, Teleostei and Chimaeroidei. It was in this paper too that the terms "hyostylic" and "amphistylic" were first introduced to describe different kinds of jaw suspension in fishes.

Huxley's classification was criticized by Günther in his book *An introduction to the study of fishes* (1880). Huxley replied to Günther through comments appended to a paper on the oviducts of the smelt (*Osmerus*), published in the *Proceedings* for 1883. In this paper Huxley reiterated his views on the soundness of his proposed scheme of relationships, and berated Günther not only for so summarily disposing of his ideas, but also doing so ". . . without the least qualification".

Huxley's second major anatomical paper (PZS, 1876) dealt with the anterior nasal apertures of another lungfish, the African *Protopterus annectens*; it provoked no response at the time.

Also in the *Proceedings* of this era was a paper by W. N. Parker (PZS, 1888) who described in detail the poison organs of weever fishes (*Trachinus* spp.), and established for the first time that definite glandular tissue is associated with the stinging spines of the operculum and dorsal fin.

Two papers by W. G. Ridewood, possibly some of the most frequently quoted by modern workers on the phylogeny of lower teleost fishes, were published in the *Proceedings* for 1904. The first dealt with the cranial osteology of elopid fishes and with the skull of lower teleosts generally, and the second with the skull in clupeoids.

The *Transactions* carried several longer anatomical papers: W. K. Parker (TZS **10**, 1878) on the structure and development of the skull in sharks and skates; St. G. Mivart on the fins of elasmobranchs, and the nature and homology of vertebrate limbs (TZS **10**, 1879); Ray Lankester on the hearts of *Neoceratodus, Protopterus* and *Chimaera* (TZS **10**, 1879) and on the external morphology of *Lepidosiren* and *Protopterus* (TZS **14**, 1896); and J. Parker on the skeleton of the oarfish *Regalecus* (TZS **12**, 1886).

When considering this period of the Society's history particular reference should be made to J. S. Budgett (Fellow 1898–1904), an outstanding Cambridge zoologist who died, aged 32, from blackwater fever shortly after returning from one of his field trips to Africa. The full magnitude of Budgett's contributions, both directly and indirectly, to ichthyology and herpetology can be gauged best from a memorial volume *The work of John Samuel Budgett* [J. Graham Kerr (ed.)] published by the Cambridge University Press in 1907.

Three of Budgett's longer ichthyological papers were published in the *Transactions* and dealt with certain aspects of the anatomy of *Polypterus* (TZS **16**, 1901), the breeding habits of some West African fishes, with an account of the external features in developing *Protopterus annectens* and *Polypterus senegalus* (TZS **16**, 1901), and a detailed osteological and anatomical study of a larval *Polypterus senegalus* (TZS **16**, 1902). One shorter paper appeared in the *Proceedings* of 1899, and two in the volume for 1903.

Budgett had particularly close associations with the Zoological Society. Indeed, his major expedition to Uganda was partly financed by funds from the Society and was directed by its Council; on another occasion, the Secretary, Dr P. C. Sclater, guided Budgett in his choice of localities for carrying out field research into the breeding habits of *Polypterus*, suggesting that the Gambia would be a most profitable locality for the work. In 1902, after returning from one of his Africa visits, Budgett applied for the Secretaryship of the Society, a post just vacated by Sclater, but he was unsuccessful (see introductory chapter in the memorial volume).

From the viewpoint of a taxonomic ichthyologist, the period between 1870 and 1910 could well be called the "African years". Some of the most significant papers published then dealt with the newly discovered freshwater fishes of the African Great Lakes, especially Lake Tanganyika. The full complexity of the large cichlid species flocks in that lake, and their evolutionary implications, gradually emerged in a series of papers published by G. A. Boulenger in the *Transactions* (1898, 1899, 1901 and 1906). These various works can be considered as the initiators of the numerous studies that, later in the century, were to be carried out on the evolutionary aspects of African lake fishes, and which have contributed much, through example and controversy, to our understanding of speciation and adaptive radiation in tropical freshwater fishes (see summaries in Fryer & Iles, 1972, and Greenwood, 1974).

Partly as a result of all this new material coming from Africa, Boulenger was obliged to revise the taxonomy of the African and Syrian Cichlidae, a work carried out in two parts and published in the *Proceedings* for 1898 and 1899.

In addition to his longer papers in the *Transactions,* Boulenger published several shorter papers on African freshwater fishes, covering for example species from the Niger delta (PZS, 1901), Kenya (PZS, 1905), the Congo river (PZS, 1902), Ethiopia (PZS,

1903), and the Transvaal (PZS, 1903). In a paper on the fishes of south Cameroons (PZS, 1903), Boulenger included some notes, the most detailed yet, on the "flight" of the little freshwater butterfly fish, *Pantodon buchholzi,* and thereby introduced a controversy that is still unsettled: does the species have a true, flapping flight or does it merely glide? (Greenwood & Thomson, 1960.)

The first decade of the twentieth century saw the publication of several papers by C. Tate Regan, a giant amongst taxonomic ichthyologists and the man who laid the foundations for all future work on the higher classification of fishes. These early papers included a classification of selachians (PZS, 1906) a revision of the orectolobid sharks (PZS, 1908), his monograph on the loricariid catfishes (TZS **17**, 1904), a revision of certain South American cichlid genera (PZS, 1905), and a paper on the fishes of Trinidad, illustrated by colour plates prepared, from water colour sketches, by James Green (PZS, 1906). Green, like G. H. Ford some 40 years earlier (see p. 90 above), was an outstanding ichthyological draughtsman and artist, many of whose drawings (and a few coloured lithographs) were published in the Society's journals.

Most of Regan's now classical papers on the classification of teleost fishes were published elsewhere (chiefly in the *Annals and Magazine of Natural History*), but one important contribution, on the anatomy and classification of the Allotriognathi, appeared in the *Proceedings* for 1907.

In the 20 years from 1910 to 1930, the admixture of papers published by the Society was essentially like that of the preceding decade (but with relatively more appearing in the *Proceedings*). Two important taxonomic works by Regan were published. One (PZS, 1913), a revision of poeciliine toothcarps, demonstrated for the first time the importance of gonopodial structure in the taxonomy of these fishes and thus once again firmly laid the foundations for future research (see Rosen & Bailey, 1963). The second paper (PZS, 1916), dealt with the morphology and classification of the related phallostethine fishes.

Later on, Regan was to publish revisions of the cichlid fishes of Lakes Malawi (PZS, 1921) and Victoria (PZS, 1922), papers which are landmarks in the taxonomy of African Cichlidae and which prepared the ground for many subsequent studies on the taxonomy and evolution of these fishes, the "Darwin's finches" of ichthyology (see summaries in Fryer & Iles, 1972, and Greenwood, 1974).

Many other taxonomic papers were, of course, published between 1910 and 1930, including some by J. R. Norman (another noteworthy British ichthyologist), amongst which can be noted his important review of the serrasalmine fishes (PZS, 1923) and the paper in which attention was first drawn to the fact that certain fishes were migrating through the Suez Canal from the Red Sea to the Mediterranean; an observation amply confirmed by later workers (Ben-Tuvia, 1966).

In the anatomical field, special note may be made of papers by R. H. Whitehouse (PZS, 1910), E. S. Goodrich (PZS, 1913) and by C. T. Regan (PZS, 1923). Whitehouse made a study of the caudal fin skeleton in several fishes (the first really comparative one of its kind) and then attempted to use his findings taxonomically. His concluding remarks

> "... The structure of the caudal fin cannot be considered a safe criterion for taxonomic purposes, though it may be useful in this connection amongst the smaller divisions, as is well illustrated by the Acanthopterygii",

have not been borne out by time. The caudal skeleton has, in fact, provided a major tool for students of teleostean phylogeny (see papers and references in the volume edited by Greenwood, Miles & Patterson, 1973).

Goodrich's paper dealt with the structure of bone in fishes, and is an early study in palaeohistology; again, the author extended his findings into the realm of phylogeny and classification.

Problems of phylogeny and classification also take up a large part of Regan's paper, which is primarily a study of the skeleton in the garpike *Lepisosteus.* It was in this paper that Regan first defined the taxa Palaeopterygii and Neopterygii, and redefined the Protospondyli, Ginglymodi and Halecostomi, concepts that were to be used in fish classification for over 40 years.

Several other anatomical papers were published in this 20-year period, including J. H. Floyd & E. M. Sheppard on the anatomy of hammer-head sharks (PZS, 1922), F. R. Wells (PZS, 1922) on the chondrocranium of the herring *Clupea harengus,* N. A. Mackintosh on the chondrocranium of the Norway cod *Sebates* (PZS, 1923), E. P. Allis (PZS, 1917) on the embryonic skull in the holocephalian *Chimaera,* and T. Goodey (PZS, 1910) on the skeletal anatomy of the frilled shark *Chlamydoselachus anguineus.*

Throughout the century under review, relatively few palaeontological papers were published. Two of the earliest were by A.

Smith Woodward, and dealt with the jaw arches in a Cretaceous shark (PZS, 1886), and the anatomy and systematics of the Liassic shark *Squaloraja polyspondyla* (PZS, 1886).

At one of the Society's meetings in 1897, R. H. Traquair exhibited some specimens of that small and enigmatic, fish-like animal *Palaeospondylus gunni,* a Devonian fossil from the Old Red Sandstone of Caithness. An account of this exhibit, together with a short note giving Traquair's views on its affinities was published in the *Proceedings* for 1897. Traquair believed *Palaeospondylus* to be a lamprey, a view which was not shared by Bashford Dean (1895, 1896), who replied to Traquair's criticisms (PZS, 1897) in the *Proceedings* for 1898. In this paper Bashford Dean restated his arguments for believing that *Palaeospondylus* was either of indeterminable affinities or that it might be a representative of a new class or subclass, the Cycliae. Even today doubt still exists about the relationships of this small creature (Moy-Thomas, 1971), despite the ideas put forward by Agnes E. Miller (PZS, 1930). Miller, using information from her studies on the tail skeleton of the lungfish *Lepidosiren paradoxa* argued that *Palaeospondylus* was in fact a young dipnoan.

D. M. S. Watson contributed some palaeontological papers to the *Proceedings*, mostly on the subject of palaeoniscoid fishes. Another of his papers (PZS, 1927), on probable viviparity in the Jurassic coelacanth *Undina* (= *Holophagus*) has recently been much quoted in connection with the discovery that the only extant coelacanth, *Latimeria chalumnae,* is indeed ovoviviparous (Smith, Rand, Schaeffer & Atz, 1975).

Apart from purely taxonomic and anatomical papers, fishes have also featured in a number of contributions dealing with such subjects as zoogeography and general evolutionary problems. But, as far as I can determine, no paper specifically concerned with the experimental aspects of fish physiology was published by the Society between 1830 and 1930, a distinct contrast with the contents of recent issues of the *Proceedings* and its successor, the *Journal of Zoology*.

Finally, one must pay tribute to the Society for its part in organizing and arranging the *Zoological Record.* Without the "Zoo Record" any ichthyologist's work would be badly hampered and be much more imprecise. The latest, 604-page, computer-assisted edition for 1971 is a far cry from Günther's highly personal, 56-page compilation for 1864. However much one may regret the passing of the Güntherian-style product, one's indebtedness to the

Society for providing an essential research tool remains a fact, and the use of that tool a reminder of what is possibly the Society's greatest direct contribution to ichthyology.

CONCLUSIONS

To sum up. Through its publications (including the *Zoological Record*), its scientific meetings and, despite its relatively short life, its Museum, the Society has made substantial contributions to the development of international ichthyology.

ACKNOWLEDGEMENTS

I am particularly indebted to Dr H. G. Vevers (Assistant Director of Science and Curator of the Aquarium), to Mr R. A. Fish (the Society's Librarian) and to my colleagues in the Museum. Mr A. C. Wheeler and Dr P. J. P. Whitehead for the information and help they have given me in preparing this paper, and especially for their contribution of unpublished information on the characters and incidents that form the background to this essay.

REFERENCES

Atz, J. W. (1971). *Aquarium fishes.* New York: Viking Press.

Ben-Tuvia, A. (1966). Red sea fishes recently found in the Mediterranean. *Copeia* **1966**: 254–275.

Darwin, F. (1887). *The life and letters of Charles Darwin*, **1.** London: John Murray.

Davy, Sir Humphry (1820). *Salmonia.* London.

Dean, B. (1895). *Fishes, living and fossil.* New York: Macmillan & Co.

Dean, B. (1896). Is *Palaeospondylus* a cyclostome? *Trans. N.Y. Acad. Sci.* **15**: 100–104.

Fryer, G. & Iles, T. D. (1972). *The cichlid fishes of the Great Lakes of Africa.* Edinburgh: Oliver and Boyd.

Greenwood, P. H. (1974). The cichlid fishes of Lake Victoria, East Africa: the biology and evolution of a species flock. *Bull. Br. Mus. nat. Hist.* (Zool.) Suppl. No. 6: 1–134.

Greenwood, P. H., Miles, R. S. & Patterson, C. (Eds) (1973). The interrelationships of fishes. *Zool. J. Linn. Soc.* **53**, Suppl. No. 1: i–xvi + 1–536.

Greenwood, P. H. & Thomson, K. S. (1960). The pectoral anatomy of *Pantodon buchholzi* Peters (a freshwater flying fish) and the related Osteoglossidae. *Proc. zool. Soc. Lond.* **135**: 283–301.

Günther, A. C. L. G. (1880). *An introduction to the study of fishes.* Edinburgh: A. & C. Black.

Gunther, A. E. (1975). *A century of zoology at the British Museum through the lives of two keepers 1815–1914.* London: Dawsons of Pall Mall.

Holdsworth, E. W. H. (1860). *Handbook to the fish house in the gardens of the Zoological Society of London.* London: Bradbury & Evans.
Mitchell, P. C. (1929). *Centenary history of the Zoological Society of London.* London: Zoological Society.
Moy-Thomas, J. A. (1971). *Palaeozoic fishes,* 2nd edn., by R. S. Miles. London: Chapman & Hall.
Raffles, S. (1830). *Memoir of Sir Thomas Stamford Raffles.* London: John Murray.
Rosen, D. E. & Bailey, R. M. (1963). The poeciliid fishes (Cyprinodontiformes), their structure, zoogeography, and systematics. *Bull. Am. Mus. nat. Hist.* **126**: 1–176.
Scherren, H. (1905). *The Zoological Society of London.* London: Cassell.
Schultz, L. P. (1957). The generic names *Barbus* and *Puntius. Trop. Fish Hobby* **5**: 14–15, and 29–31.
Smith, C. L., Rand, C. S., Schaeffer, B. & Atz, J. W. (1975). *Latimeria,* the living coelacanth, is ovoviviparous. *Science, N.Y.* **190**: 1105–1106.
Whitehead, P. J. P. & Talwar, P. K. (1976). Francis Day (1829–1889) and his collections of Indian fishes. *Bull. Br. Mus. nat. Hist.* (Hist.) **5**: 1–100.

Symp. zool. Soc. Lond. (1976) No. 40, 105–118.

MANAGEMENT OF A PUBLIC AQUARIUM

H. G. VEVERS

The Zoological Society of London, Regent's Park, London, England

INTRODUCTION

There were probably no true aquaria in the Ancient World, although Roman hosts kept ponds for oysters and they even led living red mullet in open ducts to the dinner table so that their guests might admire the brilliant colours of this noble fish before they ate its rich and succulent flesh.

In the Middle Ages when private menageries were kept by the nobles and princes of Europe, more for amusement than education, there appears to have been no attempt made to keep aquatic animals in such a way that they could be observed closely, and certainly nothing comparable with mammal cages or pits and bird aviaries. It is true that the Chinese kept goldfish in ponds and dishes—but these came to be domestic animals and a prime example of artificial selection by man—and of course monks in Europe cultivated carp and similar table fish in ponds; but a fish pond is not an aquarium.

It is a little surprising that we have few records of attempts to keep aquatic animals during the two or three centuries before 1850. In the sixteenth and seventeenth centuries there was, of course, still a good bit of copying going on, without much original observation on living animals. But in the eighteenth century, the century of experimental enquiry and industrial progress, surely then we might have expected naturalists and others to attempt aquaria on a small scale. There was, of course, no plate glass, which was first made at St. Helens in Lancashire in 1773.

Small dishes were used to keep certain aquatic animals for research purposes, notably by Abraham Trembley, one of the first of modern experimental biologists. In 1790 Sir John Dalyell of the Binns began keeping marine animals for observation in Edinburgh and continued doing so until 1850. One of his specimens of the beadlet anemone, *Actinia equina,* was collected at North Berwick in 1827, and its biography has been charmingly recorded by D'Arcy Thompson in an essay entitled "Granny" (Thompson, 1940). Sir John Dalyell used to change its water every day and to feed it occasionally on small bits of mussel or oyster. When Sir John died it

FIG. 1. P. H. Gosse.

passed to Professor Fleming and when he died it went to Dr MacBain, an Edinburgh surgeon. He died too and Granny went to an old lady who ran a girls' school, and from her it passed to a botanist, in whose hands Granny pined and finally died in 1887—60 years after she was first taken from the sea at North Berwick.

The story of true aquaria may be traced back to the thirties and forties of the last century. In 1833 at the 3rd meeting of the British Association at Cambridge, Professor Daubeny spoke on the action of light on plants. In 1837 Ward addressed this Association on the growth of plants in closed cases. By 1842 Dr James Johnstone, author of *British sponges and lithophytes,* was using small jars to keep a few red and green seaweeds, a starfish and some small molluscs and worms, and in 1846 a Mrs Thynne was keeping a few shore animals in small vessels in her house in London. In March 1850 Robert Warrington told the Chemical Society how he kept goldfish, snails and freshwater plants successfully in a tank.

The first real attempt to keep a marine aquarium came in January 1852, when both Warrington and Philip Henry Gosse were independently starting to stock small vessels with marine algae and plants. On 18th February 1852 the Council of the Zoological Society agreed that work should start immediately on the building of an Aquatic Vivarium. By the end of that year Gosse was able to take a further step and in his book *The Aquarium* he wrote:

> "Early in December, 1852, I put myself into communication with the Secretary of the Zoological Society, and the result was the transfer of a small collection of Zoophytes and Annelides, which I had brought up from Ilfracombe and which I had kept for two months in cases in London,—to one of the tanks in the new Fish House just erected in the Society's Gardens in Regent's Park. This little collection thus became the nucleus and the commencement of the Marine Aquarium afterwards exhibited there." (Gosse, 1854.)

By the end of 1853 these relatively small tanks had been used to house a varied selection of marine animals. Of these, 58 were species of fish and no fewer than 200 invertebrates. This figure included 76 species of mollusc, 41 crustaceans, 27 coelenterates, 15 echinoderms, 14 annelids as well as a sprinkling from the small invertebrate groups.

Among the seaweeds, all green, were *Ulva, Enteromorpha* and *Bryopsis.*

By 1854 some of the tanks had not been changed for more than seven months and several animals had survived for nearly a year.

Such was the first public marine aquarium (Fig. 2).

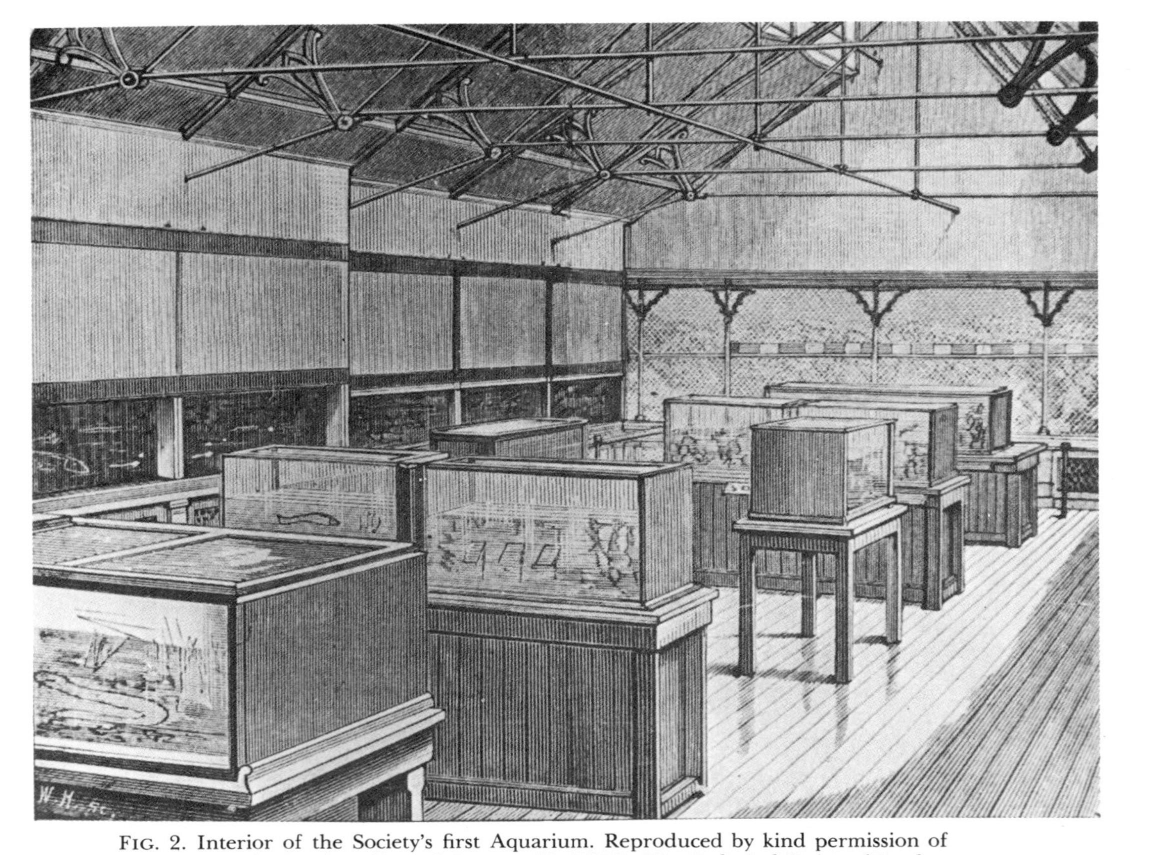

FIG. 2. Interior of the Society's first Aquarium. Reproduced by kind permission of Cassell & Co. Ltd., London. From Scherren, H. (1905). *The Zoological Society of London.* London: Cassell.

Gosse continued to supply regular consignments of marine animals to the Zoological Society during the first half of 1853, but in July of that year a misunderstanding appears to have arisen between him and the Council of the Society. On the Society's side the only inkling of this is a record of his account being questioned. The matter was put in the hands of Professor Bell and Mr Bowerbank and at the end of September they recommended a payment of £150 for the specimens provided. Gosse himself never refers to any disagreement with the Society in any of his written works, but his son, Sir Edmund Gosse, records (E. Gosse, 1890) that the misunderstanding arose because his father was getting more specimens than he needed and was sending the excess to other aquaria established as a result of the successful experiments at Regent's Park. In doing this Philip Gosse was not breaking any written or verbal agreement. Relations were also strained by the fact that the Society had allowed no mention of Gosse's name to appear in the numerous press reports describing the new aquarium. At any rate Gosse stopped supplying the Zoological Society which then obtained its specimens from August 1853 onwards from William Thompson of Weymouth, and paid enormous sums for them. For the 12 months from August 1853 Council Minutes recorded that Thompson was paid £184 for marine animals, which converted to present rates represents more than the total annual outlay on all specimens for the Society's large aquarium today.

In spite of this misunderstanding Gosse and the Zoo had taken a great step forward in presenting marine life to the public. He had successfully introduced the word *aquarium* in its modern sense, in spite of attempts by others to use *marine vivarium* or *aquavivarium.* In classical Latin an aquarium refers to a watering place for cattle.

The immediate post-Gosse period saw a flood of books on aquaria and the sea and especially on the seashore; Charles Kingsley's *Glaucus* was first published in 1855, and greatly improved in 1858 by the addition of the beautifully illustrated companion by George Brettingham Sowerby. An elegant fountain aquarium is shown in Fig. 3 (from Gosse, 1854). There were also books by Edwin Lankester in 1856 who used the term aquavivarium and by Noel Humphreys who favoured the more romantic title of *Ocean and river gardens.*

There was a gap of some years before new large aquaria were established. In August 1871 Crystal Palace Aquarium was opened and its first guide by W. Alford Lloyd was published in December

FIG. 3. The Fountain Aquarium. From Gosse (1854).

of that year. Others in this period were Brighton, Westminster and Manchester. The achievements of Lloyd as Superintendent and designer of the Crystal Palace Aquarium and later of the Aquarium in Berlin have been discussed by Wilson (1962). The other outstanding aquarium man of this period was Henry Lee of Brighton.

It was at this time that marine aquaria were made with circulating water systems, mostly on the principles laid down by Lloyd. As a result there was a great increase in the number of marine fishes which could be kept in good condition.

Aquaria became widespread too on the Continent and in America, in Paris, Antwerp, Amsterdam, Berlin, New York, and, of course, from the 1870s onwards came the spate of marine biological stations with aquaria attached, notably Naples, Plymouth, Millport and Roscoff.

THE PRESENT AQUARIUM

The Society's present Aquarium was completed and opened in May 1924, in space which had been left for that purpose beneath the Mappin Terraces when these were built in *c.* 1912.

Public aquaria all rely on some form of water circulation and there are numerous variants. In an *open* circulation, as at Naples and Millport, water is pumped direct from the sea into a high-level tank whence it flows through the exhibition tanks and then straight out to sea. In a semi-closed circulation as used at the Plymouth Laboratory, a large volume of water is taken in direct from the sea and used for a few months, usually only three or four, and then the greater part of it is replaced by new water. In a *closed* circulation the same water is used over and over again for several years, with only occasional refreshers of new water.

The Zoological Society's Aquarium uses natural sea-water in a closed circulation (Fig. 4). The sea-water is brought from the Bay of Biscay as ship's ballast and transferred to stainless steel tanker lorries at the London Docks. These vehicles bring the sea-water in a series of journeys, usually at night. The supplies of fresh sea-water are stored in underground reservoirs and allowed to rest before being used.

For some 50 years the Aquarium has used a closed circulation, and on a reduced scale, this system has also been used in several biological laboratories in Britain. The water is pumped up from a reservoir to a gravity tank situated high above the aquarium, whence it falls to the exhibition and reserve tanks. The water enters each tank at the bottom, leaves through an outlet at the top, and is then conveyed by piping to the filters, which will be described below. After having passed through the filters it flows, again by gravity, back to the reservoir.

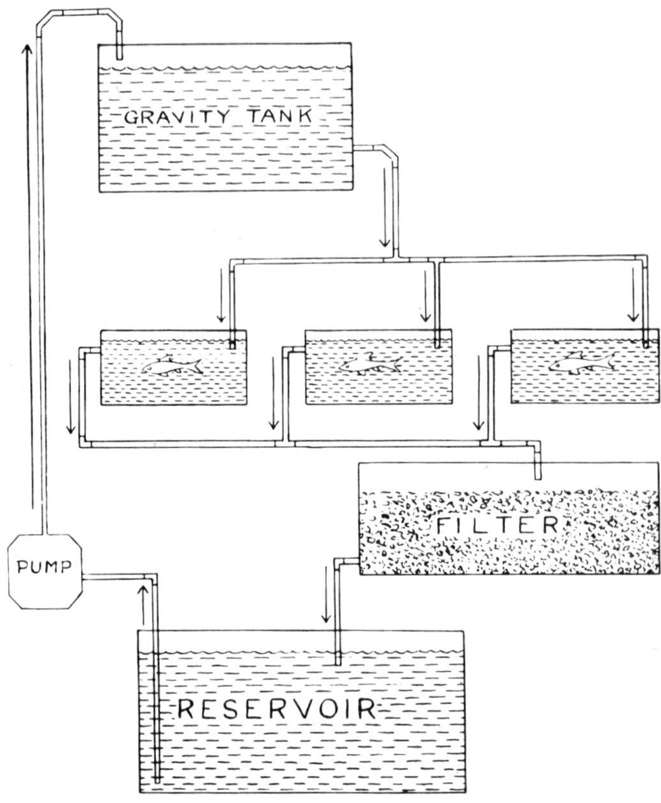

FIG. 4. Aquarium circulation system. Reproduced with permission from Vevers, H. G. (1972). Marine aquaria. In *UFAW handbook on the care and management of laboratory animals*: 537–542. 4th edn. UFAW (eds). London & Edinburgh: Churchill Livingstone.

For systems with less than 4500 litres an air-lift filter is usually sufficient. In such cases the filter is of nylon wool and is really only a sieve straining out the coarser particles. The finer material left in the water is an advantage, provided there is not too much of it, as it supplies food for filter-feeders. As a variant, to suit space requirements, there is no reason why the filter should not be built between the gravity tank and the aquarium tanks.

In such a closed circulation it is usually necessary to add about 10 to 20% of new sea-water per annum to make good losses caused by siphoning, tank-cleaning and other maintenance operations.

The policy of the Aquarium is to exhibit a wide variety of aquatic animals from the fresh and salt waters of temperate and tropical regions of the world. This entails the maintenance of a number of different circulations, of which the principal are:

Freshwater tanks
  Temperate, at 16°C
  Temperate for salmonids, at 8–10°C
  Tropical, at 25°C
Sea-water tanks
  Temperate, at 16°C
  Tropical, at 25–26°C

In addition a number of separate circulations have been established in recent years for certain of the sea-water tanks. This has been done as a result of the work by the late Dr J. H. Oliver and me in the years 1955–1960.

Before this period is was found that certain echinoderms and octopus almost invariably died within a few weeks of arrival. The sea-water circulation was then a single massive unit, filled originally with natural sea-water from the Bay of Biscay. Throughout the years there had been a slight wastage of water from small leaks and from the process of tank cleaning. This loss was made good by annual refreshers of sea-water. At the beginning of the Aquarium's life, around 1930, we have records which show that the sea-water in the circulation was still relatively close in composition to natural sea-water. But in 1956 Dr Oliver's analyses showed some astounding changes. Calcium had increased to a figure three times the normal in the sea, phosphate by 90 times and nitrate by over 1000 times (Oliver, 1957).

We immediately installed a separate self-contained circulation system—a miniature of the main system—and filled it with new sea-water which had a nitrate content of less than 1 part per million (p.p.m). In this system common octopus have lived for 8–10 months, which compares very favourably with the records of coastal aquaria. The nitrate content of the water has been kept at a low level by renewing 80% of it once a week, the discarded water being passed to the main circulation. At the same time weekly samples have been analysed and we can now say that a large octopus, when feeding well, causes the nitrate content of the water in this closed circulation of 2700 litres to rise at the rate of approximately 1 p.p.m. per day. Octopus have even laid and hatched eggs in this controlled circulation (Vevers, 1961).

The same tank has also contained a few dozen tropical sea-urchins and up to a dozen starfishes. These too have lived and thrived as they have never done before, lasting well over a year. If specimens of the tropical urchin *Diadema antillarum* are moved from this special tank to a neighbouring tank supplied with the high-nitrate water they die within three weeks. Other tanks with self-contained circulations have been set up, with water at lower temperatures, and these have been successful in keeping brittle-stars (*Ophiocomina nigra*), plumose anemones (*Metridium senile*) and other so-called difficult invertebrates. This is only one example of the type of work which is still needed on the requirements of individual invertebrate species in aquaria. There are many other animals, still difficult to keep, whose needs present a challenge to future research.

At this stage it is not suggested that nitrate itself is necessarily the damaging factor in this polluted sea-water but there is no doubt that it provides a most useful indicator of conditions which are unsuitable for many marine invertebrates.

The removal of the excess nitrate and other ions has often been suggested. However, there is no chemical method of precipitating nitrate. In theory it would be possible to remove some of the excess ions by exchange resins but this would involve a great deal of expense, and it would be difficult to remove the different ions in the correct proportions. In the sea the nitrate is, of course, used up by plants and this could probably be done, at any rate on a small scale, under aquarium conditions. It is doubtful whether this would really be economical in a large public aquarium with a quarter of a million gallons of sea-water and virtually no natural light. The amount of artificial light needed and the difficulty of removing the plant crop from the water would make it a very costly process.

The practical solution seems to be flexibility in the use of the sea-water. Where a large volume of sea-water is available its movements can be so arranged that new natural sea-water is treated as first class water and supplied only to delicate and difficult animals such as octopus, some sea-anemones and small crustaceans, and nearly all the echinoderms. After this water has been used and polluted to a known degree it can be reclassified as second class water and diverted to one of the other circulations for use by teleost and selachian fishes.

Here I should like to say a word about the danger of relying too much on hydrometer readings. These give an indication of the gravity but they do not tell you what is really happening to the sea-water. Evaporation can be made good by the addition of distilled or soft water, but Dr Oliver's work on the Zoo Aquarium circulations shows quite clearly that the processes going on in a closed circulatory system are far more complicated than mere evaporation. The changes in the chemical composition of aquarium tank water are of such a type that hydrometer readings are of little real value, and may in fact, be dangerously misleading.

MATERIALS

In addition to the provision of suitable water at the correct temperatures and with the correct composition, there is also the very practical problem of how to contain it. Care must be taken to avoid substances which are either unstable in sea-water or injurious to the animals. No copper (including brass, gun-metal and bronze) or zinc must be used in any part of the aquarium, and it is also advisable not to use aluminium or lead, nor cast iron which quickly rusts, however carefully it is painted. In fact there is no need to use these metals, as glass, polythene, slate, asbestos, porcelain, fibreglass, polyvinyl chloride or waterproof cement can all be used for tanks and piping.

In the early days of the present aquarium the main structure of the large tanks, excluding the front plate glass, was usually of slate, a material which is quite safe for living animals, but which is costly and very liable to split without warning. Moulded glass tanks are satisfactory for certain small animals, but they are fragile and do not lend themselves to the installation of water-circulation equipment. Asbestos tanks, now marketed for domestic purposes, and ordinary porcelain sinks are perfectly satisfactory for laboratory use and for aquarium reserves when there is no need to view the animals from the side through glass.

Apart from small tanks used for tropical fishes, any new tanks are now made of reinforced waterproofed cement. These do not require a metal frame to hold the plate glass, for this can be easily bedded with non-setting mastic against a true cement surface. In Britain the product Glasticord marketed by Industrial Engineering Ltd, Vogue House, Hanover Square, W.1., has been satisfactorily used by laboratories and large public aquaria as a mastic for bedding the plate glass.

A cement mix can be safely and satisfactorily waterproofed by dissolving ordinary commercial soap powder (not liquid soap or any form of detergent) in the mixing water at the rate of 1 cup to the gallon. On no account should proprietary sealing or waterproofing compounds be used unless they are accompanied by a detailed and reliable chemical analysis.

Angle-iron tanks are still used but it is difficult to protect the frame when used with sea-water, even when coated with nylon. These are being gradually replaced with tanks constructed of glass sheets cemented together with silicon-rubber glue.

The original Aquarium piping was of cast iron lined with glass or porcelain, a very expensive product, and of course all joints and pumping machinery have to be constructed in such a way as to avoid the use of copper and zinc in any form. This has now been largely replaced by relatively inexpensive plastics. Various types of polythene were used in the 1950s, but we now use rigid polyvinyl chloride which is more easily worked.

FILTRATION

In an aquarium circulation some of the larger particles of waste can be removed by passing the water through a layer of nylon wool. This is a process of sieving. Glass wool is not suitable as it fragments leaving tiny particles of glass which become lodged in the gills of fishes. Except from the aesthetic viewpoint, the visible particles of waste are, however, relatively unimportant. On the other hand, waste matter in solution can be extremely dangerous, and this applies particularly to the ammonia excreted by fishes.

In the filtration system used at Regent's Park the water is passed through a bed of coarse grit supported on pebbles (*c.* 1–2 cm diameter). The grit particles have a diameter of *c.* 2 mm; if a finer grade of sand or grit is used there is a risk that the filter will become clogged and cease to function efficiently. The grit becomes coated with bacteria, heterotrich ciliates, harpacticid copepods and other organisms which break down organic debris and wastes. When the filter is working efficiently it has been found that the water enters the bed with an oxygen content of *c.* 7 mg/litre; the exact figure depends, of course, upon the temperature. In passing through the filter the water loses *c.* 3 mg/litre of this oxygen, which has been used by the micro-organisms on the grit, mainly in the process of oxidizing ammonia to nitrite and nitrate.

The surface area of the filter should be large in relation to its depth.

The water is distributed over the surface of the grit and allowed to descend through the filter bed. It is then collected in a perforated hard polythene pipe lying 2 cm above the bottom, which leads it to the underground reservoir.

MAINTENANCE OF pH

The pH of the water should be kept at approximately 7·8 to 8·2. In nearly every case the pH of aquarium tanks decreases owing to the metabolic processes of the fishes and invertebrates. Some aquaria add calcium oxide to correct the pH, but this eventually leads to an increase in the relative proportion of calcium. On the whole, sodium hydroxide, added gradually and well-diluted, is probably the most satisfactory additive for the correction of low pH values in a closed circulation.

RANGE OF EXHIBITS

A public aquarium should, of course, aim to have exhibits that are of educational and recreational value. The vast majority of visitors will be there for recreational purposes and for them the range of exhibits will not be so critical, as long as these are displayed with imagination. Ideally the educational aims should be very broadly based, so that young schoolchildren as well as university students will find something of value.

Most of the exhibits will, of course, be fishes. Here a wide variety of species is available, covering many, but by no means all the fish families. It is always advisable to have some exhibits that illustrate biological principles, such as adaptive colouration; flatfishes to show rapid colour change, anabantids that breathe air, mormyrids and others with electric organs, cichlids that show highly advanced methods of brood protection, developing dogfish eggs, and so on.

The Zoological Society's Aquarium has always made a point of exhibiting what may be colloquially termed the primitive fishes for the benefit of the advanced students. These include garpike (*Lepisosteus*), bowfin (*Amia*), sturgeon (*Acipenser*), lungfishes (*Protopterus, Lepidosiren* and *Neoceratodus*), as well as various species of *Polypterus*.

In addition, it is essential to have a reasonably representative collection of invertebrates to include crustaceans, molluscs, echinoderms, polychaetes and sea-anemones.

References

Gosse, E. (1890). *The life of Philip Henry Gosse F.R.S.* London: Kegan Paul, Trench, Trubner & Co. Ltd.

Gosse, P. H. (1854). *The aquarium: an unveiling of the wonders of the deep sea.* London: John Van Voorst.

Oliver, J. H. (1957). The chemical composition of the sea water in the aquarium. *Proc. zool. Soc. Lond.* **129**: 137–145.

Thompson, D'Arcy, W. (1940). *Science and the classics.* London: Oxford University Press.

Vevers, H. G. (1961). Observations on the laying and hatching of octopus eggs in the Society's Aquarium. *Proc. zool. Soc. Lond.* **137**: 311–315.

Wilson, D. P. (1962). The semi-closed circulation system at the Plymouth Laboratory. *Bull. Inst. océanogr. Monaco.* Spec. No. 1B: 13–27.

Symp. zool. Soc. Lond. (1976) No. 40, 119–132.

REPTILES

A. d'A. BELLAIRS and D. J. BALL

The Zoological Society of London
Regent's Park, London, England

SYNOPSIS

The history of keeping reptiles in the collection of the Zoological Society of London is briefly reviewed, and present methods for their husbandry are described, with special reference to accommodation, feeding, breeding and precautions against snake-bite. Some possible ideas for the future are suggested.

INTRODUCTION

Opinion polls conducted some years ago among nearly 12 000 children showed that "the snake" easily topped the list as the most unpopular animal in the Zoological Gardens. A tiny proportion (0·9% of the boys, 0·3% of the girls), however, listed it as the animal they liked the best (Morris & Morris, 1965). Whether hated or loved, snakes, and to a lesser extent other reptiles, exercise a singular fascination for many people, as is evident from the large numbers of visitors to the Reptile House.

It is therefore hardly surprising that the first Reptile House in the Society's Menagerie should have been built as early as 1849, little more than 20 years after the Society was founded (see Mitchell, 1929). Much information about its denizens can be gleaned from a charming and informative book by Catherine C. Hopley (1882), and it is amusing to learn that in those days the snakes were provided with woollen blankets as a means of shelter. On one occasion a large boa is said to have swallowed its blanket by mistake, but was able to disgorge it without coming to harm. A more modern Reptile House, heated by steam pipes, was erected in 1889 and still stands as the present Bird House. The smaller building alongside it was made into a house for tortoises and has been subsequently converted into the house for humming-birds. The first Curator of Reptiles, E. G. Boulenger, was appointed in 1911 and later took charge of the Aquarium. He was the son of G. A. Boulenger, the renowned herpetologist at the British Museum (Natural History), and wrote *The Zoo and Aquarium book* (1932) and other popular books.

The present Reptile House was planned largely by its second Curator, Dr Joan B. Procter; E. G. Dawber was the architect. When completed in 1927 it was probably the most advanced establishment of its kind in the world; indeed, it still functions very efficiently today. Perhaps its most famous early inhabitants were a pair of Komodo dragons (*Varanus komodoensis*), the biggest of existing lizards, which were discovered in Indonesia as recently as 1912. One of these seven-foot reptiles soon became tame and could be trusted with small children (Procter, n.d.). This is all the more surprising since it has been learnt that these lizards are formidable predators as well as scavengers and can kill big mammals, even buffaloes, by biting through the tendons of their legs and then attacking the belly (Auffenberg, 1972).

Joan Procter, who did so much to establish the scientific reputation of the Reptile House, was a fine herpetologist, and contributed some excellent research publications; perhaps the most important was her study of the remarkable East African tortoise, now called *Malacochersus tornieri,* the flat flexible shell of which is adapted for concealment in rocky crevices (Procter, 1922). A bust of her is to be seen in the entrance lobby of the House. Subsequent Curators have been Dr Burgess Barnett who was interested in the therapeutic uses of snake venom (see Macfarlane & Barnett, 1934), and Mr Jack Lester (see Smith, 1957, for Obituary). In 1964 the new post of Herpetologist was created for Mr R. A. Lanworn, who as Head Keeper and Overseer had played an important part in managing the collection for many years. In his retirement he has written an attractive illustrated book on reptiles (1972).

Dr H. G. Vevers is at present Acting Curator of the Reptile House; there is an Overseer (Mr D. J. Ball), a Head Keeper (Mr Brian Savage) and three Keepers. There is also an Honorary Herpetologist (Prof. A. d'A. Bellairs) whose duties are mainly advisory, though he participates in an annual course of lectures given for zoology students of the University of London.

At the time of writing (January, 1976) the Reptile House contains some 327 individual animals, exclusive of sea turtles which are housed in the Aquarium. Of these there are about 112 chelonians, 23 crocodilians, 83 lizards and 109 snakes. The collection of amphibians was transferred from the Reptile House to the Aquarium in 1968, and the cooler temperatures prevailing there have been found more suitable for their maintenance.

REPTILE HUSBANDRY TODAY

During recent years there have been many advances in reptile husbandry (see Ball & Bellairs, 1972) and these are increasingly reflected in current practice at the Zoo. Much remains to be learnt, however, and it must be remembered that the maintenance of reptiles for public exhibition presents many problems not encountered by the private vivarium keeper or the laboratory worker. Thus it is necessary to consider both the optimum living conditions for the various species and the obligation to render the animals visible (or often visible) to the public gaze.

Plan of Reptile House and cages

The exhibition cages are arranged in two concentric series with a public gallery between them running round the whole building (Fig. 1). The inner cages surround a central hall which acts as a service area and a place where new arrivals can be unpacked in safety, while the outer cages are backed by service corridors (Fig.

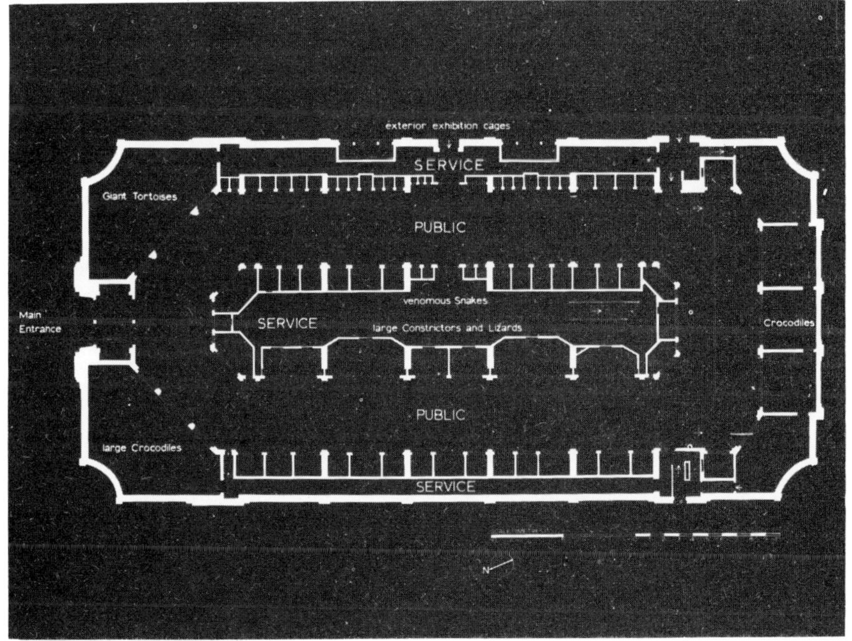

FIG. 1. Plan of Reptile House.

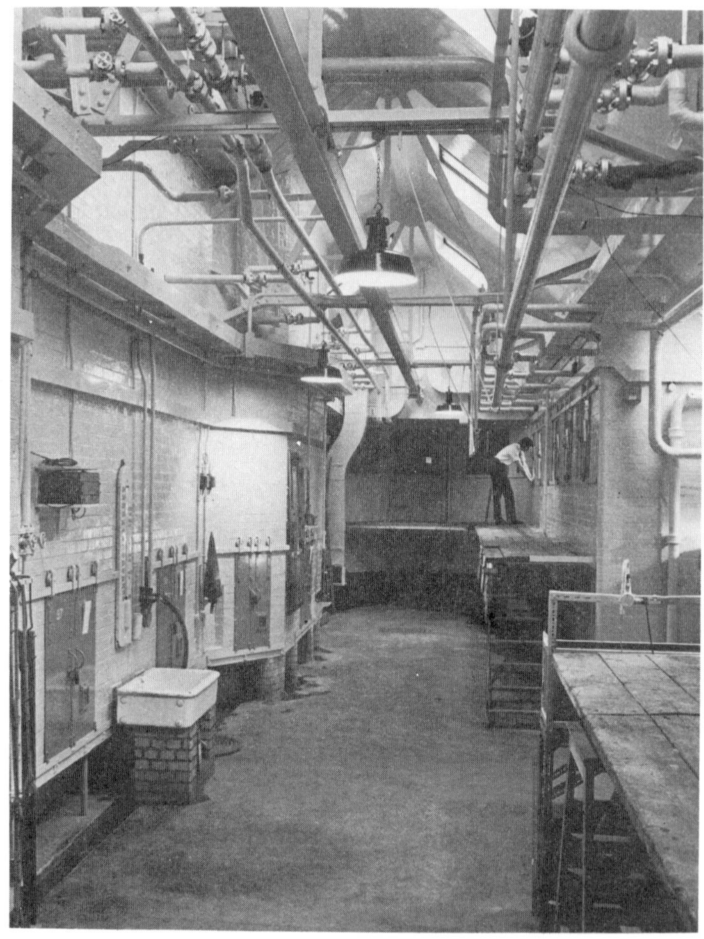

FIG. 2. Centre service area behind the cages for venomous snakes and large constrictors.

2). All the cages are illuminated, and the public gallery is kept comparatively dark. The Overseer's office and Keepers' quarters are in the basement; the first floor is used partly for accommodating animals under special observation, while sheds on the flat roof are used for cultures of locusts, an invaluable source of reptile food. There is a raised bay for crocodile pools at one end of the building and another large crocodile pool and a pen for giant tortoises near the entrance (Fig. 3). For reasons of security all poisonous snakes are restricted to the central group of cages and

FIG. 3. Giant tortoise enclosure.

the doors into them from the central hall are placed high up in the back wall, thus minimizing the chances of escape—which in fact has hardly ever occurred.

It has been shown that many species of reptiles can survive well in laboratory cages when no attempt is made to reproduce a semblance of natural conditions. Such cages are unattractive, however, from the standpoint of public exhibition and in many cases suitable "furniture" has been provided in the form of rockwork, bark, branches and growing plants. Sand, gravel, peat or loam are used as floor covering and there is always a pond of a size appropriate to the life-style of the inhabitants. Comparatively bare cages, furnished only with platforms, ponds and a few large branches are still thought most suitable for big pythons however; these seemingly sparse conditions do not appear to affect the health

FIG. 4. Prototype reptile cage constructed from plastic and glass fibre.

of the reptiles and certainly facilitate the task of cleaning. It is possible that in the future the concrete floors and walls could be advantageously replaced by some form of plastic which could be moulded into artificial rockwork. Experiments with prefabricated plastic cages (Fig. 4) for the smaller reptiles are in progress. These cages could be "furnished" in the service areas and then fitted into place. Such portable accommodation would be more adaptable than the fixed cages at present in use and might be particularly suitable for special exhibits illustrating, for example, protective colouration.

Heat, light and air

During recent years a great deal of interesting work has been carried out on the temperature relations of reptiles (see Cloudsley-Thompson, 1972; Bellairs & Attridge, 1975). In the wild these animals do not passively assume the temperature of their surroundings, but control their body temperature to a considerable extent, mainly by means of their behaviour. They make themselves warmer by seeking sources of heat (as by basking in the sun, or on a warm substratum) and cool themselves by taking shelter under rocks or plants, in holes or in the water. Many species spend a good deal of their active lives shuttling in and out of the sun and are thus able to keep themselves within a "preferred" range of body temperatures; in many diurnal lizards from subtropical or tropical countries these may be remarkably high (33–38°C), much higher than the temperature of the surrounding air. Such methods cannot operate, of course, during a northern winter, and reptiles which live in temperate climes spend about half their lives or more in an inactive state of hibernation.

The great majority of the occupants of the Reptile House require some form of artificial heating. In many cages this is achieved by means of electric elements beneath the floors, supplemented by steam panels heated from a boiler house in the Aquarium. Background temperatures of some 25–30°C, corresponding with the preferred body temperatures of many snakes, are normally maintained; higher temperatures, if required, are provided by electric light bulbs, supplemented in some cases by infrared lamps. Where additional heat rather than illumination is required special infrared heaters are used. Gradients which enable reptiles to regulate their body temperatures are produced by asbestos sheets laid in corners of the cages to cut off sub-floor heating. These sheets are surrounded by rockwork, the cavities so formed being filled with soil and stocked with plants, thus providing cool retreats for reptiles.

Although some of the larger cages receive natural daylight through the roof, electric bulbs are the main sources of illumination. Artificial light very similar to natural daylight is supplied by special lamps (Duro-Test Internationals True-Lite; called Vitalite in the USA) in cages containing desert lizards, such as certain agamid and iguanid species. Light is turned off at night and the background heating lowered.

Ventilation is provided by a system of ducts through which the air is circulated by fans.

Feeding

In the wild most species of snakes (at least when adult) feed exclusively on living vertebrate animals. In the nineteenth century live birds and mammals were fed to the snakes at the Zoological Gardens but the procedure aroused public controversy and was eventually stopped by order of the Secretary, Sir Peter Chalmers Mitchell.

In fact, pythons and other snakes which normally feed on warm-blooded prey usually accept freshly killed animals such as rats, mice and fowls quite readily and may even eat carcases which have been kept frozen and then warmed up. Some individuals, however, can only be induced to feed with difficulty, and forcible feeding is occasionally employed as a last resort to prevent the snake dying from self-induced starvation. A method commonly used is to place the bodies of freshly killed rats or mice in a liquidizer and administer the fluid through a tube (Fig. 5).

Since snakes, crocodilians and certain carnivorous lizards such as monitors can be given food which, nutritionally speaking, closely resembles their natural prey, they should seldom suffer from deficiency diseases in captivity. This is not so, however, with many chelonians and lizards, which in nature consume a more varied diet. Under Zoo conditions these animals are regrettably prone to a form of deficiency which results in softening of the shell and bones so that pathological fractures may occur. The condition may be due to shortage of vitamin D, or calcium, or to an imbalance of the calcium: phosphorus ratio in the diet (see Frye, 1973). In fact there is little scientific information about the precise dietary requirements of reptiles and further research on the subject is needed.

The practical policy is to provide as varied a diet as possible. Insectivorous lizards are fed on locusts, crickets, cockroaches, flies

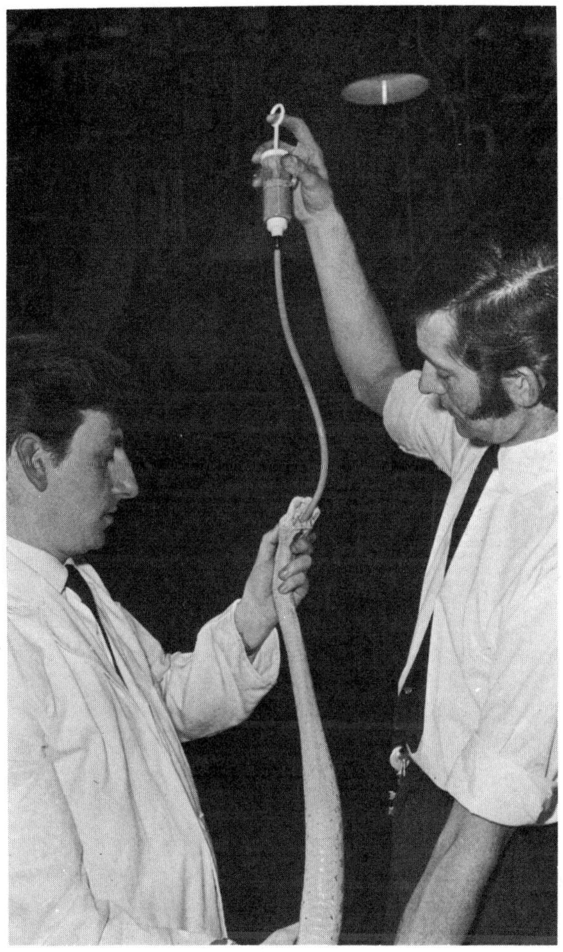

FIG. 5. Green tree-boa (*Corallus caninus*) being fed on liquidized rats.

and mealworms with multivitamin and mineral supplements such as Vionate or Beta Mix 221. Unfortunately it is not practicable to provide spiders which are known to form much of the food of many lizard species in the wild.

A number of lizards such as iguanas and the agamid *Uromastyx* are more or less herbivorous and are given mixtures of vegetables and fruit, supplemented with animal food such as locusts, chopped liver, eggs, bonemeal and dog food. Tortoises, which are often regarded as purely herbivorous animals, also appreciate a certain amount of animal protein in the form of dog food, sprats and dead mice, with suitable supplements. It cannot yet be claimed, however, that an ideal diet for many species of captive reptiles has been discovered.

Breeding

Successful mating and breeding, especially of snakes, sometimes occurs, and of course females impregnated before capture quite often lay eggs or produce young. Eggs are usually removed from the cage and incubated artificially in one of the upstairs rooms; this avoids the risk of damage to them by the movements of parents and other occupants of the cage. Young lizards can often be fed on aphids and fruit flies, and baby snakes of many species will accept freshly killed new-born mice. The successful rearing of young reptiles is, however, a difficult matter, requiring much patience and skill.

Handling and precautions against bites

Many of the larger reptiles, and of course all poisonous species, are potentially dangerous to handle, and various precautions must be taken during the unpacking of arrivals or when animals have to be transferred from one cage to another. Perhaps the best way of dealing with large crocodilians is to pass a canvas muzzle supported by a pole over the animal's head (Fig. 6); the limbs and tail can then be grasped without much difficulty.

Poisonous snakes are handled as little as possible. The more sluggish types such as vipers can be lifted by means of a simple rod with a metal loop at one end; this is slipped under the middle of the snake's body so that it balances when raised. With more active species such as cobras it may be necessary to use a special type of snake-stick with hinged jaws padded with rubber and operated by a lever at the handle (Fig. 7). For close examination the head of the snake may be pinned on a sheet of foam rubber with a T-ended rod. The snake is then grasped behind the head. If it is necessary to immobilize a snake, as for close inspection, the animal can be manoeuvred into a tube of transparent plastic; holes can be made in this to allow access to the snake's body.

Only one human death from snake-bite has occurred during the Zoo's history. This was in 1852, when an inebriated keeper swung a cobra round his head and was bitten on the nose (Scherren, 1905). There have, however, been several subsequent

FIG. 6. Securing the jaws of a Mississippi alligator. The pole may be removed and the muzzle tied to the body.

FIG. 7. Snake-stick being used to restrain a puff adder (*Bitis arietans*).

occasions when members of the staff of the Reptile House have received poisonous bites and it is clearly essential that adequate precautions should be taken against this hazard. It is no longer thought desirable that antiserum should be given as a routine measure, and treatment at the Reptile House is confined to first aid: reassurance of the victim, who should be advised to move as little as possible, wiping off excess venom from the wound, and the application of a tourniquet in certain cases. Such directions for first aid are shown on printed cards (Fig. 8), and special arrangements have been made to admit all cases of snake-bite to the Middlesex Hospital, London, where they can be kept under observation. Stocks of antiserum are held in the Reptile House and sent with the patient to the hospital for administration (if necessary). All members of the Reptile House Staff are tested for sensitivity to horse serum, since allergic reactions which often follow the use of antiserum (prepared from horses) may be more dangerous than the bite itself. Tests for blood grouping are also made in case transfusion should be necessary.

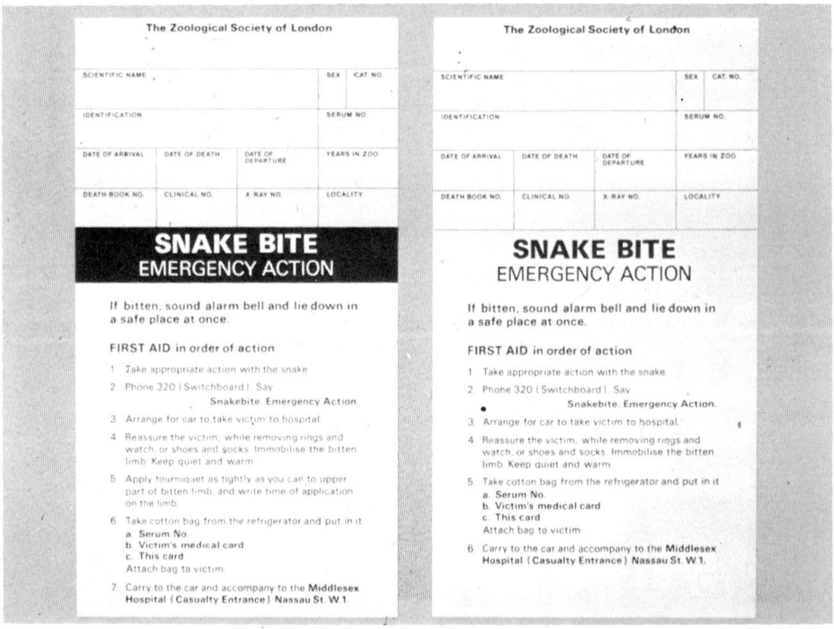

The Zoological Society of London

SCIENTIFIC NAME		SEX	CAT. NO.
IDENTIFICATION		SERUM NO.	
DATE OF ARRIVAL	DATE OF DEATH	DATE OF DEPARTURE	YEARS IN ZOO
DEATH BOOK NO.	CLINICAL NO.	X-RAY NO.	LOCALITY

SNAKE BITE
EMERGENCY ACTION

If bitten, sound alarm bell and lie down in a safe place at once.

FIRST AID in order of action

1. Take appropriate action with the snake
2. Phone 320 (Switchboard). Say **Snakebite. Emergency Action.**
3. Arrange for car to take victim to hospital.
4. Reassure the victim, while removing rings and watch, or shoes and socks. Immobilise the bitten limb. Keep quiet and warm.
5. Apply tourniquet as tightly as you can to upper part of bitten limb, and write time of application on the limb.
6. Take cotton bag from the refrigerator and put in it
   a. **Serum No.**
   b. **Victim's medical card**
   c. **This card**
   Attach bag to victim.
7. Carry to the car and accompany to the **Middlesex Hospital (Casualty Entrance) Nassau St. W.1.**

The Zoological Society of London

SCIENTIFIC NAME		SEX	CAT. NO.
IDENTIFICATION		SERUM NO.	
DATE OF ARRIVAL	DATE OF DEATH	DATE OF DEPARTURE	YEARS IN ZOO
DEATH BOOK NO.	CLINICAL NO.	X-RAY NO.	LOCALITY

SNAKE BITE
EMERGENCY ACTION

If bitten, sound alarm bell and lie down in a safe place at once.

FIRST AID in order of action

1. Take appropriate action with the snake.
2. Phone 320 (Switchboard). Say **Snakebite. Emergency Action.**
3. Arrange for car to take victim to hospital.
4. Reassure the victim, while removing rings and watch, or shoes and socks. Immobilise the bitten limb. Keep quiet and warm.
5. Take cotton bag from the refrigerator and put in it
   a. **Serum No.**
   b. **Victim's medical card**
   c. **This card**
   Attach bag to victim.
6. Carry to the car and accompany to the **Middlesex Hospital (Casualty Entrance) Nassau St. W.1.**

FIG. 8. First aid instruction cards.

IDEAS FOR THE FUTURE

The lives of reptiles, in particular their breeding behaviour, are greatly influenced by environmental factors such as temperature and probably also light. Unfortunately, little is known in detail about the ways in which these factors operate. It is also known that many reptiles, in particular certain lizards, are highly territorial and have elaborate patterns of courtship and display comparable with those of many birds. Some species are unlikely to go through normal reproductive cycles unless they have more space than can usually be given under zoo conditions.

Investigation of these interesting problems is a matter for the research laboratory and the field worker rather than the menagerie. It would be interesting, however, to set up cages for suitable species in which a greater attempt to simulate natural conditions was made. Thus lighting and temperature could be individually controlled and, for temperate-zone species, varied according to season. In this way it might be ascertained whether hibernation has any great effect on reproduction and survival, a matter about which comparatively little is known (see Peaker, 1969). Most important of all, some cages might be allotted to small potential breeding groups, with sufficient space for males to establish territories and for females to find suitable nesting sites. The practice of keeping European reptiles in an outdoor reptiliary was abandoned some years ago for various reasons. However, it might be possible to try outdoor husbandry on a more limited scale, and to maintain small colonies of such attractive species as sand, wall and green lizards both at Regent's Park and at Whipsnade. Some amphibians might also be kept in this way at comparatively little expense and special priority could be given to the maintenance of threatened British forms such as the sand lizard and the natterjack toad.

These are some of the ways in which the Zoological Society might enhance its already considerable contribution to the husbandry of cold-blooded animals.

REFERENCES

Auffenberg, W. (1972). Komodo dragons, *Nat. Hist., N.Y.* **81**: 53–59.

Ball, D. J. & Bellairs, A. d'A. (1972). Reptiles. In *The UFAW Handbook on the care and management of laboratory animals*: 490–510. (4th edn). Edinburgh & London: Churchill, Livingstone.

Bellairs, A. d'A. & Attridge, J. (1975). *Reptiles.* (4th edn). London: Hutchinson University Library.

Boulenger, E. G. (1932). *The Zoo and Aquarium book.* London: Duckworth.
Cloudsley-Thompson, J. L. (1972). Temperature regulation in desert reptiles. *Symp. zool. Soc. London.* No. 31: 39–59.
Frye, F. L. (1973). *Husbandry, medicine and surgery in captive reptiles.* Bonner Springs, Kansas: VM Publishing.
Hopley, C. C. (1882). *Snakes: curiosities and wonders of serpent life.* London: Griffith & Farran.
Lanworn, R. A. (1972). *The book of reptiles.* Hamlyn: London.
Macfarlane, R. G. & Barnett, B. (1934). The haemostatic possibilities of snake venom. *Lancet* **1934**. Nov. 3: 985–987.
Mitchell, P. Chalmers (1929). *Centenary history of the Zoological Society of London.* London: Zoological Society.
Morris, R. & Morris, D. (1965). *Men and snakes.* London: Hutchinson.
Peaker, M. (1969). Some aspects of the thermal requirements of reptiles in captivity. *Int. Zoo Yb.* **9**: 3–8.
Procter, J. B. (1922). A study of the remarkable tortoise, *Testudo loveridgii* Blgr., and the morphogeny of the chelonian carapace. *Proc. zool. Soc. Lond.* **1922**: 483–526.
Procter, J. B. (n.d., *c.* 1928). Dragons that are alive today. In *Wonders of animal life* **1**: 33–41. Hammerton, J. A. (Ed.), London: Amalgamated Press.
Scherren, H. (1905). *The Zoological Society of London.* London: Cassell.
Smith, M. A. (1957). Obituary. Jack Lester. *Br. J. Herpet.* **2**: 79.

Symp. zool. Soc. Lond. (1976) No. 40, 133–145.

THE POLICY OF KEEPING BIRDS IN THE SOCIETY'S COLLECTIONS 1826–1976

P. J. S. OLNEY

The Zoological Society of London, Regent's Park, London, England

My approach to this subject—the policy of keeping birds in the Society's collections from 1826 to 1976—has been to relate the reasons why birds have been kept, and to some extent the methods by which they have been kept, to changes which have taken place in zoological knowledge and fashion.

To give some historical perspective to the subject I will refer briefly to man's association with birds, which goes back many centuries. Symbolic figurines of pigeon-like birds have been found in northern Iraq which date from about 4500 B.C. By 3000 B.C., when written records first appeared, pigeons in Egypt had been domesticated for food and as message carriers. The domestic hen was part of the everyday scene in China and Egypt by 1400 B.C., though it is likely that it had been utilized long before then in the early civilizations of India. Birds may have been kept mainly as food, or in the case of hawks and cormorants, as food-hunters, but they have also been used for a variety of other purposes and have played a part in many aspects of human activity ranging from warfare to religious ritual. By Roman times birds figured prominently in paintings, sculptures, mosaics and in writings. They were used as watchdogs, as sources of adornment, but significantly they were also kept for the pleasure and interest they provided. It is obvious from contemporary writings that many species were kept as pets and that "aviculture" flourished. Parrots and peacocks from India, starlings, nightingales and finches were kept in private houses and parks much as they are today.

However, the *reasons* for keeping birds in captivity have changed (though they may still be kept at least partially for pleasure). In a scientific society such as this with its avowed intention of "the advancement of Zoology and Animal Physiology and the introduction of new and curious subjects of the Animal Kingdom", birds have provided a vast amount of original data of benefit to ornithology and to general biological knowledge. To be as objective as possible in my premise that the main motives for having birds in the Society's collections can be related to contemporary ornithological interests, I have referred to the most important books and journals of the period. In particular I have looked at the *Proceedings* and *Transactions* of the Zoological Society, and those of the Linnean Society, also *Ibis*, the journal of the British Ornithologists' Union, and the *Avicultural Magazine*, the journal of the Avicultural Society. This survey has shown that changes have taken place in knowledge and fashion, and to some extent this is reflected in the numbers and species kept in the collection, the methods of housing, and the type of study made on them.

The formation of the Society came at a period when ornithology was a respectable member of the sciences. It was essentially a period of classification and anatomy, a period of meticulous and eminent work from zoologists in Britain, France and Germany. It was the time of travel and exploration which produced specimens from all over the world; often rarely seen, and often species new to science. Species brought to this country for the Society's collection formed the groundwork for new and involved classification systems, based primarily on anatomical and morphological characteristics. It was a period when more emphasis was given to procuring new species than to keeping alive ones already here. For example, between 1831 and 1900, nearly 1500 new species or subspecies were brought to the collection (Table I), but a dead bird

TABLE I

Number of species or subspecies brought to the collection for the first time

Year		Number	
1831–1840		182	
1842–1846		132	incomplete records
1851–1860		209	
1861–1867		215	incomplete records
1871–1880		360	
1881–1890		188	
1891–1900		210	
	Total:	1496	

Adapted from Scherren, H. (1905). *The Zoological Society of London*. London: Cassell & Co. Ltd.

was often of more immediate value than a living one. At this time the emphasis in aviary and cage design appeared to be more on decoration and shape than on providing the more fundamental necessities for the inmates, and they came and went rather quickly (Figs 1–3).

It was in the mid 1800s that a number of far-reaching theories were promulgated. It was Charles Darwin's observations on the finches of the Galapagos Islands in 1835 that provided the stimulus to his speculations on the origin of species— speculations that led to the theory of evolution by natural selection. With A. R. Wallace in 1858 he presented his ideas to the scientific world at a Linnean Society meeting, and a year later in his own book *The origin of species.* His views changed the whole outlook of biology; at once a series of disconnected facts of natural history formed a coherent pattern. At the same time P. L. Sclater, Secretary of the Society from 1859 to 1902, published his papers on the geographical distribution of birds, and the zoogeographical regions into which he divided the world were later largely adopted by A. R. Wallace, for animals in general. New impetus was given to questions of affinity and distribution and at least until the end of the century it was a time of a number of profound papers, many of them based on birds in this collection. Fortunately, it was also the period when A. D. Bartlett was Superintendent of the Gardens (1859–1897) and

FIG. 1. Aviary for small birds, Zoological Gardens, about 1830. From *The zoological keepsake* (1830). London: Marsh & Miller.

FIG. 2. Old macaw cage, Zoological Gardens, 1831. From a lithograph by James Hakewill.

FIG. 3. General view across waterfowl lawn and pelican enclosure to an aviary containing an unlikely assortment of birds; Zoological Gardens, 1835. From a lithograph by George Scharf.

to some extent his own individual approach to his charges counteracted the rather objective uses that Sclater put the collection to. However, though much useful data was obtained from birds in the collection, it was not a period of any great advances in their management.

The *scientific* repute of ornithology began to decline markedly at the turn of the century, The over-emphasis before 1900 on classification and anatomy resulted in the belief that the classification of birds was more nearly complete than that of any other group of animals. This, coupled with the fact that birds as a class

FIG. 4. Small cages which usually contained one or at most two birds, not now in use.

FIG. 5. New aviaries which replaced small cages, and are designed to hold breeding pairs of small birds.

appeared to have rather uniform anatomy, and few fossil remains, pushed zoologists into working on mammals and reptiles. Zoology students learnt only of the anatomy of the pigeon, and initial embryology—other aspects were ignored and birds, in T. H. Huxley's phrase, were regarded merely as "glorified reptiles". Zoology departments appeared to adhere to a statement attributed to the great Professor Alfred Newton in 1900 that everything important about birds had been discovered, and between 1900 and 1940 few outstanding scientists, in this country, undertook research with birds. The most notable exception to this generalization was, of course, the Secretary of this Society from 1935 to 1942, Julian Huxley. Abroad, however, after 1920 there were a number of quite dramatic pioneering works, especially those of Rowan in North America, who initiated modern studies of breeding seasons and reproductive physiology. Rensch in Germany, whose work led to the approach to systematics that was later followed by Mayr and others, Lorenz in Austria and Tinbergen in Holland, the founders of modern work on animal behaviour. Yet when the *scientific* repute of ornithology was at its lowest the foundations for its resurgence were being laid—not by trained scientists, but by gifted and dedicated amateurs. It was during the period 1900 to 1930 that considerable progress was made in studies of the *living* bird. Though not always appreciated in their own lifetimes, the works of such people as Selous, Eliot Howard, Chance, Eagle Clarke and Burkitt, provided the basis for later studies by professional zoologists. At the same time in an equally quiet fashion the avicultural world and the Society's own bird collection flourished, especially after the appointment of David Seth-Smith as Curator of Birds. Though few radical advances were made on this front it was a period of progress with a number of important first breedings.

It was not until after the last war that ornithology gradually came back into the circle of respectability. New work on the *living* bird is now actively encouraged in zoology departments of a number of British universities and research institutes. This resurgence has been due, in part at least, to the emergence of two disciplines; ethology—the study of behaviour, and ecology—the study of the living organism in relation to its environment, animate and inanimate. Ornithologists, especially Lorenz and Tinbergen, have been the pioneers of modern behavioural studies. Their work has led to enormous advances in our knowledge and understanding of such subjects as instinctive behaviour, learning ability, vocalization, imprinting, courtship and displacement activities. At

FIG. 6. General view of the Snowdon aviary in winter.

the same time many ornithologists, especially such people as Lack, Wynne-Edwards, and Nicholson, contributed to studies on the factors which regulate animal populations. Coupled with more recent advances in our knowledge of breeding biology, the general policy of keeping birds in captivity has gradually changed.

Greater stress is now laid on keeping birds in natural surroundings similar to those they have evolved to live in. More emphasis is given to keeping species in breeding groups or pairs (Figs 4–8). More importance is given to specializing in certain families or groups particularly in those which the Society already has a good reputation for keeping and studying. The foundation of this general policy was carefully laid by John Yealland during his successful curatorship from 1951 to 1969.

FIG. 7. Cantilevered bridge inside the Snowdon aviary, overlooking nesting sites.

FIG. 8. View from bridge of sacred ibis and two chicks. The nest has been built on a wire platform.

Within the last few years a sense of urgency has been engendered by the now obvious need to *conserve* certain species, especially those particularly vulnerable to man's interference, and to produce multiple generations from a self-sustaining group. In some areas the ultimate aim may well be to reintroduce captive-bred stock into their natural habitats. The problems of breeding generation after generation, often from an original stock of a few individuals, are only now being realized, and highlight the need for careful observations and careful record keeping. Such problems as the adaptation to abnormal diets, the loss of aggression, the loss of fear of predators especially man, the lack of immunity to diseases, and the concentration of possible deleterious genetical effects, will be brought to light and solved only when a carefully considered programme is undertaken. Particular importance is now given to

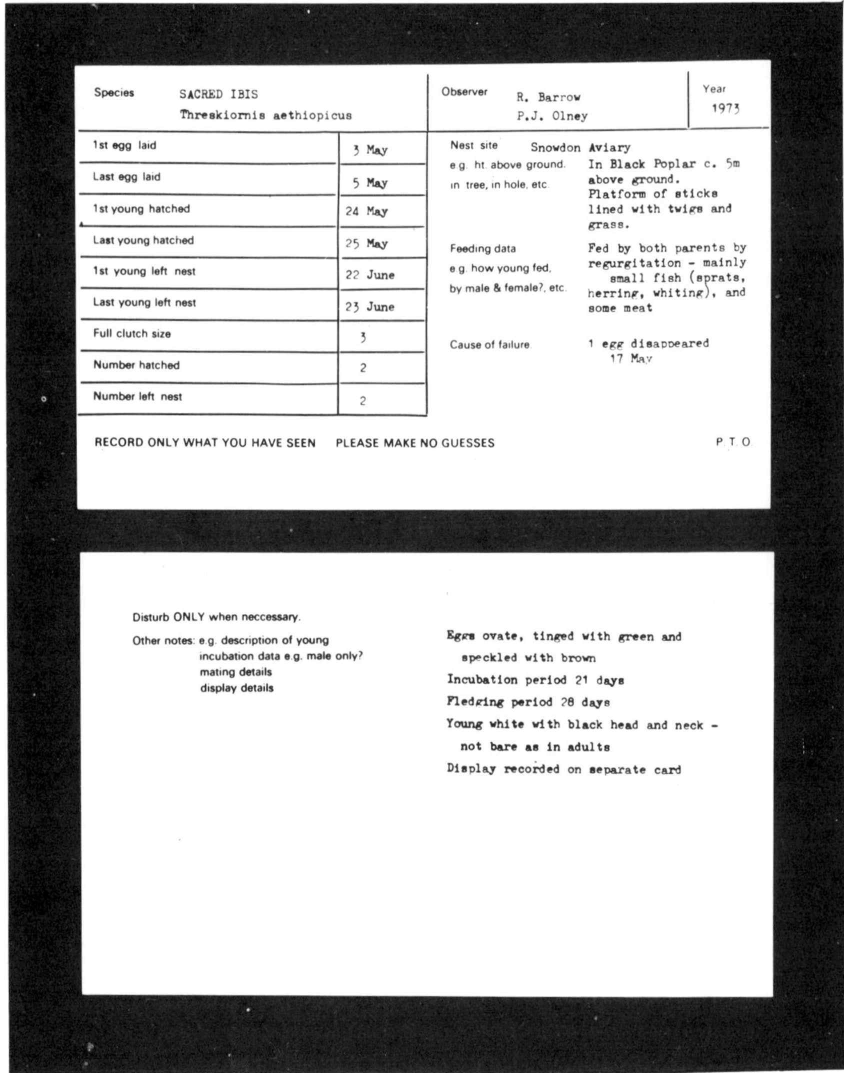

Species SACRED IBIS *Threskiornis aethiopicus*	Observer R. Barrow P.J. Olney	Year 1973

1st egg laid	3 May	Nest site Snowdon Aviary
Last egg laid	5 May	e.g. ht. above ground, in tree, in hole, etc. In Black Poplar c. 5m above ground. Platform of sticks lined with twigs and grass.
1st young hatched	24 May	
Last young hatched	25 May	Feeding data e.g. how young fed, by male & female?, etc. Fed by both parents by regurgitation - mainly small fish (sprats, herring, whiting), and some meat
1st young left nest	22 June	
Last young left nest	23 June	
Full clutch size	3	Cause of failure 1 egg disappeared 17 May
Number hatched	2	
Number left nest	2	

RECORD ONLY WHAT YOU HAVE SEEN PLEASE MAKE NO GUESSES P.T.O.

Disturb ONLY when neccessary.

Other notes: e.g. description of young
incubation data e.g. male only?
mating details
display details

Eggs ovate, tinged with green and speckled with brown
Incubation period 21 days
Fledging period 28 days
Young white with black head and neck - not bare as in adults
Display recorded on separate card

FIG. 9. Record cards for individual birds, showing the type of information collected. Data on behaviour are kept on another card.

recent advances in breeding biology, especially behavioural mechanisms. We now realize that the factors which influence breeding success in birds are complex; perhaps, because of their structure and habits, *more* complex than in any other class of animals. It is obvious that the pre-requisite of any breeding is mating, and in birds this is brought about by a complicated *sequence* of events. In an animal which has evolved to fly, whose forelimbs have been modified to act as wings and are incapable of grasping, and which has an external pregnancy (i.e. the embryo develops within a shell *outside* the body), mating is likely to be difficult. To be successful it demands the absolute synchronization of behaviour and the co-operation of both sexes, and this is brought about by a sequential system of signals and responses. A wide range of visual and vocal language has been evolved, so that individuals by their calls, postures and movements can influence and elicit responses from another individual who knows the same code. Thus courtship's primary function is to synchronize behaviour of both sexes for the purpose of breeding. In birds the sequence of events leading to successful breeding can be divided into: territory selection and defence, attraction and choice of mate, maintenance of the pair bond and nest building, copulation, egg-laying, incubation of eggs, and care of young. Any interruption or deficiency at any point may prevent the occurrence of all subsequent events. It is this *fundamental* fact, probably rarely realized, that we need to keep in mind in our breeding programmes with the birds in this collection. Of course, identification of the factors which cause the sequence to occur is often difficult. Though once identified, it may be simple—e.g. the provision of the right type of nesting area or nesting material, it may also be complicated and inter-related to other factors, e.g. a change in seasonal diet. Because we need to know in detail, it is vital that full records are kept of any breeding, whether in the wild or in captivity. From birds in captivity it is much easier to obtain information, and this in itself is a reason for having birds in our collections. In recent years, in this and in other major collections, extensive records are kept on individuals (Fig. 9), and it is the dissemination of this information which is the backbone of all future breeding programmes.

Hand in hand with observation and record keeping goes the need to provide more areas of natural surroundings for potential breeding species, and recent new aviaries—and changes in old aviaries—provide just that.

Throughout the Society's history the general policies underlying the keeping of birds have changed, and will go on changing as our knowledge increases. In no way have I intended my remarks to be a criticism of past policies, but I hope I have shown that they have been a reflection of current ornithological and zoological thinking.

Symp. zool. Soc. Lond. (1976) No. 40, 147–165.

PRIMATES AND CARNIVORES AT REGENT'S PARK

M. R. BRAMBELL and SUE J. MATHEWS

The Zoological Society of London, Regent's Park, London, England

INTRODUCTION

The primates and carnivores at Regent's Park have attracted a great amount of interest ever since the founding of the Society 150 years ago. The sesquicentennial year provides an opportunity to assess progress and to speculate on future trends, especially as the Michael Sobell Pavilions for Apes and Monkeys, opened in 1972, are now just coming into their own, and as the New Lion Terraces for the large felids are about to be completed.

The primates and carnivores, containing as they do our own species' closest relatives and the largest of the terrestrial predators, are especially important to the zoo. Not only are they better known and better understood by the visiting public, but they make high demands on the resources of manpower, housing and feeding. Furthermore, both groups contain many species which are now vulnerable in the wild. The future of these may well depend on establishments such as the Society's collections if they are to survive at all.

In the last few years the zoo has been forced to review its management policies in the light of new knowledge on the physiological and behavioural needs of the animals. In addition, it is no longer easy, nor is it desirable, to be forever replacing stock with wild captures. The breeding of young stock is desirable not only as a means of attracting public interest; it is essential for the zoo to be able to provide its own followers, and in some cases it contributes significantly to the preservation of species.

Thus the zoo is at the beginning, the rather shaky beginning, of a new era of management, in which the continuous species group is the unit to be managed. The beginning is shaky because sustained breeding successes have previously been few. There have been periods of great success with particular species, but more often than not these successes fizzled out without leaving a continuing healthy stock.

The oldest groups of mammals in the Society's collections are the three well-established "regulars": Przewalski's horses, Père David's deer and European bison; and the Barbary sheep which had been obtained by 1913 for the Mappin Terraces. Other stock cannot be traced back far before coming to a wild import. It will be many more years before the majority of species are breeding more than enough young to provide replacement breeding stock, but the signs are that there has been a significant turn in events.

The Society has broken out of the "picture gallery" concept of the zoo where the good collection exhibited as many kinds of animals as it could obtain. Now it is content to exhibit far fewer kinds in order to have the space and the resources to maintain those kinds in more realistic groups. Only ten years ago the zoo had well over 50 species of larger monkeys and apes on exhibit, but only about 120 individual animals. Now we have 16 species, but there are 72 individuals. There were 17 species of cats; now there are eight.

With a great many species the nutritional barrier has been broken and the zoo is grappling with the social and behavioural barriers. It is beginning to think in much broader terms than its own management. There is more and more discussion (and not an insignificant amount of action) on the management of species in national and even international terms. Not only are there stud-books (which are aids to management, rather than the actual means of management), but there are several zoos which are working together to manage their stocks of particular species. In time the need to do this will become so demanding that it will be a moral obligation for a particular zoo to surrender its sovereignty over particular stock in order that a species can be kept in a viable group.

The world herd of Arabian oryx is the most spectacular current example of this, but the process was foreshadowed by the Duke of Bedford's purchase of scattered individuals of Père David's deer towards the end of the last century. It is now developing in other groups, especially among the larger primates and the carnivores. The Society has played a significant part in these developments. It brought the concept of interzoo co-operation into the limelight with its efforts to bring the male giant panda at Moscow Zoo together with the female at London Zoo. It contributed the most prolific of the founding female Arabian oryx to the world herd. It has joint breeding ventures for orang-utans, Californian sealions, scimitar horned oryx and gorillas with the Jersey, Edinburgh, Marwell Park and Bristol Zoos respectively, to mention just a few examples.

Efficient management and housing must take into account common themes as well as the many differences between groups

FIG. 1. Scimitar horned oryx. This herd is owned jointly by the Zoological Society of London and Marwell Zoological Park.

and individual species. The primates contain 145 species in ten families. They can be regarded as a group of animals which are agile, alert and intelligent; most are generalized feeders which usually live in trees in the tropics and subtropics, and which reproduce rather slowly and in low numbers.

The carnivores are an even more diverse group. There are at least 250 species arranged in seven families, ranging from dogs and bears, through the raccoons, weasels and civets, to the hyaenas and cats. In general they can be described as alert, agile and quite intelligent animals which feed by capturing other animals, which usually live on the ground, and which reproduce fairly quickly and in large numbers. Unlike the primates, they occur at all latitudes, from the equator to the subpolar regions. Of course there are many exceptions: otters are aquatic animals using the land as a refuge; giant pandas get most of their nutrition from plant foods, while the aardwolf has become an anteater, and has fewer and less significant teeth than the other carnivores.

PAST HOUSING

Over the last 150 years there has been a certain amount of equivocation between indoor, semi-outdoor and, in some cases, wholly outdoor housing of the larger primates, and indeed of zoo animals in general.

The 1829 zoo had a Monkey House (somewhere to the south of the present Parrot House) which was small and had no outside cages, though the 1829 guide says that it was intended to add wired cages. At first this building held other mammals as well as monkeys and was called the Repository, the second building to have this name (see p. 154). The monkeys were kept in ones and twos in low, over-heated, badly ventilated, and probably badly drained dens. The public crowded up to the bars and the monkeys could grab at the visitors—there was a complaint of the tearing of the veils and dresses of ladies pushed too close to the bars by the crowds behind.

At the same time, there was a row of three monkey poles which appear to have been about 3·5 m high and to have been topped by a small platform with a shelter. The monkeys, one to each pole, were tethered by a chain to the waist and were able to come down to the ground. They were taken in at night. The public were not kept back from the bottom of the pole and feeding these monkeys must have been the highlight of many young visitors' day at the zoo.

In 1839 the monkeys were moved to a new house which was the building shown as the "First Monkey House" in Scherren (1905). At first it had only indoor dens with an area for the public and a service area. A year later three large outdoor cages were added. The animals were kept in larger groups than before, but the indoor part of the house was still kept very hot.

In 1864 a new monkey house was built, more or less on the site of the 1927 house. It did not have any outdoor cages, although the baboons and hardier monkeys were kept elsewhere in the open. It was "fitted up in the style of a conservatory, so as to be as light and sunny as possible, and to reproduce as nearly as maybe the circumstances under which its occupants live in their native haunts" (unascribed quote in Mitchell, 1929). It had a large central cage into which all compatible monkeys were put, and the walls were lined with smaller cages. Behind the scenes was a poorly lit, over-hot and insufficiently ventilated sick bay. In 1906, after there had been a severe outbreak of tuberculosis, this house was cleared of its stock and completely disinfected, with new cages being built.

The apes kept in the last century were housed in various parts of the zoo. By the end of the century most were in the Monkey House, though the gorilla which arrived and died in 1887 had been housed in the Sloth's House on the site of the present main offices. An Ape House had been built by 1902 on the site of the present Reptile House; indeed the white tiled walls of the apes' cages still exist as part of the keepers' area. The public were separated from the apes by glass and the animals were kept indoors in heated enclosures. The orang-utans and chimpanzees fared much better in it than before.

In the 1920s the Monkey Hill was built on the site of the present Animal Hospital. It was designed by Miss Procter, Curator of Reptiles and Amphibians. Eighty sacred baboons were put on the hill and were to achieve immortality only a few years later with the publication by our Secretary of *The social life of monkeys and apes* (Zuckerman, 1932). The baboon colony was a success and lasted at least until the war.

In 1952 the hill was restocked with 252 Rhesus monkeys. These had been obtained in India at just the time when the trade in Rhesus for research began. At that time the easiest Rhesus to catch were those that lived in close proximity to man in the villages. There was a very high incidence of tuberculosis and a pathetic list of deaths which were wrongly blamed on the hill. The concept was therefore abandoned.

Also in the 1920s steps were taken to replace the 1864 Monkey House and the 1902 Ape House. A plan for an Ape and Monkey House was drawn up, and an experimental house, consisting of three cages, was built on the site of the present Supplies Building. A young orang-utan, some African monkeys and a group of marmosets were put into these cages. The plan was adopted and the new Monkey House was opened at the end of 1927. It was a long hall with indoor cages along both of the side walls, attached to two rows of outdoor dens. Most of the apes were moved into this house, but a few lived on in the experimental house where the chimpanzee Jubilee, the zoo's first great ape baby to be reared, was born.

FIG. 2. Inside cage in the 1927 Monkey House.

However the house was not a success. The animals lived in cramped cages, separated from those on either side only by a single screen of open wire mesh (Fig. 2). They were unable to live in social groups, and were unable to isolate themselves from the threats and jealousies of their neighbours. Mating behaviour was frequently interrupted and maternal behaviour was often inhibited. Further troubles occurred when the electric hotplates, put into the cages to provide warm resting places, began to short circuit because of the action of the monkeys' urine.

By the end of the 1950s it was clear that the 1927 house must be replaced. A spectacular building was designed by

FIG. 3. Pigtail macaques in the Sobell Pavilions. The animals live in groups and normal activity is encouraged.

F. P. Stengelhofen which showed many improvements on what had gone before. A prototype section of the building was erected inside the empty South Mammal House in 1968, but the design had two serious drawbacks; not all the species groups would have had access to outdoors, and the cost was more than the Society could afford. So it was that the stage was set to design the buildings now known as the Sobell Pavilions which cater for the larger monkeys and the apes (Fig. 3).

The smaller primates were first kept in various houses scattered all over the zoo. They were later concentrated into the North, Central and South Mammal Houses until 1967 when the Clore Pavilion opened. Here the primates have done well and in the ensuing nine years about 100 have been born (including many marmosets), the majority reared by the mothers.

The housing of carnivores over the last 150 years has been easier to follow for the larger species, though it is still rather obscure for the smaller animals.

The 1827 plans for the Zoological Gardens drawn up by Decimus Burton included a house for carnivores which had sleeping dens where the animals were off public view, and outside yards. However, the idea of keeping exotic animals in fresh air was strongly contested and the plan was dropped. Instead a temporary Repository was built not far from the site of the present Zoo Shop. In it many of the carnivores were kept in closed and heated conditions. At the same time there were also built a Bears' Pit, an Otter Pool, cages for dogs and foxes, and a den for wolves, allowing all these animals to be kept in the open air.

By 1842 the fresh air school was ready to make another attempt and the Burton Terrace was built. The last dens and paddocks of this survived until 1971 when they were pulled down to make room for the east block of the Sobell Pavilions. Large cats and bears lived in the cages under the terraces until the Lion House was finished in 1876. The north side was then used for hyaenas and large dogs, the south side for bears.

The 1876 Lion House had 14 airy indoor exhibition cages, 28 sleeping and cubbing dens, and four (later increased to seven) outdoor cages. It was a great advance on what had gone before, and it lasted exactly 100 years. When it opened it held six lions, seven tigers, two jaguars, two leopards, three pumas and a clouded leopard. Three of these had been born in the zoo, a litter of one male and two female lions. Cats had been breeding in the zoo as early as 1838 (three pumas and eight Persian cats), but longevities were poor and the zoo was encountering great problems in maintaining the cats.

Meanwhile the housing of the other carnivores also developed. In 1913 the Mappin Terraces were built, with the middle terrace designed for bears. The six bear paddocks, though in effect still the conventional bear pit with one side replaced by a dry moat instead of a wall, were landscaped and were much larger than what had gone before. Not until the brown bear paddock at Whipsnade was built did the Society break away from the "bear pit" concept, in essence a relic from the renaissance menageries of Europe. Nevertheless the bears have thrived on the Mappin Terraces, with a good health record and good breeding results.

There were no special provisions for housing most of the other carnivores. The only exceptions were the otters, which right from the beginning have usually had purpose-designed housing; and the wolves, which in 1956 had a large wooded paddock along the eastern boundary made over to them. Most of the small carnivores were housed in a variety of buildings including the North Mammal House (originally a Lemur House), the 1913 Central House alongside the present Insect House, and the south Mammal House (once the Tortoise House) on the site of the gorilla and chimpanzee outdoor enclosure in the Sobell Pavilions. The small cats were housed in the building which later became the Education Centre in the north-east corner of the main gardens. In 1967 the small carnivores, like the small primates, were concentrated in the Clore Pavilion for Small Mammals.

In conclusion of this review of past housing, it should be remembered that the indoor/outdoor argument of the last century involved not only temperature, but the quality of the air. Several animals died from respiratory complaints aggravated by severe fogs. This could only be resolved when the clean air legislation of the 1950s rid London of its fogs.

PRESENT AND FUTURE HOUSING

The planning of the Sobell Pavilions and the New Lion Terraces have had a lot in common. Both were designed by the same architect, working to a brief prepared by the same curator, and both were designed with the enclosures at Whipsnade in mind.

The Sobell Pavilions were designed in the period immediately following the work done on the Stengelhofen plan. Thus the Society's staff were better prepared to meet the demands for information put on them by the architect. The most important considerations of the briefs were, first, the requirements for keeping breeding groups of healthy, active animals, and, second, the needs of the keepers. The needs of the public were kept in mind, but not allowed to override the other criteria. Perhaps the most significant features of the buildings are the free access in and out at all times of the year, and the space frame climbing area over the whole of the outside area (see Toovey & Brambell, 1976, for further details).

In planning the replacement of the 1876 Lion House the Society had the advantage of having the highly satisfactory exhibits for lions and tigers at Whipsnade. However it was not clear how small an open, planted area could be before severe poaching of the ground surface would take place. In addition the large cats cannot be expected to be active for very long, unlike the larger primates who are active if given the opportunity.

Thus in the period before the formal opening, there is some trepidation about how the animals and the enclosures will treat each other. If there is one lesson that can be learned from the many zoo buildings that have been erected all over the world, it is that it is impossible to predict the interaction of animals and architects until the two have come face to face, faeces to facia (Fig. 4).

Each new building is planned and heralded against the background of its predecessor, but is eventually judged, and condemned if necessary, against the background of itself. It is clear that the Mappin Terraces must be modified or replaced as the home for bears. However, the Sobell Pavilions have settled in well and we expect them and the New Lion Terraces to be satisfactory for some time to come. The Charles Clore Pavilion for Small Mammals, the place where most of the small primates and carnivores have been housed, is basically a good building. While there is room for improvement in the light of experience, the building can be modified without major constructional problems. It is likely that this building too will be able to keep up with current thinking for many years to come.

MANAGEMENT

The day-to-day management of the animals has probably not changed very much in the years which have passed since the zoo first opened. The shovel, brush and hosepipe are still the tools for cleaning, the bucket still the means of putting food into the cages.

FIG. 4. Cheetahs in the open enclosure.

Although curators may dream of automated cleaning, such dreams are soon dashed when new gadgets break down.

For example, hydraulic sliding doors have many attractions until they break down. Yet they take just as long to operate, and a keeper is more likely to be distracted when he only has to push a button than if he is manually shutting a slide. Manual operations are capable of considerable improvement by design, and it is our belief that machinery should only be introduced when its need has been properly established and the equipment itself has been proved. Too often have animal keepers found themselves saddled with prematurely introduced gadgets which have failed under the pressures of day-in, day-out animal keeping.

FEEDING

Probably the biggest advances in the zoo, especially in the last two decades, have been in feeding. Outwardly many of the diets appear to be very similar to those fed in the last century, and the differences are ones of detail, albeit essential detail. There is now a far better understanding of which materials, and in what amounts, need to be absorbed by an animal for it to meet all its bodily requirements. There is more understanding of the contents of foods and of the digestive processes by which the animals render the essential nutrients contained in the food available for digestion. A recent survey (Brambell, in press) of mammal diets at Regent's Park has shown that there are now only about 20 dietary types among the 185 species in the collection.

Many of the problems were recognized quite early and at least partial remedies were incorporated into early diets. However over the years diets became less and less nutritious. Passed on from generation to generation of keeper without a real understanding of what the food was expected to provide, they shed first one and then another aspect.

Although the first gorilla in 1887 was fed only on fruit and bread, Bartlett had recognized that the larger primates would benefit from a little animal protein (Bartlett, 1899). This was gradually dropped from the diet until in the immediate post-war years the primates were relying for most of their protein and much of their minerals on the food being thrown into their cages by the visitors. Virtually nothing but fruit and cabbage was being provided by the Society and the diet lacked many essential elements.

The trouble was a fundamental one. The keeper staff knew from their own eyes that the animals would eat the foods they offered, but would not eat the fully balanced diets that were being pushed by the Society's advisers. The advisers knew from experience of single animals kept in laboratory cages that the primates should eat the balanced diets. It was dietetics versus nutrition. The impasse was not broken until the advent of the expanded pellet. At last it was possible to offer the animals something they *should* eat which they *would* eat. Nutritional needs had been wedded to dietetic possibility. Both nutrition and dietetics probably have a long way to go, but at least today we can call upon a vast amount of knowledge in assessing what are the likely nutritional needs of an animal. Moreover we have many useful indices which can be used while the animal is still alive to tell us whether or not it is in a state of incipient deficiency of the major nutrients (Fig. 5).

The big cats have had a rather similar dietetic history. As early as 1864 the high incidence of malformed palates in zoo-born lion cubs (most of which did not survive) was noted. In that year three cubs born at the zoo to two different mothers had thyroid glands enlarged to 20 times their usual size. The cubs were thought to have died from pressure on the recurrent laryngeal nerve at the time of birth. As a result of this, Crisp (1864) attributed London Zoo's poor lion breeding record to the lack of secluded indoor cubbing dens.

The three $3\frac{1}{2}$-year-old lions moved into the 1876 house at its opening were then longer lived than any other lions born at the Gardens in the nineteenth century (Scherren, 1905). Flower (1929) reports about 60 lions being born alive in the gardens between 1863 and 1881, but there were not enough to prevent continued intakes of lions. Thus there was a serious management problem which remained unsolved until this century, in spite of the acknowledged success of Wombwell's Menagerie and the fame of the Dublin Zoo which was breeding lions to maturity by the 1860s.

This situation occurred although Bartlett recognized a difference in the feeding regimes of the different establishments. He pointed out that captive animals were fed only muscle and bone, whereas the natural diet included skin, intestines and blood as well (Bartlett, 1899).

The animal nutritionist of today recognizes the problem as being one of a mineral and/or vitamin deficiency. The main shortcomings of an all-meat diet for large cats are the low levels of vitamin A, and the imbalances of, and absolute deficiencies in, the

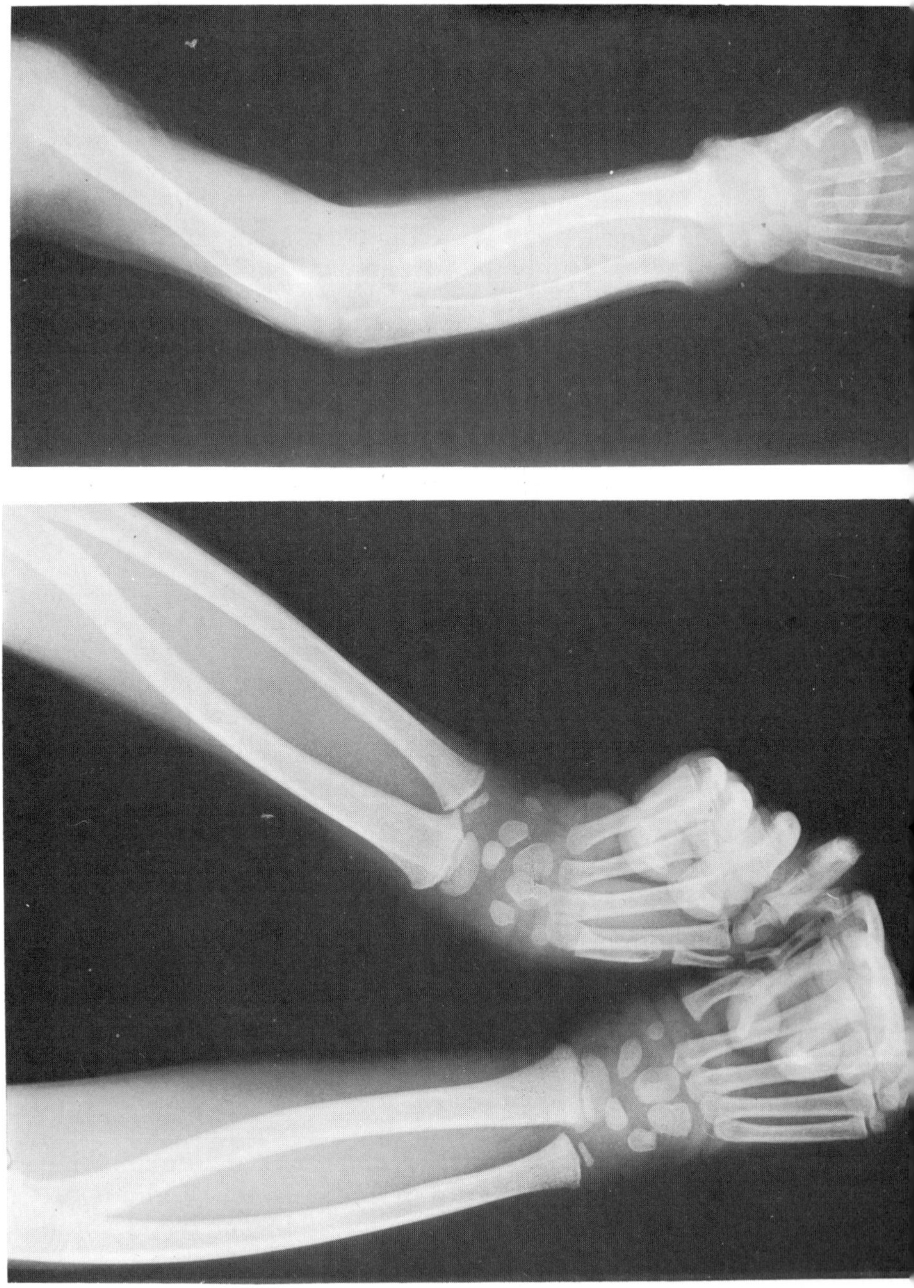

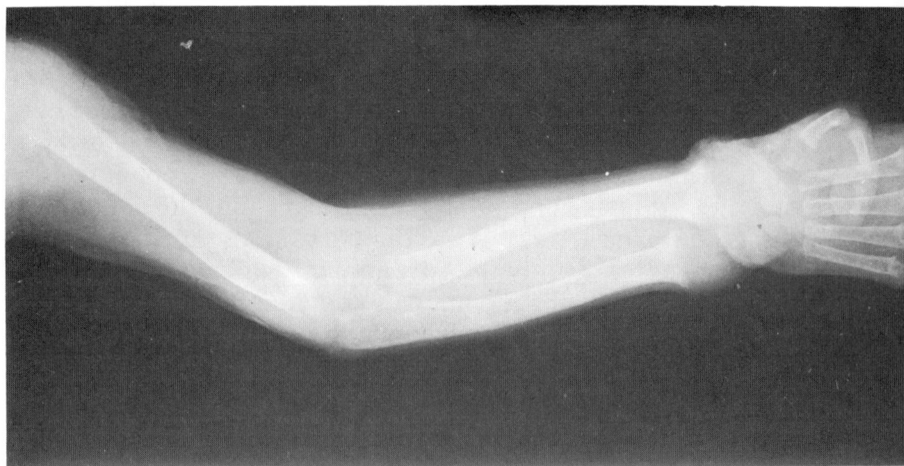

FIG. 5a. Nine-month-old male chimpanzee. Radiograph of forearm showing rickets and osteomalacia due to deficiencies in Vitamin D, protein and minerals in the former poorly balanced diet. b. Eleven-month-old female chimpanzee. Radiograph of forearms showing normal development and calcification. This animal was reared by the mother and received only the standard balanced diet now fed to chimpanzees.

amount of calcium and phosphorus in the meat. (Cats have a very high requirement for vitamin A.) It was the continuing failure to totally solve the problem of diet supplementation with this vitamin which occupied, and partly led to the establishment of, the Society's Animal Welfare and Husbandry Committee in 1958. We still feed meat to our large cats, but it is supplemented with high levels of vitamin A and with both calcium and phosphate. Other zoos have abandoned meat in favour of a manufactured sausage, but it is quite likely that an expanded biscuit, similar to that used for domestic cats, will be the next stage in ensuring adequate nutrition for the larger carnivores.

PERFORMANCE

In the final analysis the success of an animal house and its management is judged by its inhabitants and manifested in breeding success. Table I shows birth numbers for the different families

TABLE I

Survey of births of primates and carnivores at Regent's Park between 1828 and 1974

	1828–1924		1925–1964		1965–1974		Total 1828–1974	
	No.	% of total	No.	% of total	No.	% of total	No.	% of total
Lemuridae	67	(55)	36	(30)	18	(15)	121	(100)
Lorisidae	9	(12)	28	(37)	39	(51)	76	(100)
Cebidae	2	(10)	4	(19)	15	(71)	21	(100)
Callithricidae	4	(5)	26	(34)	47	(61)	77	(100)
Cercopithecidae	51	(34)	53	(35)	47	(31)	151	(100)
Pongidae	0	(0)	12	(32)	25	(68)	37	(100)
Total primates	133	(28)	159	(33)	191	(40)	483	(100)
Births per year	1·4		4·0		19·1		3·3	
Canidae	307	(61)	159	(31)	39	(8)	505	(100)
Ursidae	35	(37)	41	(43)	19	(20)	95	(100)
Procyonidae	26	(54)	15	(31)	7	(15)	48	(100)
Mustelidae	26	(27)	57	(60)	12	(13)	95	(100)
Viverridae	40	(45)	47	(53)	2	(2)	89	(100)
Hyaenidae	28	(100)	0	(0)	0	(0)	28	(100)
Felidae	156	(32)	241	(49)	92	(19)	489	(100)
Total carnivores	618	(46)	560	(42)	171	(13)	1349	(100)
Births per year	6·4		14·0		17·1		9·2	

TABLE II

Acquisitions of primates and carnivores at Regent's Park 1925–1974

	1925–1964		1965–1974	
	Total acquired	Births (%)	Total acquired	Births (%)
Lemuridae	145	24·8	34	52·9
Lorisidae	293	9·6	117	33·3
Cebidae	319	1·3	53	28·3
Callithricidae	329	7·9	110	42·7
Cercopithecidae	1632	3·2	88	53·4
Pongidae	282	4·3	44	56·8
Total	3000	5·3	446	42·8
Canidae	450	35·3	62	62·9
Ursidae	133	30·8	29	65·5
Procyonidae	204	7·4	27	25·9
Mustelidae	339	16·8	48	25·0
Viverridae	398	11·8	39	5·1
Hyaenidae	21	0	2	0
Felidae	667	36·1	109	84·4
Total	2212	25·3	316	54·1

of primates and carnivores in the years from 1828 (when records were first kept by the Society) to 1974. In Table II breeding success is compared with total numbers of animals acquired, over the last 50 years from 1925 to 1974.

Over 3400 primates were acquired by London Zoo between 1925 and 1974; 350 of these (10·2%) were born in the zoo, a few were born in other zoos, and the remainder came from the wild. However if the last ten years of this period are compared with the preceding 40 years it can be seen that births rise from 5·3% of the total to 42·8% of the total. Moreover a great many recent acquisitions which were not born at London were born in other zoos, particularly Jersey.

The lemurs have the highest percentage of births (31%). Lemurs have always bred well at London Zoo: 67 births occurred before 1925. A Lemur House was in existence throughout much of the last century, but did not have any special features other than provision for enough space to allow social groups to be kept (c.f. the larger monkeys and apes, pp. 150–151).

In view of the relative success of lemurs and the popularity of bushbabies as pets, it is rather surprising to find that in the Lorisidae, births comprise only 16% of the total number acquired in the years 1925–1974. However, over half of these births have occurred in the last ten years.

The marmoset and tamarin births are 17% of total numbers acquired between 1925 and 1974, with well over half occurring since 1965. These animals have a high requirement for vitamin D_3, a fact foreshadowed as long ago as the 1920s when ultraviolet light was provided and "Vita-glass" roof lights were installed (Mitchell, 1929). These practices have since been discontinued and the recent successes are largely attributable to supplementation of the diets with vitamin D_3 and to increased consideration of the protein levels in the food.

The gibbons (Hylobatidae) and apes (Pongidae) have been treated together for the purposes of this survey. Births are 11% of total acquisitions, but sustained breeding has only occurred in the last few years—69% of births since 1965, none before 1925. This success can be attributed to better facilities for maintaining the animals in social groups and to improvements in the diets.

Neither the New World monkeys (Cebidae) nor the Old World monkeys (Cercopithecidae) have had a very good breeding record. Births are 5% and 6% of all acquisitions respectively, and again there has been considerable improvement in breeding success in recent years. This is attributed not only to better housing and diet, but also to changes in the origin of the acquired animals. In past years the zoo received many unwanted monkey pets. Most had not been fed correctly and were not very healthy on arrival. Many had been kept away from their own kind since very young, and were unable to mix. In several cases there were no others of their kind to mix with. In more recent years the zoo has avoided adding such animals to its collections. The number of cebid births is still very small, and only the capuchins and douroucoulis have any sustained breeding success. Many of the cebids are delicate in captivity and may not be able to tolerate as wide a range of environmental conditions (especially of temperature and humidity) as other monkeys. Certainly they are still imperfectly understood and they are difficult monkeys to keep.

The carnivores have been relatively more successful in the zoo than the primates. About 2500 animals were added to the collection from 1925 to 1974, of which 731 (29%) were born. Of course it should be remembered that carnivore births are usually multiple, while most primates have a single young. Therefore the number of occasions that carnivore births took place is relatively much less than the number of animals born implies.

The breeding success of the carnivores is due largely to the births among the dogs, bears and cats (39%, 37% and 43% of acquisitions respectively). The dogs and cats have always bred well and a high proportion of young have been reared to maturity, but although 35 bears were born in the years up to 1925, none were reared. It was not until after World War II that the bears really began to rear their young consistently.

On the other hand, 28 hyaenas were born in the years before 1925, but none since, although 23 have been kept in that period. Procyonids, viverrids and mustelids have all had periods of success, but social, behavioural and environmental needs for these animals to breed in captivity appear to be difficult to meet.

In the primates in the last 50 years there has been a transformation from a state of almost nil breeding, with tremendous reliance on acquisitions from outside the Society, to one in which importing from the wild has fallen to almost zero and breeding is the only practical source of replacement stocks. In the carnivores the situation has been a little different, particularly with the cats. There have always been plenty of births, but the survival rate has increased. In the past the offspring were sold off and the breeding stock was retained until breeding stopped because of old age or the death of a partner. There were periods of high breeding followed by periods of no breeding until new stock was brought in and the cycle began again. Large cats are costly to keep, and a continuous process of breeding is difficult to achieve within the limits imposed by housing and finance. The solution for maintaining viable self-perpetuating groups in perpetuity is to manage the stock in each species on an interzoo basis.

CONCLUSION

Ahead lies a much more detailed attempt to meet the behavioural and ecological requirements of the animals. The zoo of the future must be an establishment which breeds its own followers. Good management is the process of using our knowledge of anatomy, physiology, behaviour and ecology to determine how best to place the animals within their normal adaptive range for as many variables as we know to be relevant.

When all the relevant variables are eventually understood it will be possible to keep the animals in conditions in which breeding is the inevitable result. Considerable advances have been made among the carnivores and primates, and among several other groups of mammals. However there are still some groups, notably the edentates and the various types of anteaters, which do not breed well in the zoo and about which we must learn a lot more. The same critical approach to management when applied to these groups should lead to similar results.

Again and again we have found ourselves saying: we have been here before. Many of the ideas which are contributing to the success of present-day management were thought up a long time ago. We would like to think that our successors will regard 1976 as a milestone marking the time when we were beginning to think in terms of managing populations, rather than being satisfied with success with individuals; the time when we began to recognize that even the largest collections have to pool their resources to maintain viable breeding stocks over long periods.

References

Bartlett, A. D. (1899). *Wild animals in captivity.* London: Chapman & Hall.

Brambell, M. R. (In press). Diets of mammals at the London Zoo. In *Handbook of nutrition and food.* Rechcigl, M. (Ed.), Cleveland, Ohio: CRC Press.

Crisp, E. (1864). Report to meeting on 12.4.1864. *Proc. zool. Soc. Lond.* **1864**: 158–159.

Flower, S. S. (1929). *List of the vertebrated animals exhibited in the gardens of the Zoological Society of London 1828–1927.* **I**. *Mammals.* London: The Zoological Society of London.

Mitchell, P. Chalmers (1929). *Centenary history of the Zoological Society of London.* London: The Zoological Society of London.

Scherren, H. (1905). *The Zoological Society of London.* London: Cassell & Co.

Toovey, J. & Brambell, M. R. (1976). The Michael Sobell Pavilions for apes and monkeys. *Int. Zoo Yb.* **16**: 212–217.

Zuckerman, S. (1932). *The social life of monkeys and apes.* London: Kegan Paul.

Symp. zool. Soc. Lond. (1976) No. 40, 167–178.

MAMMALS AT WHIPSNADE

V. J. A. MANTON

Whipsnade Park (Zoological Society of London), Dunstable, England

In the Centenary History of the Zoological Society of London, Sir Peter Chalmers Mitchell, then Secretary of the Society, wrote that he could

> "foresee in the future of this and other zoological societies a linkage of the intimate city 'zoo', where the maximum number of species will be shown with a few individuals of each" and "the country park where the number of the species will be more limited and the number of individuals greater" (Mitchell, 1929).

How does Whipsnade, the "country park" of the Society, purchased in 1927, match these ideas today? Perhaps an examination of some of the mammals kept there can illustrate this.

As at 31st December 1975 there were 934 specimens recorded in the Park and these covered 79 species. Although this means an average group or herd of 11·8 per species, it should be remembered that in a number of species, such as elephant, dolphin, common hippopotamus (*Hippopotamus amphibius*) (Fig. 1), giraffe and sealion, it is neither economically viable nor a practical possibility to keep more than a few. Indeed, if we ignore those 21 species represented by four or fewer specimens, the average rises to 15·2—surely an indication of the intention to show a "greater number of individuals" per species.

As to breeding successes, a check last year showed that 792 or 84·8% of the mammals present at the end of the year were born in the Park. This total, of course, does not include the many "home bred" specimens exchanged or deposited with other collections. For example, 714 Bennett's wallabies (*Macropus rufogriseus*) (Fig. 2) were sent away in the last 13 years—over 50 per year. Those who are not familiar with Whipsnade perhaps are unaware of the unique situation whereby eight species have complete freedom to roam round the 500 acres of the Park. It is probably because of the size of the area in which they find themselves that we have such great success in breeding them. The species concerned are the Bennett's wallaby (*Macropus rufogriseus*), Chinese water deer (*Hydropotes inermis*), muntjac (*Muntiacus* sp.), mara (*Dolichotis*

FIG. 1. Family of common hippopotamus, 1975: male "Henry" arrived 1950, female "Belinda" arrived 1950 and her male calf born October 1974, female "Wendy" born at Whipsnade in 1967 and her female calf also born there in February 1974.

FIG. 2. Bennett's wallaby free-roaming at Whipsnade.

FIG. 3. Part of the home-bred Thomson's gazelle herd at Whipsnade.

FIG. 6. Père David stag at Regent's Park.

FIG. 4. (*above, opposite*) American bison herd on Bison Hill at Whipsnade.

FIG. 5. (*below, opposite*) Part of breeding herd of Père David's deer at Whipsnade.

FIG. 7. Part of imported white rhinoceros herd at Whipsnade.

FIG. 8. White rhinoceros calf born at Whipsnade.

patagonum), common peafowl (*Pavo cristatus*), North American turkey (*Meleagris gallopavo*), jungle fowl (*Gallus gallus*) and helmeted guineafowl (*Numida meleagris*).

This percentage means that many species are captive breeding to three, four, five or more generations and certain species are represented entirely by Whipsnade bred stock. These include the wolf pack (*Canis lupus*) (last import in 1953), the herd of Thomson's gazelle (*Gazella thomsoni*) (last import in 1954) (Fig. 3), the American bison (*Bison bison*) (last import before 1954 except for a new bull in 1963) (Fig. 4), the Père David's deer (*Elaphurus davidianus*) (Figs 5 and 6), the llama (*Lama glama*) (last import before 1950), the peccaries *Tayassu tajacu*) (last import before 1953), the axis deer (*Axis axis*) (last import before 1959) and the guanaco (*Lama guanicoe*) (last import in 1960). Thus, in terms of the total world population or the world captive population, the Society has at Whipsnade an important reserve bank of captive-bred stock from which surplus may be drawn to meet the needs of other national collections: Père David's deer have been sent to Peking; white rhinoceros (*Ceratotherium simum*) (Figs 7 and 8) to Leningrad, Copenhagen and Alexandria; cheetah (*Acinonyx jubatus*) to Melbourne, Montpellier, Paris and Jersey; and Sumatran tiger (*Panthera tigris*) to Wassenaar and Rotterdam. It is even more important to stress the value of keeping some species in viable numbers since the bulk of world captive breeding takes place in only a few herds. Take for example the Przewalski horse (*Equus przewalskii*) (Figs 9 and 10). On 1st January 1974, 219 Przewalski horses were in captivity in 54 collections and 30 births recorded in 14 of them, but 24 of these took place in only eight collections which had 102 specimens between them (Volf, 1975). Here the Society owns the fifth largest group in captivity. Again with Père David's deer, as at 31st December 1974, 726 animals were registered in 74 collections and 153 births took place in 27 of them. However, 101 of these were at Woburn and 10 at Whipsnade—the next highest number (Manton, 1976). The last edition of the European bison (*Bison bonasus*) herd book (Zabinski, 1973) lists 833 animals in captivity throughout the world in 162 collections of which 213 occur in seven collections. However, only 167 births were recorded in 68 collections. Whipsnade's herd consists of one male and 10 females and 27 live births have occurred since it was started in 1964 (Fig. 11).

The breeding successes at Whipsnade are not entirely confined to herd species. In 1966 a female cheetah was presented to

FIG. 9. Przewalski stallion at Whipsnade.

FIG. 10. Part of the breeding herd of Przewalski horses at Whipsnade.

FIG. 11. European bison cow and calf at Whipsnade.

Whipsnade and began a saga which to date has resulted in 28 births, of which six were second generation captive-bred animals (Figs 12 and 13). My paper to the first Symposium on Breeding Endangered Species in Captivity, held in Jersey in 1972, summarized the information then available from those few collections where births had taken place (Manton, 1975). The conclusions reached then appear to be still valid today and breeding is continuing—the last litter was born on 29th March 1976 and is again an F2 captive-bred generation one. Even the births in safari parks since have not caused me to doubt my views and indeed have tended to strengthen them.

In the wild state cheetah tend to be solitary animals or at most to live in family groups. Myers (1976) quotes one per 150 square miles to be a not unusual stocking rate in East Africa. Occasionally a number of males may be seen consorting with a female in oestrus. It appears though that the members of a group seldom remain

together from youth through to sexual maturity. In captivity, at least until recently, the practice was to obtain youngsters, rear them artificially for a pets' corner and keep them together for as long as possible to make a good show. In all the collections whose management I have investigated and in which breeding has taken place, sexually mature animals have been introduced to one another after separate travel or rearing on their own. The domestic cat has been shown to be an induced ovulator and it is my belief that it is not too unreasonable to consider the cheetah to be in the same category. Evidence at Whipsnade indicates that oestrus may follow introduction or re-introduction of a "strange" male to the female. Indeed, on four occasions, oestrus, with or without mating, was recorded between six and 14 days after this (Table I). On the other two

TABLE I

Length of gestation and time of mating in cheetahs at Whipsnade

Time between introduction and mating(s) (days)		Gestation (days)
14	+	91
13	+	91
15 + 57 +	96 + 94 +	94
? 49	?	93
6	+	93
7	+	93

occasions the intervals between matings and parturition varied greatly. If the mating dates were correct, foetal resorption could have accounted for the returns to oestrus and repeated mating. I should explain that normally the young are left with the dam for the best part of the year. Those born late one year or early the next one are retained until the end of the visitor season. No relationship is shown when comparing the dates of removal of the young and the onset of oestrus.

FIG. 12. (*above, opposite*) Imported cheetah with 1970 litter.

FIG. 13. (*below, opposite*) Female cheetah born at Whipsnade in 1970 with second generation captive-bred litter.

Finally, I would like to reinforce the emphasis placed by the Curator of Mammals on the importance of the supra-national management policy for captive-bred stocks. I am glad to say that at Whipsnade we have certainly been instrumental in operating this policy for the Hartmann's mountain zebra, Przewalski wild horse and Sumatran tiger.

References

Manton, V. J. A. (1975). Captive breeding of cheetahs. In *Breeding endangered species in captivity*: 337–344. Martin, R. D. (Ed.) London, New York & San Francisco: Academic Press.

Manton, V. J. A. (1976). 1974 world register of Père David's deer (*Elaphurus davidianus*). *Int. Zoo Yb.* **16**: 227–229.

Mitchell, P. C. (1929). *Centenary history of the Zoological Society of London.* London: Zoological Society of London.

Myers, N. (1976). The status of the leopard and cheetah in Africa. In *The world's cats* **3**(1): 53–69. Eaton, R. L. (Ed.) Seattle: Department of Zoology, University of Washington.

Volf, J. (1975). 1973 world register of Przewalski horses (*Equus przewalskii*). *Int. Zoo Yb.* **15**: 300–301.

Zabinski, J. (Ed.) (1973). *European bison pedigree book.* Warsaw: National Council for Nature Conservation.

Symp. zool. Soc. Lond. (1976) No. 40, 179–202.

150 YEARS OF BUILDING AT LONDON ZOO

J. W. TOOVEY

The Zoological Society of London, Regent's Park, London, England

> "Lions at play, free as their own jungles at home; tigers crouching, springing, gambolling, with as little restraint as on the hot plains of their native India—such is the dream of everyone interested in Zoology. We are all tired of the dismal menagerie cages. The cramped walk, the weary restless movement of the head, . . . the bored look, the artificial habits . . . . Thousands upon thousands will be gratified to learn that a method of displaying lions and tigers, in what may be called by comparison, a state of nature, is seriously contemplated at last".[1]

Not an enthusiastic description of the New Lion Terraces, but of the Old Lion House from the *Daily News* of 1869. It is perhaps important to realize that throughout its history the Zoological Society has striven for the same ideals, and when they have not been achieved it has only been through lack of knowledge, or the means, at the time. This continuous search for perfection was well expressed by the President, Professor Flower, in his Jubilee Address of 1887—

> "We have a responsibility to our captive animals, brought from their native wilds to minister to our pleasure and instruction, beyond that of merely supplying them with food and shelter. The more their comfort can be studied, the roomier their place of captivity, the more they are surrounded by conditions reproducing those of their native haunts, the happier they will be. . . . Much as has been done in this direction, we must all admit that there is still more required. The buildings of today will, we may even hope, some day seem to our successors what the former ones seem to us".[2]

The Society has a history of continuous building that must be remarkable for any institution, and for the last 150 years there have been holes in the ground, loud noises, and buildings going up and coming down; all very restless, but an absolute boon to builders and architects. All the way through architects have been involved,[3] and after each one has proved himself the Society has tended to settle down with him for several years. As a result only nine known architects were employed in the 114 years up to 1940, and in the great rebuilding since the 1939 war only a further seven firms or

[1] Numbered notes at end of chapter.

individuals have been responsible (Appendix A). From the beginning the buildings have echoed contemporary style, and so the Zoo's history presents a fascinating microcosm of 150 years of British architecture, from Regency and Classical styles to Neo-Gothic Revival, from Victorian Functional to Neo-Renaissance, through to Italianate simplicity (Appendix B). Then, mercifully, the Zoo was spared the worst of the 1930s Functional style, and jumped straight into the forefront of the real Modern Movement, and since then, apart from perhaps just after the war, a remarkably high standard of contemporary architecture and engineering has been maintained.

On 3rd June 1826 the Society's Council decided ". . . to proceed upon the Plan to the amount of £1000. The President was requested to communicate with Mr Decimus Burton on the subject, respecting terms etc. . . .". So it is clear that right from the start Burton was closely involved in the planning and detailed development of the Zoo. He prepared layout plans, supervised the drains, and designed most of the buildings, ranging from a coal cellar to the great Carnivora Terrace that, sadly, was never built to his design. On 2nd June 1830 Burton was formally appointed Architect to the Society on an annual retainer of £150, rather than on a percentage fee as before.[4] The earliest structures were small, Follies set in an elegant garden for entertainment and curiosity, in the tradition of Nash's villas in Park Village East (Fig. 1). They went long ago, but fragments of Burton's work survive today in the Raven's Cage, the Clock Tower, and the Giraffe House of 1836. Fortunately his East Tunnel of 1829 has hardly been altered, and today looks much as it did when built (Fig. 2). On the whole, Burton's influence has been through his sensitivity and concern with planning, and the high quality of his design.

A Mr Elmslie followed Burton and it was he, not Burton, who designed the Carnivora Terrace built in 1843 (Fig. 3). The Terrace was really only a form of superior menagerie, but all the same it remained in use up to 1919. There was little building between 1851 and 1860; and from then until 1880 the scene was dominated by Mr Anthony Salvin jun. Most of his buildings were Classical in feeling, but he seems to have coped with the Gothic Revival style just as ably. He completed the Eastern Aviary in 1864, and it is still in use today. The new Monkey House of the same year (Fig. 4) continued in use for 63 years until 1926; it was an elegant Classical conservatory, later considered unsuitable as there were no outside cages. His Elephant House of 1869 (Fig. 5) lasted until 1938, and

FIG. 1. The early structures were Follies set in an elegant garden (*c.*1830—Burton?). From Bishop (1832).

FIG. 2. The south end of the East Tunnel, built in 1830 (Burton), remains almost the same today. From a lithograph by James Hakewill.

FIG. 3. The New Carnivora Terrace of 1843 (Elmslie). From a scrapbook dated 1843–1853 held in the Library of the Zoological Society of London, inscribed *London*, by C. Knight, No. 117 24th June, 1843, and subtitled CXVII. The Gardens of the Zoological Society. Pages 257–272.

FIG. 4. The third Monkey House of 1864 (Salvin). No outdoor cages, but it lasted until 1926. From Scherren (1905), by kind permission of Cassell & Co. Ltd, London.

FIG. 5. The New Elephant House of 1869 (Salvin). In use until 1938. Reproduced with permission from *The Illustrated London News*, 26th June, 1869.

FIG. 6. The New Lion House of 1876 (Salvin). The outside cages were not completed until 1877. From Scherren (1905), by kind permission of Cassell & Co. Ltd, London.

was a competent exercise in the Gothic style. But Salvin's major work was the new Lion House of 1876, again in his Neo-Classical style (Fig. 6); it was carefully thought out after Mr Bartlett the Superintendent had toured throughout Europe in 1868 inspecting lion houses, and so the plans were not ready for approval until 1874. The animals started to move in on 15th January 1876, and the building had been in use for almost 100 years when the last inhabitant left in December 1975. Whatever we may have thought of it today, the house did provide space for visitors, and did represent an advance in animal husbandry on the grand scale, providing facilities for handling and moving animals, and, most important of all, the outside enclosures of 1877. Frank Buckland enthusiastically described them as "Playgrounds, in which the idea of letting the large Carnivora roam completely loose at their own free will, has at last been realised".[5]

Mr C. B. Trollope followed. In fact he was a structural engineer, and not an architect, which explains the design of his

FIG. 7. The New Reptile House of 1883 (Trollope). Converted into the Bird House in 1927. Reproduced with permission from *The Illustrated London News*, 8th September, 1883.

Reptile House of 1883, which still stands today (Fig. 7). Although Neo-Renaissance outside, the House was firmly Functional Victorian inside, using a prefabricated wide-span structure with cast-iron columns, giving plenty of room to visitors and enabling it to cope easily with the transformation in 1927 into today's Bird House. In 1896 Trollope also designed the Ostrich and Crane House, again on simple practical lines, so that it too is in use today.

At this point it is interesting to reflect on building costs.[6] The 1839 Monkey House cost £1688, the Carnivora Terrace of 1843 cost about £4300. Salvin's Eastern Aviary was only £2688, his Monkey House £5095, while his 1869 Elephant House cost £6355, and the Lion House the vast sum of £11 643. Trollope's buildings were of the same order of cost, the Ostrich House costing only £3400. So it can be seen that building costs remained fairly static for 60 years, and the situation remained much as this up to the last war. Lubetkin's 1933 Gorilla House cost all of £4060, while the Penguin Pool cost only £1700. After that inflation took over until, today, the New Lion Terraces are costing over £900 000. This may seem an incredibly large sum, but the 1876 Lion House would set the Society back £420 000 today, not £11 600, an inflation rate of 3720%. Maintenance costs too are a regular burden, and the old so-called "well built" buildings seem to have performed no better, or worse, than today's structures. The old Lion House alone needed major attention in 14 of the 50 years prior to the last war.

After 1887 the mood turned towards "reproducing the native haunts" aspired to by Professor Flower, and in the following year the Great Aviary was constructed so that the birds might lead a more natural life. From then on things moved fairly rapidly and progressively, especially with the advent of Sir Peter Chalmers Mitchell as Secretary in 1903. Another large aviary, now the Southern Aviary, was opened in 1905, together with the Sealion Pond, which was the Zoo's first attempt at a natural enclosure for mammals, although interestingly enough it began as a mixed exhibit of sealions and penguins.

Soon after, Mr J. J. Joass, with Sir Peter Chalmers Mitchell, designed the Mappin Terraces (Fig. 8), and they were opened in 1914. A romantic concept based on Carl Hagenbeck's scenic panorama at Hamburg in 1907, the Terraces are most successful, though perhaps hardly "native haunts" for most of the animals. Joass drew up a master plan in 1909 which met with opposition. The problem was thought to be resolved when a Captain Swinton was brought in on the basis of his experience with the London

FIG. 8. An idealized preview of the Mappin Terraces, which were completed in 1913 (Joass). The Aquarium was inserted underneath in 1924. Drawing by Douglas Macpherson in *The Graphic*, 15th February, 1913.

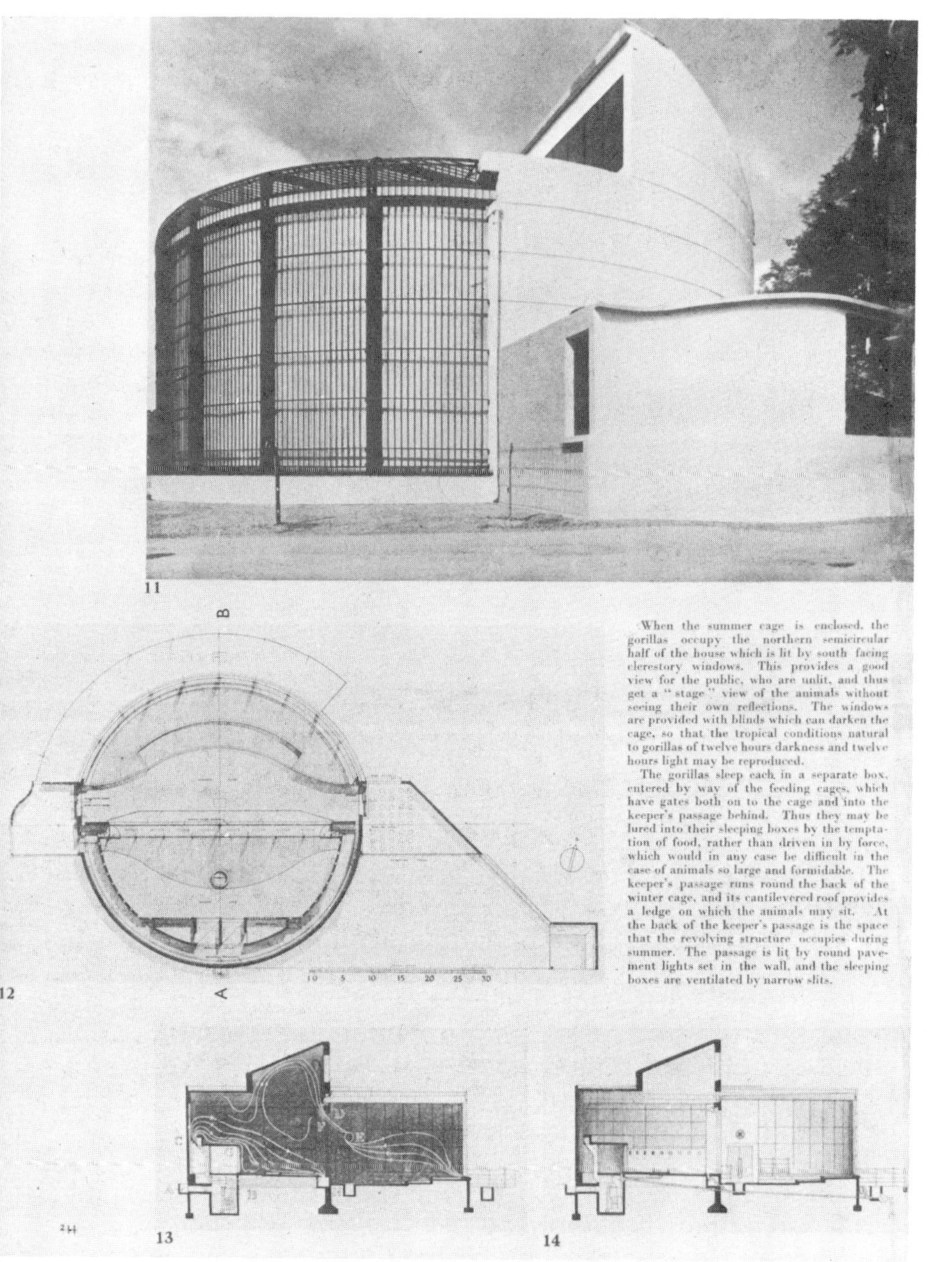

When the summer cage is enclosed, the gorillas occupy the northern semicircular half of the house which is lit by south facing clerestory windows. This provides a good view for the public, who are unlit, and thus get a "stage" view of the animals without seeing their own reflections. The windows are provided with blinds which can darken the cage, so that the tropical conditions natural to gorillas of twelve hours darkness and twelve hours light may be reproduced.

The gorillas sleep each in a separate box, entered by way of the feeding cages, which have gates both on to the cage and into the keeper's passage behind. Thus they may be lured into their sleeping boxes by the temptation of food, rather than driven in by force, which would in any case be difficult in the case of animals so large and formidable. The keeper's passage runs round the back of the winter cage, and its cantilevered roof provides a ledge on which the animals may sit. At the back of the keeper's passage is the space that the revolving structure occupies during summer. The passage is lit by round pavement lights set in the wall, and the sleeping boxes are ventilated by narrow slits.

FIG. 9. The New Gorilla House of 1933 introduced the Zoo to Modern Architecture (Lubetkin). Reproduced with permission from *The Architectural Review* Vol. 73 p. 244.

FIG. 10. The New Elephant House. A model of the scheme stopped by the 1939 war (Lubetkin). Photograph by John Havinden.

County Council and New Delhi, but in 1913 he produced a grandiose axial plan that was most unsuitable for Regent's Park, or any Zoo. Luckily, Joass was still in favour as an architect and he designed several fine buildings, of which only the 1924 Aquarium involved animals. His main works are the 1910 Main Offices, and the Pavilion and Regent Buildings of 1922 and 1929, both sited according to the Swinton grand axis. Only by luck, or foresight, was the Regent Building not repeated on the site of the Parrot House, where the great plan would have had it filled by munching elephants, instead of Fellows having an elegant luncheon.

There was little new building for animals during this period, other than the Reptile and Monkey Houses by Sir Guy Dawber, who also designed the present Main Gate, all in his pleasant and unobtrusive Italianate style. The 1927 Reptile House was a great success, largely through the efforts of the Curator, Miss Joan Proctor, who incorporated many new technological ideas. But,

FIG. 11. The Cotton Terraces. The first of the major post-war developments, completed in 1963, it centres around the 1836 Giraffe House by Decimus Burton (Shepheard and Stengelhofen).

although in the same year the newspapers heralded the new Monkey House as "the Monkey Palace", it was not a satisfactory building. In spite of a prototype cage unit having been tried out for three years, it seems to have relied too much on weak technology and not enough on zoology and physiology.

This fascination with evolving technology shows in the 1933 Gorilla House by Mr Berthold Lubetkin of the Tecton group (Fig. 9). Based on functional ideas such as improving the air flow, and the use of moving walls, it was a fine forward-thinking building which has proved adaptable enough since to accommodate bears, chimpanzees and orang-utans at different times. His Penguin Pool of 1934 symbolizes London Zoo to many people, and was a pioneering structure of the Modern Movement, even if it is not ideal for penguins. Lubetkin's North Gate Kiosk of 1936 still

FIG. 12. The Elephant and Rhino Pavilion of 1965 (Casson and Conder).

stands, but the Art Studio of the same year was demolished in 1962. His Elephant House of 1938 (Fig. 10) could have been the most important building of its era, but work was stopped because of the 1939 war, and afterwards the scheme was abandoned.

The war and its aftermath occupied the next 10 years. In 1947 Mr Stengelhofen was appointed as the first Staff Architect, to start repairing and reconstructing the Zoo and to look ahead to the future. He prepared a Development Plan in 1950, but no action was taken. Then in 1955 Lord Zuckerman became Secretary, and in the following year Sir Hugh Casson was invited to prepare a further plan, which was ready by 1959. The ground was cleared by starting with the various Service Buildings and the Central

FIG. 13. The Charles Clore Pavilion for Small Mammals was completed in 1967 (Black and Bayes).

Boiler House, all designed by Stengelhofen. Then the next 15 years saw as much major development as did the corresponding years in the nineteenth century, with the emphasis on re-housing the animals.

Access to the North Bank was improved by Sir Hugh Casson and Mr Neville Conder's new bridge in 1961. In 1963 the Cotton Terraces (Fig. 11), designed by Mr Peter Shepheard and Stengelhofen, were completed. They show a return to the ideas in favour at the beginning of the century, with a landscaped setting and straightforward buildings; practical for the animals and clearly

FIG. 14. The Michael Sobell Pavilions for Apes and Monkeys (Toovey). Below: a close-up view of the orang-utan enclosure.

FIG. 15. The New Lion Terraces. Model of the scheme opened in 1976 (Toovey).

attractive for visitors. This involvement with landscape is taken further with the 1965 Northern Aviary by Lord Snowdon and Mr Frank Newby; the problem of a difficult site was imaginatively solved by using the new computer technology becoming widely available in the early 1960s. The dramatic Elephant House of the same year (Fig. 12) by Casson and Conder is much less involved with landscape, perhaps because of the nature of its inhabitants, and is interesting to compare with Lubetkin's 1938 solution. Both are sculptural, strong, rather urban buildings.

After that, landscape takes over. Although the 1967 Clore Pavilion by Sir Misha Black and Mr Kenneth Bayes is an enclosed building, it has a strong outdoor feeling (Fig. 13). It is a highly technological solution relying on air conditioning and artificial lighting, but not letting them intrude on its neat and reticent design, which leaves visitors with a memory of animals and planting rather than of architecture.

Our design for the 1972 Sobell Pavilions (Fig. 14) derives from the thinking behind the Clore Pavilion, but takes the concept out into the open air, providing the maximum involvement of animal and visitor with landscape. The sleeping dens and handling facilities are no longer on view, so could be entirely designed to meet the animals' needs and simplify management. Out of doors, the activities of the animals determine the design, and they can walk and climb, or run about on the ground and live in family groups, or just sit up in the "branches"—which have proved flexible enough to cope with unexpected inhabitants such as the giant pandas. Our design (Fig. 15) for the New Lion Terraces, which also incorporates a new Aviary for Water Birds, has a much looser approach, but otherwise it follows the same general principles as the Sobell Pavilions. Landscaping is an essential part of this scheme, and after a few years the plants should take over the architecture, making the Terraces into a true "Zoological Garden".

During this period the scientific and educational aspects of the Society's activities were not overlooked. A Veterinary Hospital with full operating facilities opened in 1955; it was designed by Stengelhofen, who was also responsible for the Wellcome Institute of Comparative Physiology which was built beside it in 1963. Then in 1965 a fine concrete building by Lord Llewelyn-Davis solved some difficult problems by placing the Nuffield Institute of Comparative Medicine above new Meeting Rooms for the Society, while leaving a way through from the old East Tunnel; this then allowed the old Meeting Rooms in the Main Offices to be converted to an

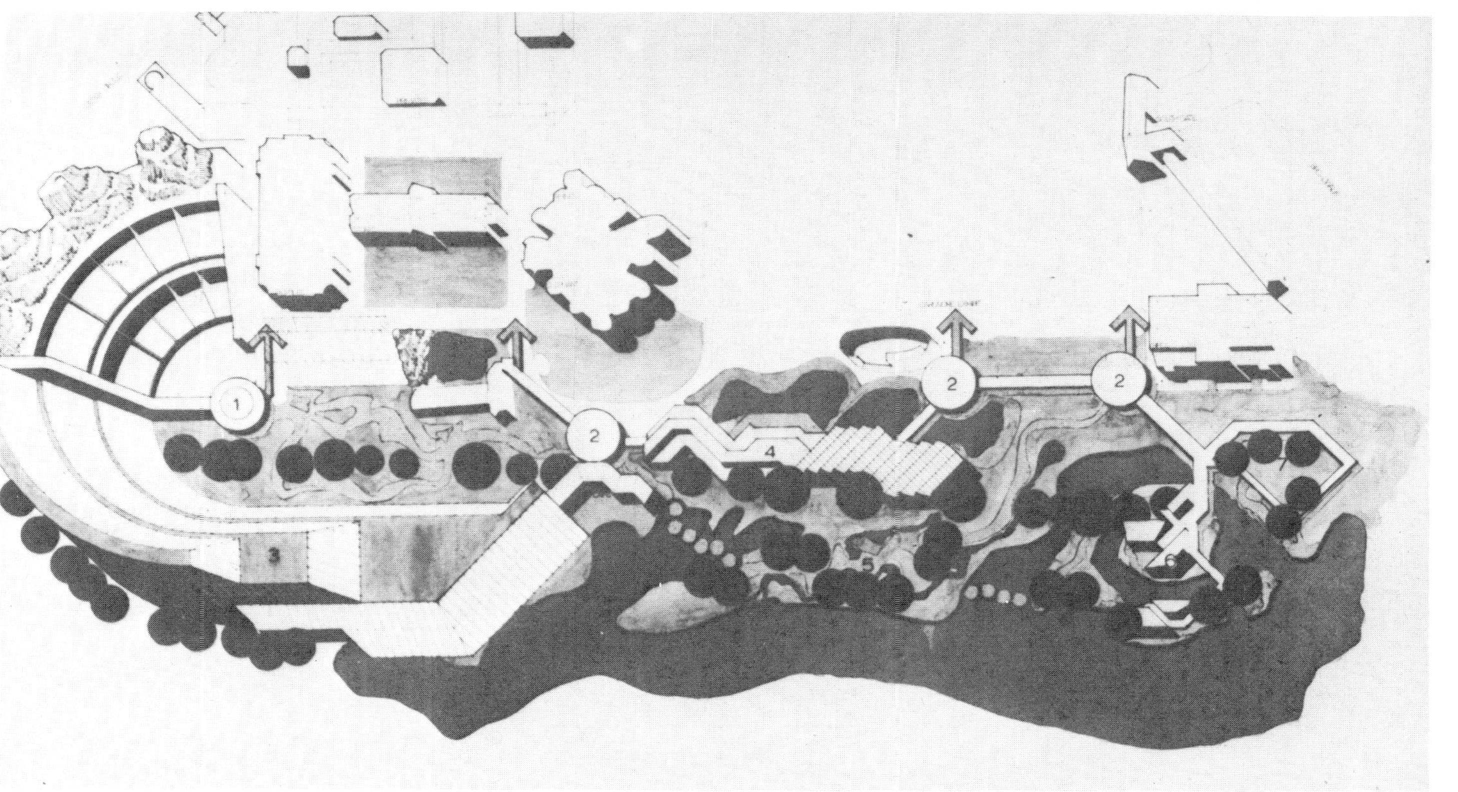

FIG. 16. Regent's Park Extension. Project on ten acres of the Park (Toovey). 1, Ecology exhibit & ecosphere; 2, orientation points; 3, tropical jungle exhibit; 4, temperate deep-sea exhibit; 5, temperate river-life exhibit; 6, children's islands; 7, family area; 8, perimeter lake.

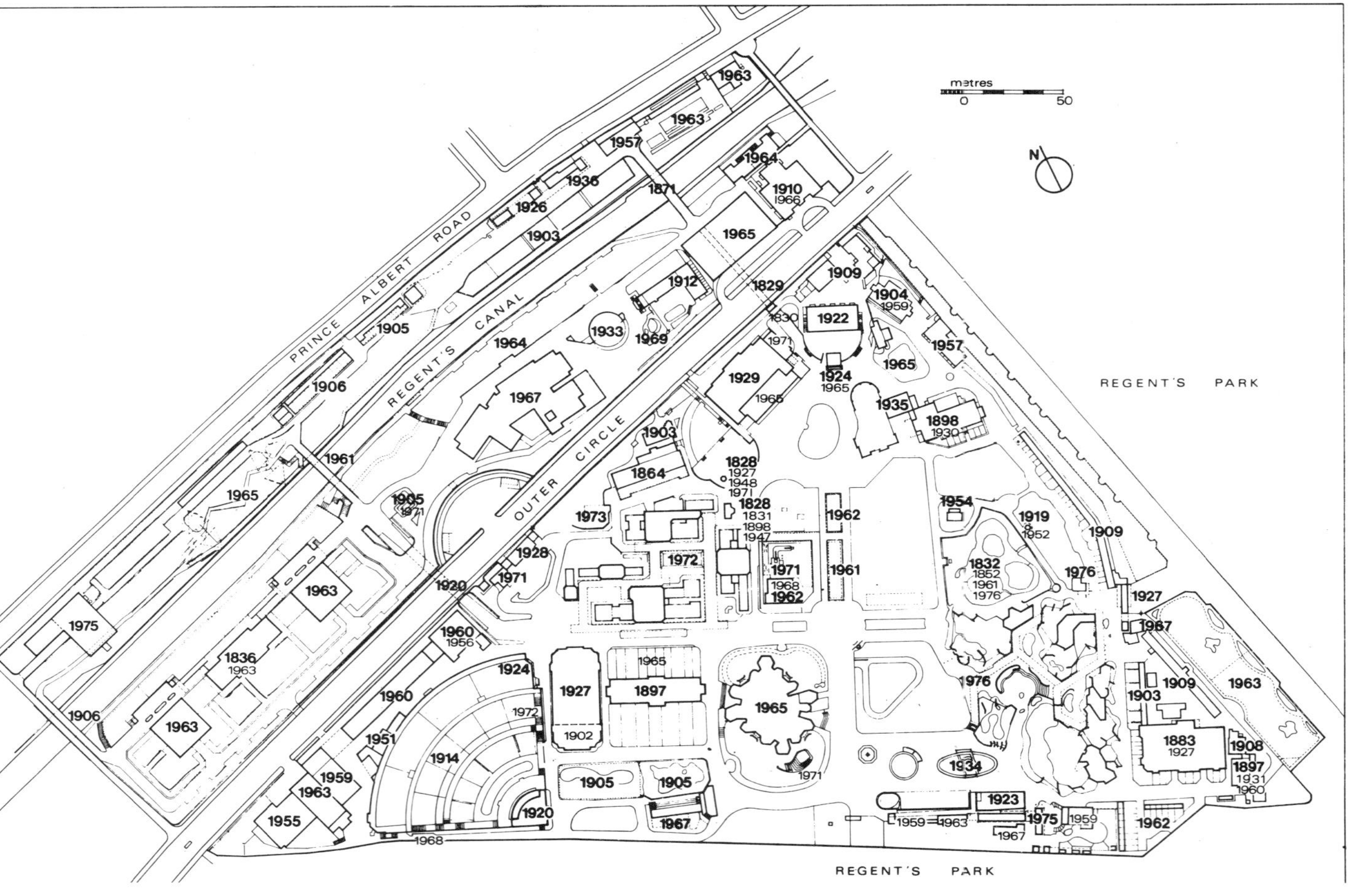

FIG. 17. Plan of London Zoo showing the buildings and landscaping in 1976, with the year in which they were completed superimposed

enlarged Library in 1966. The Education Department had worked in converted accommodation until the Zoo Studies Centre, designed by Casson and Conder, was opened in 1975. Placed on the sloping North Bank of the Regent's Canal this strongly modelled building contrasts well with the rural nature of the site.

This great 15 or more years of rebuilding has benefited the conditions and health of the animals, and has led to greatly improved management. It has certainly made the Zoo a more attractive and interesting place for visitors. But there is still a lot to be done and a lot to learn. At present the Society does not have a formal development plan on paper, and in many ways this can be considered to be a good thing, since it will not have to be saddled with a preconception when new projects do come forward. However, it does have a concept, in which science, education and information figure strongly. Sometimes an idea does get put down, such as the one recently prepared by us (Fig. 16), that attempts to fill in some gaps in the present exhibition, and to use the 10 acres of Regent's Park to which the Society is entitled one day. A great Tropical House would be grafted on to the Mappin Terraces, where it could be entered separately from outside the Zoo. The animal exhibits and landscape would be interwoven along the boundary, following on from the concept of the adjoining New Lion Terraces, but having an emphasis on water, with marine and freshwater exhibits, a children's zoo on islands, and more bird and ungulate paddocks. No doubt very, very expensive. It would be just one way forward; there are many others; but whichever way it follows the Zoo will never fail to be exciting, or to continue to be "new and curious" as its founders hoped.

References and Sources

The Architects, Engineers, and Building-Trades Directory (1868). London: Wyman and Sons.

The Architectural Review (various for Lubetkin, Tecton 1930–1940).

Avery Obituary Index of Architects and Artists (1963). Columbia University: G. K. Hall & Co.

Bishop, J. (1832). *Henry and Emma's visit to the Zoological Gardens in the Regent's Park.* London: A. K. Newman & Co.

Buckland, F. (1882). *Notes and jottings from animal life.* London: Smith Elder & Co.

The Builder (various for obituaries).

Chalmers Mitchell, Sir P. (1929). *Centenary history of the Zoological Society of London.* London: Zoological Society of London.

The Dictionary of Architecture (1887). The Architectural Publication Society. Whiting & Co.

The Dictionary of National Biography (various for obituaries).
Royal Institute of British Architects—"Biography File Index": RIBA Journal (various for obituaries): Members Directories: "Names Index".
Scherren, H. (1905). *The Zoological Society of London.* London: Cassell & Co.
The Zoological Society of London— Archives: Annual Reports for the years 1828–1975: Council Minutes: Guide Books for the years 1858–1975: Press Cuttings for the years 1843–1936: "Survey of Existing Structures on the Society's premises at Regent's Park", Architects Department December 1948 (Stengelhofen).
Toovey, J. & Brambell, M. (1976). The Michael Sobell Pavilions for Apes and Monkeys. *Int. Zoo Ybk* **16**: 212–217, plates 36–41.

Notes

1. From the *Daily News* for 23rd January 1869.
2. Annual Report for 1887: extract from the Jubilee Address made on 16th June, by the President, Professor Flower. Similar sentiments were expressed by the Secretary, Sir Peter Chalmers Mitchell, in his Centenary Address, recorded in the Annual Report for 1929.
3. Annual Report for 1830: Auditor's recommendation reported "no Surveyor or Architect on any occasion be paid by a percentage charge on disbursements or tradesmen's bills".
4. Annual Report for 1830: . . . "The Council applied to Mr Decimus Burton to become Architect of the Society at a fixed salary; and they are happy to announce their having secured the valuable services of that gentleman on the following terms. His salary to be £150 per annum. He will have the sole direction of all future buildings and alterations; the bills and accounts of which are to be examined by him previously to their payment. . . ."
5. *Notes and jottings from animal life* by Frank Buckland MA. Whole chapter headed "Playground for the Lions and Tigers at the Zoological Gardens". The dates he gives are not correct.
6. Annual Report for 1828: "The nature of the soil, which consists of a thick ungrateful clay, increases the cost of every work." Annual Report for 1883: List of ten principal New Works since 1860:

		£
1860–62	New wing of Antilope-house	2339
1862–78	New Refreshment-rooms	8193
1863–65	Eastern Aviary	2688
1863–65	New Monkey-house	5095
1863–65	New Entrance-lodges	1402

1867–68	Wapiti-house	2421
1868–70	Elephant-house	6355
1872–73	New North Lodge and Entrance Gate	1244
1875–77	Lion-house and Outer Cages	11643
1882–83	Reptile-house	9174

Appendix A

Architects and Engineers, listed approximately in the order in which they worked at London Zoo

Decimus Burton FRIBA, FRS, FSA, *b.*1800 *d.*1881 (in 1868 used offices at 6 Spring Gardens SW).

Elmslie (in 1868 firm known as Elmslie and Franey used offices at 43 Parliament Street, SW).

Thomas Bellamy FRIBA, *b.*1798 *d.*1876.

Anthony Salvin, jun. FRIBA, *d.*1881 (in 1868 used offices at 4 Adam Street, Adelphi, WC. Not to be confused with his father of the same name, who died in the same year).

Charles Brown Trollope AICE (in 1868 used offices at 2 Upper Charles Street, SW).

A. Flower (no record at RIBA, so may not have been an Architect or Engineer).

John James Joass FRIBA, *b.*1868 *d.*1952.

Sir E. Guy Dawber FRIBA, *b.*1862 *d.*1938.

Edward Salter, ARIBA.

Berthold Lubetkin (a founder member of the Tecton firm of Architects).

Sir Hugh Casson MA, PRA, RDI, FRIBA, FSIA, *b.*1910
Neville Conder AA.Dipl.(Hons), FRIBA, FSIA
} Casson Conder and Partners.

Peter Shepheard CBE, B.Arch, PPRIBA, MRTPI, PPILA, *b.*1913.

Lord Richard Llewelyn-Davies MA, AA.Dipl., FRIBA, FRTPI, *b.*1912.

Sir Misha Black OBE, RDI, PPSIA, *b.*1910 Kenneth Bayes FRIBA, FSIA	Black, Bayes and Gibson.
Lord Snowdon *b.*1930 Cedric Price MA, AA.Dipl., ARIBA Frank Newby MA, FIStructE, MConsE	Together for Northern or "Snowdon" Aviary.

Franz Stengelhofen LRIBA, *b.*1901 (appointed 1947, retired 1966).

John Toovey AA.Dipl.(Hons), FRIBA, *b.*1933 (appointed 1967).

Appendix B

List of buildings and landscaping at London Zoo in 1976 (Fig. 17)

Date	Present Name	Architect	Alterations or change of use
1828	Clock Tower	Burton	Turret added 1831. Rebuilt 1898, 1947
1828?	Raven's Cage	Burton	Rebuilt 1927, 1948. Moved 1971
1829	East Tunnel	Burton	North and South entrances added 1830
1832	Three Island Pond	—	Extended 1852, altered 1961, 1976
1836	Giraffe House	Burton	Altered 1963 (Stengelhofen)
1864	Eastern Aviary	Salvin	
1871	East Bridge	—	
1883	Bird House	Trollope	Reptile House until 1927
1897	Ostrich House	Trollope	North paddocks altered 1965. Tortoise House until 1931. Change of use 1960
	Hummingbird House	—	
1898	Parrot House	Flower	Fellows' Tea Pavilion until 1929 (Salter)
1902	Reptile House	Trollope	Ape House incorporated in 1927 Reptile House (Dawber)
1903	Keepers' Lodge	—	
	Peafowl Aviaries	—	
	Crane and Goose Paddocks	—	
1904	Catering Store	Trollope?	Small Mammal House until 1959, then Education Centre
1905	Sealion Pond	—	
	Southern Aviary	—	
	Owls' Aviary	—	
	Beaver Pond	—	Originally Otter Pond

Date	Present Name	Architect	Alterations or change of use
1906	Primrose Bridge	—	
	Northern Pheasantry	—	
1908	Research Unit	—	Originally Pathology laboratory
1909	TV Building	—	Sanatorium until 1957
	Birds of Prey	—	Designed by Curator Seth-Smith.
	East Service Gate	—	Lodge built before 1900
1910	Main Offices	Joass	Library inserted 1966 (Stengelhofen)
1912	Insect House	—	
1914	Mappin Terraces	Joass	
1919	War Memorial	Joass	Originally in front of Reptile House. Moved 1952
1920	West Tunnel	—	
	Mappin Cafe	Joass	
1922	Pavilion Building	Joass	
1923	Penguin Cafe	—	
1924	Aquarium	Joass	(Engineers, Alexander Gibb and Partners)
	Photoshop	Joass	Originally Bandstand, Information Kiosk 1965
1926	North Gate	—	Closed 1975
1927	Reptile House	Dawber	Incorporating 1902 Ape House
	South Gate	—	
1928	Main Gate	Dawber	
1929	Regent Building	Joass	Cafeteria forecourt 1965 (Stengelhofen)
1933	Ape Colony	Lubetkin	Gorilla House, later used for bears, chimpanzees orang-utans
1934	Penguin Pool	Lubetkin	
1934	Garden Cafe	Salter	
1936	North Gate Kiosk	Lubetkin	
1951	Boiler House	Stengelhofen	
1954	Small Parrots Aviary	Stengelhofen	Originally for budgerigars
1955	Hospital	Stengelhofen	
1957	Lavatories	Stengelhofen	
1959	Workshops	Stengelhofen	
1960	Stores/Garage	Stengelhofen	
1961	Canal Bridge	Casson/Conder	
	Gibbons' Cage	Stengelhofen	
1962	Cockatoo Aviary	Stengelhofen	
	Southern Pheasantry	Stengelhofen	
	Zoo Shop	—	Extended 1968
1963	Cotton Terraces	Shepheard/ Stengelhofen	Incorporating 1836 Giraffe House (Burton)
	Wellcome Institute	Stengelhofen	

Date	Present Name	Architect	Alterations or change of use
	Social Club	Stengelhofen	
	Wolf Wood	Stengelhofen	
	Gardening Department	Stengelhofen	Rebuilt over several years
1964	South Canal Bank	Shepheard	
	Staff Flats	Llewelyn-Davies	
1965	Nuffield Institute	Llewelyn-Davies	
	Elephant and Rhino Pavilion	Casson/Conder	Pool altered 1971
	Snowdon Aviary	Snowdon/Price/ Newby	Including shelter
	Flamingo Pond	Stengelhofen	
1967	Clore Pavilion	Black/Bayes	
	Sealion Stand	Stengelhofen	Includes dog paddocks
	South Gate Lavatories	Stengelhofen	
	Children's Zoo	Stengelhofen	Rebuilt over several years
1968	Mappin West Steps	Toovey	
1969	Otter Exhibit	Toovey	
1971	Tavern Bar	Toovey	
	Playground	Toovey	
	Main Gate Kiosk	Toovey	Including terrace in front
1972	Sobell Pavilions	Toovey	
	Mappin East Steps	Toovey	
1973	Main Gate Lavatory	Toovey	
1975	Children's Zoo	Toovey	
	Education Centre	Casson/Conder	
1976	The New Lion Terraces	Toovey	
	South Gate Kiosk	Toovey	

Symp. zool. Soc. Lond. (1976) No. 40, 203–214.

THE MAINTENANCE OF ANIMAL HEALTH IN THE COLLECTIONS

D. M. JONES

The Zoological Society of London, Regent's Park, London, England

In 1829 the Society appointed a Mr Charles Spooner of Camden Town as Medical Attendant at a salary of £60 a year. His duties were to attend the Gardens three times a week and prescribe for and examine the animals and to keep a record of his observations and practice. It appears that Mr Spooner was not very satisfactory, for he soon complained about the increasing number of animals and asked for a rise. The record discretely states that soon after this, his services were dispensed with. The famous Mr Youatt, who had a veterinary practice in Oxford Street, was then appointed Medical Superintendent and his duties were to accompany the Head Keeper on Mondays and Thursdays, to inspect every animal and give directions as to their maintenance. He was also required to carry out post-mortems. After his death in 1847 the Superintendent of the collection was in overall charge of animal health, and he called largely on the services of the nearby Royal Veterinary College. In 1907 a Sanatorium was built with isolation wards, mainly in an attempt to treat tuberculosis which was then widespread in the collection, but it was found that the disturbance in moving sick animals to a strange environment often outweighed the benefit of the intensive care which they were to be given. Forty years before, the Society had instituted its prosectorial service, and the regular post-mortem of animals from the collection was beginning in the early part of this century to help those such as Peter Chalmers Mitchell and Geoffrey Vevers, who were trying to introduce new concepts in the husbandry of the stock. Chalmers Mitchell could find no evidence to support the then current theory that the British climate was largely detrimental to animal health and that the animals must therefore be kept in heated houses. In his Centenary History (Mitchell, 1929), he described the small cats as being "odorous with their own effluvia" and set about improving the standards of hygiene and increasing air flow and available light in the houses. With the help of Vevers, who was a keen parasitologist, a regular examination of the faeces of all the stock in the collection was instituted. Although tolerant of public feeding, they were well aware of its problems, and tried to channel these activities

FIG. 1. Supporting a limb fracture in an alligator using plaster of Paris, 1922.

by giving visitors specific instructions as to what they could and could not feed. In general, the 1920s and 30s probably saw the first real effort to establish a planned programme of medical care in the collection (Figs 1 and 2).

PREVENTIVE ASPECTS

The application of preventive medicine is particularly important in the management of non-domestic species. This is mainly because of the difficulties of handling such animals and also, from a clinical point of view, because they are not nearly such good patients as their domestic relatives. The three most important aspects of preventive medicine in this situation are: housing and enclosure design, correct nutrition, and efficient keepers.

The principles behind the design of animal housing have been explained in another chapter, and it has been shown there how essential it is to have some knowledge of the environmental requirements of each species. Increasingly, it is becoming more important to keep viable reproducing units of each species and this usually means larger numbers of that species than might have been

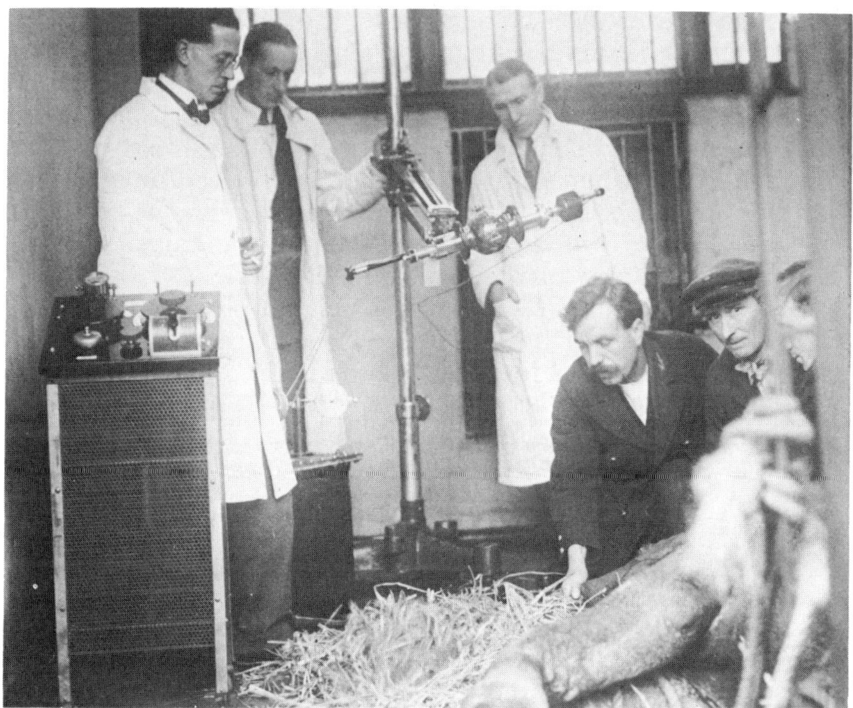

FIG. 2. Using an early design of X-ray machine to examine an injured elephant's leg (1926). The radiation scatter from such an X-ray tube was considerable.

the case previously. As a result extra facilities are needed for holding spare males, nursing mothers and groups of subadults. In order that the animals can be inspected easily and handled when necessary, facilities for doing this must be built into the house and the designer must take account of the fact that an animal under the influence of narcotic or sedative drugs will become ataxic and may collide with objects if they project from the walls or roof. Surfaces have to be easily cleaned and disinfected but at the same time the flooring may have to be rough in order to prevent an animal from slipping. A loose box with wall surfaces covered in ceramic tiles may be ideal from the medical point of view, but a compromise usually has to be reached to prevent the exhibit looking visually sterile and uninteresting as a result.

Most animals, once accustomed to an enclosure, rarely try to realize their full potential in crossing the boundaries of that enclosure. Nevertheless, particularly with species that may be dangerous to the staff or public, barriers have to be constructed to cope with all possibilities. The unpredictably nervous behaviour of some ungulates, especially when in small enclosures, usually necessitates the building of barriers which are opaque. A similar design is often advisable when aggressive males are housed next to each other. The additional expense of building this type of enclosure and the fact that the visitor has paid to see the animal, dictates against the more extensive use of the solid barrier.

Apart from a requirement in all cases for efficient ventilation and in some cases for heating, there is evidence that the maintenance of a high level of humidity is also important in the design of some exhibits, particularly of animals from tropical rain forests. Ungulates from deserts usually adapt over a 12-month period to the cooler, damper conditions in Britain but they do not rapidly adapt to new species of nematode parasites or to a diet of fast-growing grasses, low in fibre content but high in protein. For this reason such animals must only have access to perennial grasses which have been heavily grazed by other species or mown by a machine. Yards with a hard, well drained surface are needed for such animals throughout most of the year, and in the case of some species, notably the equines and the giraffes, these surfaces also provide a high degree of friction to prevent overgrowth of their hooves.

The extensive research carried out into the nutritional requirements of domestic species has helped considerably in designing diets for our own stock. All animals need amino acids, fats, carbohydrates, vitamins, minerals and water, but they differ in the amounts of each nutrient required and in the way in which they obtain it. With mammals in particular, the design of a diet begins by using the data available from the most closely related domestic species. The same approach can be used to some extent on reptiles and birds, but knowledge is inevitably more limited for these classes. The Keeper, the Curator, and the Veterinary Officer, each contributing some knowledge of a particular animal's requirements, then have to decide on the most palatable means of supplying it, within the usual economic limitations. Knowledge of a species' food preferences in the wild is useful to some extent, but one is restrained from using the information literally because these items are unlikely to be available. It is often not realized that long-term storage, particularly of meat, fish and vegetables, leads to considerable deterioration in their vitamin content. Partly because of this, and also because certain traditionally accepted diets in zoos (such as red meat for cats) are highly deficient in a variety of nutrients, a number of complete diets and supplements are made commercially to our designs. These are modified from time to time as new information comes to light. These diets are now available to other collections in the country, and this is a good example of how the Society's applied research programme helps other stockmen as well. By using accumulated information in the normal course of clinical and pathological examinations, it has been possible to show, for instance, that giraffes, camels and wallabies almost certainly have a higher requirement for vitamin E than other related species and that certain forms of vitamin A probably cannot be used by fish-eating animals. One problem which has not been solved yet, and which is particularly important at Whipsnade, is that of keeping animals which are adapted to utilize trees and shrubs on a system which is more suitable for the grazer. Usually it is not practicable to feed large quantities of leaf and twig throughout the year. It has not yet been possible to identify accurately the differences between the normal diet of the browser and that of the grazer and, unfortunately from the captive animal's point of view, much of the ecological work which has been done on these species in the wild has yielded very little information on the nutrient content of the plants being eaten. The Biochemistry Department in the Nuffield Institute, who have carried out a considerable amount of work on this subject, have shown that one of the major differences is that the content of some of the unsaturated fatty acids is higher in bark and leaf than in grasses. Attempts to correct these differences in the artificial diet are rendered useless, probably because the fatty acid does not reach the duodenum in its original form, having been hydrogenated in the rumen. This is the type of problem which remains on the high priority list for applied research.

In many ways the efficient keeper is the most important factor in the development of any programme for maintaining animal health. It is he who must make the theory work in practice and from whom a feedback of information is required to ensure continued advances. Nothing useful can be achieved without keen, interested stockmen. The formal courses which the Society holds for keeping staff provide him with a sound theoretical background, but the best means of learning is by seeing and talking about a problem as it arises. The veterinary staff have to make every effort to involve a keeper in activities which affect his stock and to make sure that they are reasonably available to the man who wants to ask questions. The clearer the idea the keeper has of the factors involved in animal husbandry, the more interested he becomes, and the more useful he is in preventing and dealing with the problems that could arise.

THE VETERINARY SERVICES

The Society runs a comprehensive veterinary service staffed by two full-time veterinary surgeons who are supported by 12 lay staff. One veterinary surgeon is based at each of the two collections and a continuous emergency service is maintained by involving three further members of staff who are also veterinary surgeons to cover for some weekends and holiday periods.

FIG. 3. Anaesthetized common zebra undergoing surgery for ruptured tendons in the Animal Hospital.

At Whipsnade, there are comprehensive facilities for post-mortem examination of carcases and a small surgery. Most of the clinical work is carried out in the paddocks and houses, and samples for laboratory examination are submitted to the technicians at Regent's Park. The facilities at Regent's Park are more comprehensive. Material from the post-mortem rooms at both collections can be processed for histopathology, and specialized technical knowledge is available for bacteriology, parasitology and to some extent mycology. The hospital section includes facilities for major surgery (Fig. 3) and comprehensive clinical investigation. A range of rooms and cages with controllable ventilation and heating can accommodate almost any animal in the collections up to about 500 kg in weight. There are special facilities for handling dangerous individuals and part of the building can be isolated for the strict quarantine of new arrivals. The staff of this section includes three registered animal nurses and a radiographer. Clinical cases, whole carcases and samples for examination can be submitted to the Department by veterinary surgeons in general practice.

Clinical and pathological records are kept of all the animals examined and summaries of the main findings are published every other year in the Society's scientific reports. The pathological records are also incorporated into the World Health Organization's co-ordinated pilot programme for international recording of zoo and wild animal disease data. The facilities of the department and the bank of data are available to visiting research workers.

Owing to the wide range of interests and equipment in the Society's research institutes (the Nuffield Institute of Comparative Medicine and the Wellcome Institute of Comparative Physiology), the work of the veterinary clinical and pathological sections is substantially backed up by other scientific staff. Much of the accumulated physiological data which is fed back into the management of the stock comes either directly from the research programmes carried out in the Institutes or as a by-product of a wider field of research. This is especially the case in the fields of haematology, mycology, reproduction, biochemistry and radiography.

The remaining gaps in the laboratory facilities available on the Society's premises are then filled by other research workers who are interested in the comparative information which material from the collections offers them. Much of the virology and histopathology, the determination of serum enzyme and electrolyte levels, vitamin and mineral estimations, and the identification of some pathogenic bacteria and parasites falls into this category. As a result, these people must take some of the credit in the formulation of the husbandry policies of the collections.

PATHOLOGY

Much of the Society's early scientific activity revolved around the prosectorial room. Not only was the latter half of the last century an era when comparative anatomy was particularly fashionable, but it became evident, as time went by, that routine systematic post-mortem examination of all animals dying in the collection yielded a wealth of information which was useful for management planning. Chalmers Mitchell used the records from the prosectorium between 1870 and 1902 in an effort to prove that the application of heat alone had very little to do with maintaining an animal's health, without considering other environmental factors. In fact the records showed that in many cases the hot, stagnant air had probably led to considerable problems. Post-mortem examinations are of immediate use in that they give an indication as to whether other animals in the group, and even people in contact, might be at risk. The finding for example of *Salmonella* at such an examination would lead to the screening of all "in contact" stock, and as a result the removal of the source or treatment of the carriers and clinically affected animals. In the long term, such examinations can also provide very useful information on features such as body weight, foetal development, disease incidence in a particular group, and the pin-pointing of sub-clinical nutritional deficiencies. Injury to particular areas of the body may indicate that a change is required in the design of the enclosure. Losses amongst males of a territorial species might indicate an overpopulation problem. The routine examination of faeces for the evidence of parasites and pathogenic bacteria is regularly undertaken and programmes of control worked out accordingly. Many of today's anthelmintics are almost too efficient, and care has to be taken, particularly in the large enclosures at Whipsnade where reinfestation with the parasite is almost inevitable, that the equilibrium between the parasite and its host is not upset to the disadvantage of the latter. An illustration of this point has been seen in a number of collections recently where Bactrian camels harbouring large numbers of nematode worms related to the genus *Ostertagia* of domestic species were treated with an anthelmintic which removed the adult worm from the abomasum. This occasion then precipitated the emergence of the fourth stage larvae from the mucosa taking the place of those removed and in so doing causing tissue damage. Where the chances of reinfestation are less likely, as in for instance a well-kept reptile exhibit or aviary, with hard cleanable surfaces, then complete eradication of the parasite may be possible and advantageous.

CLINICAL MEDICINE

The popular myth that clinical medicine in zoological gardens is a branch apart from other aspects of veterinary science has now largely been dispelled. There is no doubt that a sound knowledge of the natural history of each species being dealt with, increases the help which the veterinary services can offer the Curator and his keepers. Only a very small proportion of the clinical conditions met with in a zoological collection are peculiar to non-domestic species. There are two major problems which are different from the management of domestic stock. The first is the difficulty of handling and sedation, which has now been largely overcome. The second is the approach to diagnosis and treatment which must always take into account whether more harm may be done in examining the animal and providing "treatment" than by leaving it well alone. As more is learnt, the element of doubt in dealing with a case decreases. Previously, chemical immobilization of a non-domestic animal was carried out as a last resort. More recently, the technique is almost always used as early as possible in the course of a clinical problem. This has been partly due to the advent of a number of relatively safe but potent sedative and narcotic drugs which have largely revolutionized the handling of these species. It has been possible to evaluate extensively almost all the drugs with these properties which have come on to the market in the last ten years and, with very few exceptions, safe sedation and anaesthesia is now practicable for almost all animals. The Society's veterinary staff have shown that techniques and drugs developed for use in domestic stock can quite safely be used with the agents which are employed to capture the animal. At Whipsnade in the last ten years techniques have been developed using mixtures of narcotic and sedative drugs which, by manipulation of the dose rates, allow the operator under field conditions to carry out a variety of manoeuvres, from examining lame elephants and moving zebras from one paddock to another, to providing long-term narcosis for transport to other collections thus avoiding the damage which animals often inflict on themselves in crates.

As far as is practical, animals entering the collection from any source, but particularly from abroad, are screened for evidence of infectious disease. Many of these animals will in any case be subject to Ministry of Agriculture quarantine regulations, but regardless of this all stock are isolated for at least 28 days for the protection of the other animal residents, and in some cases for the protection of the staff and public as well. Now that safe vaccines are available for feline enteritis, canine distemper and leptospirosis, these are used regularly in the majority of carnivores in both collections. If there is any evidence in the country at a particular time of an increasing incidence of diseases such as Newcastle disease or duck virus enteritis, as has occurred in recent years, stocks of suitable vaccines are kept ready for immediate use in case of emergency. Polyvalent clostridial vaccines are used routinely in the sheep, and occasionally in other susceptible ruminants at Whipsnade. Passive protection, in the form of polyvalent antisera, against certain clostridial toxins and serotypes of *Salmonella* and coliform bacteria, are usually given to colostrum-deprived animals that have to be hand reared.

Continuous monitoring of animal health is carried out largely as part of the daily clinical activities in the collections (Fig. 4). If a healthy animal is immobilized for movement to another enclosure, blood samples will be taken in most cases for routine haematology and biochemistry, plasma samples will be prepared for vitamin A and trace element composition, and serum will be deep frozen for antibody estimations which are usually undertaken when a screen for a particular disease is carried out for the whole collection. In the case of Whipsnade, for instance, this has included recently examination of the sera of 30 species for the presence of *Brucella* and *Leptospira* antibodies. Sometimes, several years after the serum sample has been taken, a problem is encountered where that original sample becomes invaluable for background information. In addition to this, the serum bank provides valuable material for comparative studies by outside research workers.

SUMMARY

The maintenance of health in the Society's collections is based on sound preventive planning by a team of people ranging from the Curators and Veterinary Officers to the youngest Keeper. The methods are those widely accepted in agriculture, although the practical difficulties in implementation, particularly where regular catching and handling is involved, are to some extent special to the non-domestic species. It is easy to look critically at the casualty figures and disease incidence of 100 years ago, but it has to be

FIG. 4. Obtaining a blood sample as part of a routine clinical examination of a sedated tiger at Whipsnade.

remembered that in addition to the many sophisticated laboratory aids which we possess in the 1970s, each generation of animal managers has gleaned something from the experience of its predecessors.

Reference

Mitchell, P. Chalmers (1929). *Centenary history of The Zoological Society of London*. London: Zoological Society of London.

Symp. zool. Soc. Lond. (1976) No. 40, 215–222.

RESEARCH

L. G. GOODWIN

The Zoological Society of London, Regent's Park, London, England

INTRODUCTION

The papers given at this symposium will have made it clear that the Zoological Society of London, throughout its long history, has been deeply concerned with research in many fields of biological science.

The studies of the great anatomists, among whom Professor Cave himself occupies a prominent position, have laid the basis of comparative anatomy and taxonomy. The Society's curators, prosectors and pathologists, through their reports and their collaborative studies, have made important contributions to the understanding of disease processes in various species of animals. Sir John McFadyean, the founder of the *Journal of Comparative Pathology*, was closely associated with the Society's work at the beginning of the present century and Murie, Garrod, Plimmer, Scott and Hamerton, all at one time members of the Society's staff, were among the founders of the science of comparative pathology.

To choose a few subjects at random—among Hamerton's observations can be found an account of the alleviation of lower limb paralysis in monkeys with degenerative lesions of the spinal cord by giving injections of liver extract (Hamerton, 1941). This was probably among the first demonstrations of the action of vitamin B_{12} on nervous function. Snakes from the reptile house provided material for the studies of Burgess Barnett & R. G. Macfarlane (1934) on Russell viper venom and its effect on the coagulation processes of blood.

Lawrence & Rewell (1948) studied the anatomy of the arteries in the head of the giraffe, in search of a clue to a mystery: why, with a change of hydrostatic pressure of 6 metres of blood, does the brain not become heavily congested when the head is lowered? The buffering action of the rete mirabile at the termination of the carotid and the presence of a bypass in the form of an exceptionally large occipital artery offers a plausible explanation. A classical series of studies on the menstrual cycles of primates was made by Lord Zuckerman while he was Anatomical Research Fellow of the

Society; the first of these was published in 1930 (Zuckerman, 1930). With characteristic energy, he also published two important books on the *Social life of monkeys and apes*, and the *Functional affinities of man, monkeys and apes* (Zuckerman, 1932, 1933).

RESEARCH ARISING FROM THE MAINTENANCE OF THE COLLECTIONS

Animals are mortal; their remains furnish materials for study by anatomists, taxonomists, pathologists and parasitologists, and there has been a continuous outflow of these materials to research workers in museums and University departments. After the First World War, Ronald Leiper was Professor of Helminthology at the London School of Hygiene and Tropical Medicine; he helped the Society with the organization of the prosectorium and with the identification of parasitic worms. This association lasted throughout his long life, was continued by the late Professor J. J. C. Buckley, and is still maintained by his successor, George Nelson.

A similar arrangement was made with regard to protozoa with Dr C. M. Wenyon: many of the organisms described in his classical treatise (Wenyon, 1926) were derived from the Society's animals. Wenyon was still regularly collecting blood films from the Zoo when I joined his staff at the Wellcome Bureau of Scientific Research in 1939.

This kind of traffic has advantages for both sides. The external worker obtains access to unique materials; the Society obtains expert advice on the identification of organisms, or is provided with chemical data that may be of importance to the health of the collection. It does not, however, remove the need for routine laboratory diagnostic facilities at the Zoo, to identify hazards that might threaten the other animals in the collection, the staff, or the public. It is important to know, without delay, of a threat from anthrax or tuberculosis, of a *Salmonella* or *Yersinia* infection or of an outbreak of ornithosis, so that appropriate measures can be taken. It is also of great value to have a haematology service so that blood disorders, from nutritional or other causes, can be identified and an attempt made to put them right. We now have a fair idea of normal ranges for the numbers of blood cells and many other haematological parameters; these have been published so that all can refer to them (Hawkey, 1975).

Biochemical tests are often of value in diagnosing the cause of disease in a sick animal and there was a time when I made an effort to run a range of tests as part of the laboratory services in the Zoo.

But clinical chemistry in hospitals has become a highly sophisticated, automated process, and generous help is available to us; it seems to me that our limited resources are better deployed in other directions. We carry out a few specific tests such as estimations of vitamin A, and tests for rancidity of fish fed to dolphins and other animals. Such tests are of importance in monitoring current problems and the time comes when they lose their usefulness and can be replaced by others.

David Jones has dealt with another aspect of applied research—the comparative pharmacology of immobilizing drugs and anaesthetics used in the management of the collections, especially at Whipsnade. Dr Martin on pp. 283–319 discusses the value of routine chemical and other tests for monitoring pregnancy. These studies are making enormous improvements in the efficiency with which wild animals can be managed and bred in captivity.

Unless an epidemic of disease occurs—and as husbandry improves, epidemics fortunately become rarer—the accumulation of evidence on a particular problem is very slow because of the small numbers of any one species that come to hand. An interesting lesion may be detected in the liver of a crocodile, but it may be years before another crocodile with a similar lesion appears. Certain peculiarities of physiology or behaviour may occur when an orang-utan has to be anaesthetized to carry out essential surgery. It may be many months before there is sufficient reason to anaesthetize another orang. The research worker in the Zoo is surrounded by interesting material but he often has to wait for an occasion to get his hands on it, and progress may be slow.

But he has to be there. Daniel Cowan, of the Department of Pathology, Eastern Virginia Medical School, put the situation clearly at the recent Symposium to mark the one hundredth anniversary of the Zoological Society of Philadelphia:

> "How does research in comparative pathology relate to the programs of a zoo? The most obvious reason is that the zoo is where the animals are. Zoos, aquariums and game farms may be the only places left where large numbers of highly diverse, genetically intact species may be found and observed. Domestic species and most especially laboratory animals are often hardly more than bizarre artifacts, so detached from their environments and so highly modified that their existence depends on the constant intervention of man. . . . They are useful in the same way that caricatures are useful in attracting attention to peculiarities of face or expression, but like caricatures, they are not accurate reflections of reality." (Cowan, 1975.)

THE RESEARCH LABORATORIES

Two research Institutes, the WICP (Wellcome Institute of Comparative Physiology) and NICM (Nuffield Institute of Comparative Medicine), were established in Regent's Park about 12 years ago. The task laid down for the Wellcome Institute was to study reproductive physiology, and for the Nuffield Institute to study disease in animals—including the human animal— in all its aspects. These objectives were set by the Committees that planned the programmes of the Institutes, and we have gone some way to fulfilling them, as our published Scientific Reports will show (Zoological Society's Scientific Reports, 1967–69; 1969–71; 1971–73; 1973–1975).

We have studied the blood cells and the cerebral circulations of many animals, the curious activation of fibrinolytic activity by vampire bat saliva, the action of the teeth in the lower jaw of the wallaby, essential fatty acids in giraffes, rickets in mammals and reptiles and the protein that carries vitamin D in their blood, vitamin A levels in dolphins, diabetes in tuco-tucos, the massive release of ova from the ovary of the plains viscacha, and cyclical changes in the cervical mucus of the elephant. We have collected dead ducks from St James's Park, Teddington Lock and the Norfolk Broads, vampire bats from Trinidad and Peru, blood from people in malarious areas of Africa, India, Indonesia and South America, and semen from elephants in the Kruger National Park.

You may say that we have done too many things, but I do not think so. We have had the opportunities, facilities, accumulated experience and the interest to study, by modern techniques, aspects of biological science that have not been tackled before. The keepers, curators, veterinarians and research staff of the Society now form a team of a quality that has never been seen in a Zoo before, and this infusion of professional expertise into the zoo world has done much good. Handling, housing, nutrition, health and breeding records have all improved. The presence, on the spot, of competent scientists who can, without delay, take advantage of an opportunity to collect blood, tissues or recordings not usually within reach, and to deal with them at once, provides information that otherwise goes to waste. Outside collaborators are useful on many occasions but they can never be a substitute for the man on the spot.

Dr Martin gives in his volume (pp. 283–319) an account of research on reproduction in the Wellcome laboratories and I will therefore deal only with projects that are in progress at the Nuffield laboratories; three examples will serve to show how comparative studies can throw new light on problems of human and veterinary medicine.

Sickle cells

In the course of routine haematological examinations, Dr C. M. Hawkey found that the erythrocytes in the blood of some animals—deer, mongooses and genets—changed from the normal discoid form to elongated, sickle-shaped cells a few minutes after the blood had been drawn. The deformed cells resemble those found in human patients that suffer from the genetically determined sickle-cell disease, although the change in shape occurs when the pH, not the oxygen tension, alters.

The formation of a sickle cell in human blood is thought to take place because molecules of haemoglobin-S, the abnormal blood pigment associated with the sickle-cell gene, arrange themselves into long stacks within the cell membrane, pushing the cell out of shape and rendering it fragile and easily destroyed in the circulation. Anaemia follows. The cells of deer, mongooses and genets are not abnormal and are not fragile. They pose an interesting problem for the haematologist, geneticist and molecular biologist, the study of which could throw light on the pathology of the human disease.

Botulism

During the summer of 1969, several waterfowl died in St James's Park. The authorities thought that the birds had been poisoned as a result of the use of horticultural chemicals; post-mortem examinations, chemical and biological tests carried out by the Society's staff showed conclusively that the birds were victims of botulism—poisoning by the toxin of *Clostridium botulinum*. This anaerobic bacillus flourishes in the mud of lakes and ponds during warm weather when the water level is low. Exactly how the toxin reaches the birds in the first place is not clear but it is one of the most poisonous substances known and it paralyses the muscles so that the birds cannot feed, or they drown. In 1952 the disease killed 5 million ducks in the USA, in 1970 many birds died in Holland and, more recently, botulism was responsible for the death of 50 000 birds in the Coto Doñana reserve in Spain. During the summer of 1975, birds died on Regent's Park Lake, many parts of the Thames Valley and on the Norfolk Broads.

Once established, it is easy to see how the disease spreads—blowfly larvae grow in the bodies of dead birds and take up the toxin. It does not harm them but makes the maggots lethal to birds that eat them.

A sudden mortality among birds in the London Zoo (and in other collections) occurred at Christmas 1974. The cause was rapidly traced, by laboratory tests, to botulinus toxin in maggots from a commercial source that had been used as food. The outbreak was quickly brought under control.

Dr G. R. Smith and his colleagues have been studying this disease, which is of great importance to the conservation of waterfowl. The distribution of the organism has been mapped in ponds in the London area and in several reserves such as Slimbridge, Windermere, the Norfolk Broads, the Coto Doñana and the Camargue. In many locations, the organism was found in the mud and the bacteriological types responsible for poisoning man, birds and fish have all been detected. By studying the chemistry and biology of mud in which the organism fails to survive, it may be possible to devise methods of control.

This is a good example of basic field and laboratory research that may have an important bearing on conservation.

Essential fatty acids

My third example relates to the importance of polyunsaturated fatty acids in the diet. Dr M. A. Crawford and his colleagues have been studying the part these nutrients play in the structure of cell membranes, particularly in the brain. About half of the dry weight of the brain is lipid, much of it in the form of phosphatidyl choline and ethanolamine esters of fatty acids with long chains of carbon atoms, joined at strategic points by double bonds. The structure of these molecules is important in the architecture of the cell membrane. Two important series of acids are derived from the shorter chain linoleic and linolenic acids found in vegetation. Herbivores can elongate these molecules by adding carbon atoms metabolically; carnivores obtain the long-chain acids ready-made in the bodies of other animals. If insufficient quantities of essential fatty acids are obtained in the food, animals lose their fur, fail to grow and fail to breed. Cats appear to have adapted themselves so completely to obtaining their long chains ready-made that they no longer possess the enzymes necessary to build them.

This work is clearly of importance in the nutrition of breeding groups of both wild and domestic animals. It is also of importance to human mothers during pregnancy and lactation, and to the growing infant.

THE FUTURE

During the past 18 months we have been deeply introspective about what we are and what we should be doing. A Committee of the Society's Council, under the chairmanship of Dr C. E. Gordon Smith, has recommended that the staffs of the two Research Institutes, the Animal Hospital and Pathology Laboratory, together with the Curators, should all become members of a comprehensive Institute, covering comparative zoology, physiology and medicine. This procedure should help to demonstrate that we all work together as a team, and do not carry out our activities in watertight compartments.

We believe that we need to strengthen the permanent scientific staff, both in numbers and in tenure, so that longer-term programmes are not so frequently interrupted by the termination of grants. Projects grants must always be tailored to the particular interests of grant-giving bodies; there are many areas of research, of interest to us, that do not fit into the areas of biological science that they traditionally support. We have often had schemes, agreed by all to be interesting, but supported by none.

We believe that we have unique opportunities for studies in developmental biology; our expertise in reproductive physiology and nutritional biochemistry can be combined to make studies of importance to human and animal biology during pregnancy, lactation and infancy.

We also think that the Society is ideally placed to become an international centre for the scientific study of problems of conservation. Skilful management and husbandry have already put us in the forefront of the movement to breed endangered species in captivity. Our facilities for keeping animals, especially at Whipsnade, together with our scientific background, can put real knowledge, at present sadly lacking, behind attempts to breed good stocks of rare animals for posterity. Expert services and techniques in reproductive physiology and biochemistry, immunology, parasitology and, especially, cytology and genetics to determine sex and to monitor the effects of inbreeding, are all needed to make this venture a success.

We have the animals, the scientists, the facilities and the experience. But we should like the money.

References

Barnett, B. & Macfarlane, R. G. (1934). Relative potency of certain snake venoms to coagulate haemophilic blood. *Proc. zool. Soc. Lond.* **1934**: 977–978.

Cowan, D. F. (1975). Comparative pathology research in the Zoological Garden. In *Proceedings of the Centennial Symposium on Science and Research, Philadelphia Zoo*: 6-19. Snyder, R. L. (Ed.) Philadelphia: Zoological Society of Philadelphia.

Hamerton, A. E. (1941). Report on the deaths occurring in the Society's Gardens during the years 1939–1940. *Proc. zool. Soc. Lond.* **110**: 151–188.

Hawkey, C. M. (1975). *Comparative mammalian haematology*. London: Heinemann.

Lawrence, W. E. & Rewell, R. E. (1948). The cerebral blood supply in the Giraffidae. *Proc. zool. Soc. Lond.* **118**: 202–212.

Wenyon, C. M. (1926). *Protozoology*. London: Baillière, Tindall & Cox.

Zoological Society of London (1970). *Scientific Report 1967–69*. London: Zoological Society.

Zoological Society of London (1972). Scientific Report (1969–71) *J. Zool., Lond.* **166**: 499–610.

Zoological Society of London (1974). Scientific Report (1971–73). *J. Zool., Lond.* **173**: 37–158.

Zoological Society of London (1976). Scientific Report (1973–75). *J. Zool., Lond.* **178**: 443–550.

Zuckerman, S. (1930). The menstrual cycle of the Primates Part I. General nature and homology. *Proc. zool. Soc. Lond.* **1930**: 691–754.

Zuckerman, S. (1932). *The social life of monkeys and apes.* London: Kegan Paul.

Zuckerman, S. (1933). *Functional affinities of man, monkeys and apes.* London: Kegan Paul.

Symp. zool. Soc. Lond. (1976) No. 40, 223–231.

THE ROLE OF EDUCATION

W. S. BULLOUGH and FEONA HAMILTON

Birkbeck College, University of London and *The Wolfson Library, The Zoological Society of London, England*

INTRODUCTION

The intentions of the founding Fellows of the Zoological Society of London and its associated Zoological Gardens were no doubt mixed, but it is certainly clear that both the instruction of the public at large and the advancement of zoological knowledge among gifted amateurs and professional scientists formed a significant part of their thoughts. In their very different world, with its wide belief in the dogma of special creation, the Prospectus of 1825, setting out the aims of the proposed Society, considered that "zoology, which exhibits the nature and properties of animated beings, their analogies to each other, the wonderful delicacy of their structure, and the fitness of their organs to the peculiar purposes of their existence, must be regarded not only as an interesting and intellectual study, but also as an important branch of Natural Theology, teaching by the design and wonderful results of organization the wisdom and power of the Creator". Clearly their concept of education by the study of natural history had a different bias to our own.

It may be noted in passing that other people at that time were not so impressed by the serious nature of the new venture. Thus in a Foreign Office circular to British Consuls throughout the world, asking them to do anything in their power to help the procurement of animals, the hope is expressed that "even if this establishment should not materially advance the purposes of science, at least it will become an object of public interest as a place of amusement", while the Literary Gazette in an acid article commented that "there is neither wisdom nor folly new under the sun".

THE PUBLICATIONS

Nevertheless, the serious intentions of the new Society were apparent in the almost immediate establishment of a Museum, a Library, and a publication which later became well known as the *Proceedings of the Zoological Society of London.* Beginning in 1830 these *Proceedings* expanded steadily in size and importance; only recently has the name been changed to the *Journal of Zoology.* In 1833, shortly after the foundation of the *Proceedings,* the *Transactions of the Zoological Society* were also founded. These have appeared irregularly and have included "suitable memoirs deemed too important for the Proceedings" (Mitchell, 1929:117). The early volumes are particularly fine, being illustrated by magnificent full-page plates many of which were hand-coloured.

The next publication, the *Zoological Record,* began in 1864 as a joint venture with the British Museum (Natural History) "to acquaint zoologists with the progress of every branch of their science in all parts of the globe; and to form a repertory which will retain its value for the students of future years" (Mitchell, 1929:121). In 1886 the Zoological Society took sole responsibility for this publication.

Much later, in 1935, there began a fourth important publication of the Society, the *Nomenclator Zoologicus,* in which it was planned to publish the names of every zoological genus and subgenus established since 1758, the date of Linnaeus' *Systema Naturae* (10th Edn).

Finally, in 1959, the Society established two further important journals. One of these, the *Symposia of the Zoological Society of London,* is described later; the other is the unique *International Zoo Yearbook,* which provides authoritative information on developments in the zoo world for the members of staff of zoos everywhere.

All these journals have flourished to the present day and their cumulative contribution to the dissemination of zoological knowledge over the last century and a half is incalculable.

THE LIBRARY

Like the publications, the library has always been regarded as an integral part of the Society's structure. As early as 1825, Sir Stamford Raffles proposed in a circulated letter "that the [Zoological] Society shall have . . . a Library of all Books connected with the subject" (Mitchell, 1929:9).

One of the early benefactors of the Library was Mr Edward T. Bennett, Secretary to the Society from 1833 to 1836, who donated some 218 volumes from his own collection. The growth was rapid: to 20 000 volumes in 1883, to 40 000 in 1910, and so to an estimated 120 000 today. Originally the Library was accommodated at various addresses in the West End, and it was only moved to its present site in 1910, when the Society's new offices were opened. The shelves then lined the walls of the Meeting Room, which was also the Reading Room, and this arrangement continued until only recently when the new Library was constructed within the shell of the old Meeting Room. The Wolfson Foundation generously met the cost of this important redevelopment so that the Library is now known as the Wolfson Library.

The service of the Library to education and the development of zoology in Britain has been outstanding, and today in its newly-housed and newly-classified state it is one of the world's most important collections of zoological literature.

THE EARLY EDUCATIONAL EFFORTS

The various publications and the Library have, of course, always been services that were directed more towards the zoological specialist than towards the layman. The first specific efforts to promote information for the interested layman began with a legacy of £2000 received in 1870 from Mr Alfred Davis, a Fellow of the Society, to form "a Perpetual Fund for any purpose which may seem to . . . the Society most conducive to its interests". This led, in 1874, to the first public lectures being given in the Picture Gallery in the Gardens "on Tuesdays and Fridays at 5p.m. between Easter and Whitsuntide". After some success and some failure they were finally abandoned in 1902.

The next event, of considerable later significance, was in 1910 when the Society's Secretary, Dr Chalmers Mitchell, as he then was, co-operated with the Education Committee of the London County Council to establish courses of instruction for school teachers, who subsequently brought their children for educational tours of the Gardens.

This type of activity was resumed after the First World War, and in the 1930s elementary school children in groups of not less than 25 with one teacher were admitted from Tuesdays to Fridays at the cost of one penny per child and five pence for the teacher.

Also in the 1930s the first university students were admitted in groups, with a lecturer, at a reduced rate. At this time, for instance, the Sheffield University Zoology Department annually hired a passenger coach, which was slipped from the London express to deliver its complement of students for a day at the newly opened Whipsnade Zoo.

Other experimental activities of the time were the establishment of the popular magazine *Zoo* (later known as *Animal and Zoo Magazine*) which was widely used in schools; of the Zoological Film Productions Ltd, which made some 12 popular films, six educational films, and thousands of feet of film for scientific record; and of a Studio of Animal Art begun by Dr Julian Huxley with grants from County Councils, the Slade School of Art, and the Royal Academy.

These schemes had varying success but all were terminated by the onset of the Second World War.

THE PRESENT SCHOOL ACTIVITIES

After this long period of experimentation and development, the post-war years have seen the firm establishment of a wide-ranging pattern of educational activities. As Lord Zuckerman noted in 1973, the Zoo comprises a unique museum of living animals, which when properly arranged and properly cared for, provides "an indispensable tool in the teaching of biology . . . at all levels, from the ordinary visitor to school pupils, students and research workers".

The main emphasis has been on the needs of school children. Early in 1957, after consultation with the Inner London Education Authority, Lord Zuckerman arranged an experimental one-day teaching course in co-operation with Dame Mary Green, then the headmistress of Kidbrooke Comprehensive School. The outcome was the establishment in 1958 of the Education Department under the first Education Officer, Mr D. G. Lambert. The initial efforts were concentrated on three levels, non-academic, "O" level, and sixth form. In the first year the number of children attending was 15 760, which figure has expanded steadily to the present (1975) total of 56 831. Over the same period, mostly under the second Education Officer, Mr M. K. Boorer, the teaching staff has also expanded from 1·5 to 4. Simultaneously there has been a widening of activity to include the use of the Whipsnade facilities, the introduction of programmes for still younger children and for backward children, the development of courses that co-ordinate with the new Nuffield approach to biology teaching, and the special conservation lectures arranged in 1970 to coincide with the "Europe and Conservation Year".

One of the most stimulating of these new developments, begun experimentally in 1970, has been the Sixth Form Symposia. Each of these Symposia lasts for a day and brings before its audience the acknowledged experts in the chosen field. The first, organized by Professor E. J. W. Barrington, dealt with "The natural history of hormones", and its success was so outstanding as to be embarrassing. The number of children applying to attend was so great that, half in earnest, the Education Committee discussed a proposal that only the Albert Hall would be adequate to hold them. It was, however, felt essential to attempt to create an intimate atmosphere in which the children would be able to meet the eminent zoologists face to face, and consequently a system of rationing had to be devised which has subsequently been accepted with regret by the schools. These Symposia are now an annual event and they have ranged widely in subject from "The problems of pollution" to "The natural history of the dinosaurs".

Indeed, the success of the Education Department as a whole, has been so great as to create the need for the foundation of a specially-built Education Centre (Fig. 1). Planning for this began in 1971, the building being financed in part by a grant by the Wolfson

FIG. 1. The new Education Centre, opened in 1975

Foundation. At the same time the Inner London Education Authority agreed that it should establish in the Zoo its own Teachers' Centre for Life Studies, and the building was therefore expanded to accommodate both activities. This Teachers' Centre is to have its own staff employed by the ILEA, and the first Warden is Mr John Wray. Together these conjoint ventures of the Society and the ILEA now provide educational facilities that are probably unique in the world, and that must have great significance for the future.

EDUCATION AT THE UNIVERSITY LEVEL

Since the 1930s repeated attempts have been made to cater for the needs of university students. The most successful and enduring of these began in 1959, when a special two-day course was instituted mainly for the undergraduates of the various London University Colleges. Again the response was greater than the facilities allowed and a rationing scheme had to be instituted since it was felt that the class would lose its informal and intimate character if the number of students was allowed to rise much above 20. These courses, which have concentrated on the main vertebrate classes—the fishes, amphibians and reptiles, birds, and mammals—have been given by experts both from the London Colleges and from the Society's own staff. They are now an annual event and are held in the Easter vacation.

At a still higher academic level, the year 1959 saw the beginning of what has become a series of Symposia of great international significance. Designed to bring together the main workers in particular fields of research, the first Symposium of the Zoological Society of London, entitled "Hormones in fish", was organized by Professor I. Chester Jones of the University of Sheffield. Since then there have been two or three such Symposia each year dealing with the widest range of topics from evolution in cnidarians to evolution in man, and from the endocrine system of insects to the structure of mammalian skin.

The proceedings of all these Symposia have been published and they constitute the latest major scientific publication of the Society. With this present Symposium, the 40th volume has been reached.

OTHER EDUCATIONAL ACTIVITIES

In recent years a range of other varied educational activities have also been organized, initially on an experimental basis. Thus the

Christmas lectures of 1951, given for the children of Fellows, have become a popular annual event, and have led to the establishment of the XYZ (Exceptional Young Zoologists) Club for children under 18. The activities organized by the Club have ranged widely from the publication of the *Zoo Magazine* (issued thrice yearly) to tours of other zoos, lectures on keeping pets, Christmas "At Homes", and Easter "Brain Trusts". The most ambitious tour yet organized was in August 1969, when a group of young people flew to Nairobi and visted the main game reserves in Kenya and Tanzania.

Another activity has been the establishment of courses for the Keepers in the general management of animals in Zoological Gardens. The Society's own Diploma Course was started in 1956 but later this developed into a joint activity with the Paddington Technical College. The resulting course, designed for the education of professional zoo keepers, has been a pioneer of its kind in the world.

Finally, in relation to the development of television, there has been a whole series of exciting developments beginning in 1954 with an expedition, organized jointly with the British Broadcasting Corporation, to make a natural history film in Sierra Leone for showing on television. This led a year later to the first of the famous "Zoo Quest" expeditions, directed by Mr David Attenborough, in this case to British Guyana.

These operations resulted, in 1956, in the establishment in the Zoological Gardens of a special Television and Film Unit. The Society's own funds were not used in this venture which was financed by Granada Television, although the facilities of the Unit were not exclusive to them. The first director was Dr Desmond Morris and the animal programmes that were produced had audiences of millions. The BBC also used the facilities of the Unit for making films for their admirable series of schools broadcasts.

By 1961 about half a million feet of film had become available for educational and scientific purposes, and when in 1964 the Granada contract came to an end, the studio was presented to the Society, which then established a Television and Film Unit of its own. This Unit has since been incorporated into the Public Relations Department, which together with the Library is also responsible for holding the extensive collection of still photographs many of which are unique historical records (Figs 2 and 3). The Department now supplies material for educational purposes, and also makes new film records for the Society's archives.

FIG. 2. The last quagga. Photographed in 1870 in the Zoological Gardens.

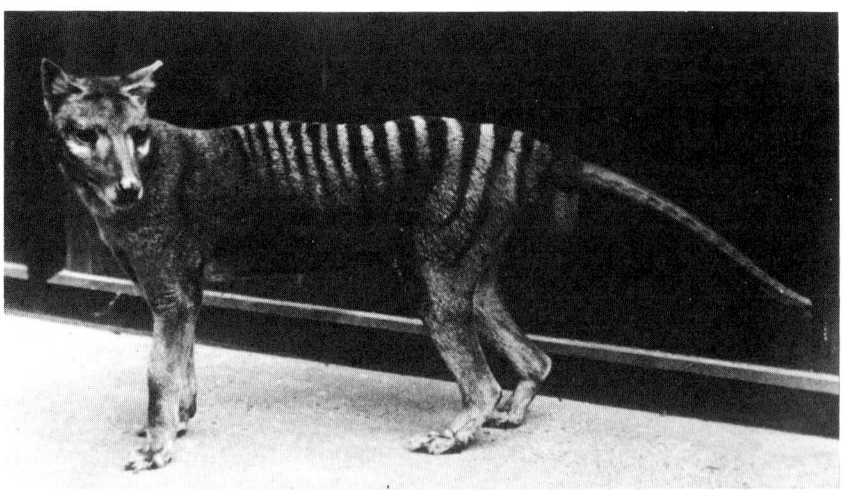

FIG. 3. The thylacine or Tasmanian wolf, now possibly extinct. Photographed in 1913 in the Zoological Gardens.

THE FUTURE

The difficult economic conditions of the present day make it particularly dangerous to attempt to forecast the shape of the future in any of its aspects, but as long as the Zoological Society of London and its Gardens exist, it is certain that demands will continue on its resources and expertise in support of zoological education in all its aspects. Indeed, as Lord Zuckerman has emphasized (this Symposium, pp. 1–16), it is these educational activities that provide the main justification for the very existence of zoos.

At the level of the layman the constant barrage of questions that arrive daily at the Society and that require all the resources of the Public Relations Department, of the Library and of the Curators' departments is likely to continue: "what do you call the noise that a camel makes?", "how and where do dolphins sleep?", "do you have a photograph of a hairy frog?", "why doesn't a whale get the bends?" and so on *ad infinitum.* It is to meet this kind of non-stop questioning that the Society produces a wide range of pamphlets describing popular groups of animals from monotremes to monkeys and suggesting techniques for the care of pets from tortoises to chipmunks.

At the next level the commitment to the education of school children in the Greater London Area must be preserved. The never-ending excitement generated by the unique collection of animals will certainly continue to ensure the demand for this service.

Finally, the most serious aspect of the Society's educational work, supported by the resources of its specialized library, must always be the wide-ranging support of zoological teaching and research at the professional and university levels and the dissemination of new zoological understanding through the various specialized publications.

Thus, in spite of all the present difficulties, there is every reason to believe that the educational activities of the Society will continue to thrive. Indeed it may be confidently expected that, as in the past, these activities will be able to meet whatever new challenges may arise.

REFERENCE

Mitchell, P. C. (1929). *Centenary history of the Zoological Society of London.* London: Zoological Society.

Symp. zool. Soc. Lond. (1976) No. 40, 233–252.

THE LIBRARY AND SCIENTIFIC PUBLICATIONS OF THE ZOOLOGICAL SOCIETY OF LONDON: PART 1

R. FISH

The Zoological Society of London, Regent's Park, London, England

SYNOPSIS

The events leading up to the establishment of the Library and the formation of the Committee of Science and Correspondence are described, and the reasons why the scientific meetings and publications began in 1830 are discussed.

This is followed by an account of the growth of the Library and its services, and of its present condition.

INTRODUCTION: THE EARLY YEARS 1825–30

Although the scientific meetings and publications of the Zoological Society have for most of its history formed a crucial part of the Society's contribution to science, it is interesting that neither these nor the Library is mentioned in the first prospectus of March 1825, almost certainly written by Sir Stamford Raffles, and indeed no mention appears to have been made of the scientific meetings or publications until 1830, four years after the Society's foundation. Although Raffles was professedly more interested in the scientific than the utilitarian aspects of the proposed Society, this interest appears to have been restricted to the more practical aspects, in particular the formation of a "zoological collection", i.e. the Museum.[1] He envisaged a Society bearing the same relation to zoology that the Horticultural Society does to botany.[2]

This is particularly striking in view of the fact that zoologists had for a number of years been extremely unhappy about the facilities available to them for communicating and publishing papers on their subject. For example N. A. Vigors, an entomologist and later first Secretary of the Zoological Society, had in 1822 proposed to the Rev. W. Kirby, the leader of entomological studies in this country, the formation of a new Entomological Society to collect and disseminate information on the subject. Kirby, who represented the views of many zoologists of his time, hesitated to adopt any course likely to harm the Linnean Society, which was the

[1] Numbered notes at end of chapter.

Society which had taken natural history as its province. He suggested instead the establishment of separate committees within the Linnean and the publication of a naturalist's journal "which would include all that intelligence which does not accord with the general plan of the Linnean Transactions".

Vigors seized on these suggestions with alacrity. He considered the present situation entirely unsatisfactory. "There is no way by which we can obtain an insight into the discoveries of our contemporaries, or ascertain the actual progress of the Science". He goes on to relate the shortcomings of the Linnean Transactions.

> "The Linnean Transactions, as you are aware, cannot embrace these objects: they come out at intervals too distant for the constant diffusion of knowledge that is necessary; they are too costly for general circulation; and are devoted to subjects too important to take in that subordinate but still valuable mass of information that is fitted only for the pages of a periodical journal. Composed, also, of original papers alone, they could not take in the already-published discoveries of others, or allude to the system or mode of arrangement established on the Continent or elsewhere, the communication of which is indispensable for the advancement of science, but not consistent with their own regulations".[3]

The Zoological Club of the Linnean Society was established in November 1822, with Kirby as its first Chairman, in an attempt to overcome these problems without hurting the interests of the Linnean. Unfortunately, the Council of the Linnean Society did not look with a favourable eye on the formation of the Club, fearful no doubt that it might mark a bid for independence by zoologists. Onerous conditions were imposed, one of which forbade the Club to publish papers read to it, but insisted these be submitted to the Linnean Society for consideration. When the Asiatic Society applied for a Charter in 1823 the Council of the Linnean again feared that their monopoly in natural history was threatened and opposed the granting of the Charter unless the study of natural history was expressly excluded from the Society's objects.[4]

There can be little doubt that this jealous attitude on the part of the Council of the Linnean Society had a profound effect on the foundation and early years of the Zoological Society. Zoologists had themselves sought to overcome some of its inconveniences by the publication of an independent journal—the *Zoological Journal*, first issued in March 1824. This was the first purely zoological journal published in this country and was "conducted by" Thomas Bell, J. G. Children, J. de C. Sowerby and G. B. Sowerby.

On the other hand, many zoologists agreed with Kirby that the future of their subject would be better served by remaining loyal to the Linnean Society and by action within that Society, than by founding a new Society to compete with it, thus introducing discord within their own ranks. Raffles himself was elected a Fellow of the Linnean Society on the 15th February 1825. He was on friendly terms with many of the members of the Zoological Club and attended its meetings on more than one occasion. He must have been abreast of the views of its members. By his own account he regarded the support of men of science as of more value than that of the country gentlemen whose support he also solicited.[5] Had he in his Prospectus proclaimed that the proposed Society would hold scientific meetings and publish journals he would have faced zoologists with a clash of loyalties. Very probably this was the reason for the great emphasis on the more practical aspects of the new Society, and the absence of any plans for publications or scientific meetings during its early years.

Even the Library was not mentioned in the first prospectus. The first recorded mention of it was in a cautious statement in the minutes of a meeting of the proposers of the Society on the 26th February 1826, that the immediate objects of the new Society should not extend beyond the introduction and domestication of new breeds of animals, and that a Museum and Library should only be added as resources admit.

The revised prospectus, issued probably in the following month, simply announced this decision in its preamble. In April 1826 when the Society was finally inaugurated the formation of a Library was firmly established as one of its objects.

On p. 35 of his *Centenary history of the Zoological Society*, published by the Society in 1929, Sir Peter Chalmers Mitchell tells how his attention was attracted to an account in *The Times* of 27th April 1827 of a lecture on the dissection of an ostrich given by Dr Joshua Brookes at the invitation of the "noble directors of the Society". He hails this as the first Scientific Meeting of the Society. Actually it was one of the weekly meetings which took place throughout the 1827 season

> "at which lectures on various branches of Zoology given by Joshua Brookes, Esq., F.R.S. and by the Secretary [N. A. Vigors], were attended by a numerous audience of scientific gentlemen and other members of the Society; the lectures by Mr. Brookes, the subject of which was chiefly the comparative anatomy of Birds, were in part

> illustrated by demonstrations of the structure of the Ostrich, performed on the specimen presented by His Majesty".[6]

There is a report of Dr Brookes' fourth lecture on the comparative anatomy of the ostrich, delivered on the 13th June 1827, in the *Weekly Review*. It says that: "His attention was on this occasion chiefly directed to the thoracic and abdominal viscera". After describing the lecture the report continues:

> "Notwithstanding the profusion of technical terms introduced by him in his anatomical demonstrations of the numerous preparations which he exhibited, the greatest interest was shown throughout the whole of a very long address by a crowded audience, consisting chiefly of ladies".

For the reason already given these are unlikely to have been scientific meetings. They had no connection with the later Scientific Meetings, but appear to have been public lectures, similar to those given at the Jardin des Plantes in Paris, and possibly even in imitation of them.

It is interesting to speculate what would have happened had there been no change in the situation. In fact a profound change was brought about by the death of James Edward Smith, President of the Linnean Society, on the 17th March 1828. In his possession were the Library and Collections of Linnaeus. Instead of bequeathing them to the Linnean Society as expected, his will directed them to be sold. His executors demanded a price of £5000. The Linnean Society, with a capital of £190 and £550 in cash, determined to acquire them. By herculean efforts they succeeded in begging and borrowing sufficient money, but the Society was doomed to penury for many years.[7]

Zoologists realized that they could now expect little from the Linnean. The Zoological Club was an early casualty. Beset by financial problems it ceased to exist in November 1829. On the other hand the Zoological Society, thanks to Raffles and Davy, had achieved what the Linnean had failed to do, that is attract the support of the wealthy country gentlemen, and in its gardens at Regent's Park it had a source of income denied to the Linnean. The way was now open for the Zoological Society to take over full responsibility for zoological science in this country. Nevertheless its Council proceeded cautiously. On the 21st July 1830 it nominated a Committee of Science and Correspondence to meet on the second and fourth Tuesday of the month. The Committee's first meeting was held on the 9th November 1830, and by the 8th February 1831 its sessions had aroused so much interest that they were thrown open to the membership of the Society at large.

All those who consider this period are intrigued by the relationship between the Zoological Club of the Linnean Society and the Zoological Society of London. Whatever that relationship, the Zoological Club was the immediate forerunner of the Committee of Science and Correspondence. Its meetings were held on the second and fourth Tuesday in the month, like those of the Committee, and many of its members were among those invited to attend the original Committee meetings. The *Zoological Journal* which ceased publication in 1835 was likewise the forerunner of the Zoological Society's *Proceedings*. Abstracts of the proceedings of the Committee were issued monthly from January 1831 and this eventually developed into the *Proceedings of the Zoological Society*.

There is no doubt at all what zoologists thought of the situation now. They were triumphant. In the jubilant rhetoric of Vigors' address on the dissolution of the Zoological Club in 1829:

> "You have seen zoology emerging from the seclusion of the closet, where, like a thing of mystery, it lay hid under the monopolising patronage of a few; you have seen it gradually passing into light, and winning its way by its own native attractions, until attaining its legitimate station in public estimation, it has become the popular and universally acknowledged favourite of the day".

THE LIBRARY

At the inaugural meeting of the Zoological Society, on the 29th April 1826, the new Society resolved to direct itself to three objects—the formation of a collection of living animals, a museum of preserved animals, with a collection of comparative anatomy, and a Library connected with the subject.

Donations to all these collections soon began to come in, and indeed a committee for the formation of a library was appointed at the first meeting of the Council of the Society, presided over by Sir Thomas Stamford Raffles, on the 5th May 1826. The Committee consisted of the Rev. Dr Edmund Goodenough, son of Bishop Goodenough, who had been one of the founders and first Treasurer of the Linnean Society, H. T. Colebrooke, Director of the Royal Asiatic Society, and N. A. Vigors, Secretary of the Zoological Society.

The first recorded list of donations to the Library was read out at a special meeting of the Society on the 7th March 1827. It consisted of two volumes of the *Zoological Journal*, presented by G. B. Sowerby, and six monographs, among which were several works by Vigors, presented by the author, and Vieillot's *Observations sur la classification méthodique des oiseaux*, presented by Temminck.

Library

Bound Volumes

	Vol.
Transactions of Zool Soc. — Parts 1&2)	2
Cuvier sur les ossements fossiles —	6
Catalogue of Contents of Hunterian Mus.	5
Hall on Medulla oblongata	1
Jardines Illustrations	1
Owen on Pearly Nautilus	1
Hamilton's Fishes	2
Abhandlungen &c. Berlin	1
Home's Lectures —	6
Memoires de la Soc: Imp. Mosc.	1
Trans: Royl. Soc: Lit.	1
Goulds Himalayan Birds	1
Lears Parrots	1
Goulds British Birds	11
Manuel des Mollusques	1
Histoire des Reptiles	4
" " Vers	3
History of the Museum &c. Paris	2
Cooks Sketch of Spain	2
Bulletin de la Soc: Imp. Mosc.	5
Zoognosia	3
Magazine of Hist Nat.	5
Maw's S: America	1
Gardens delineated	2
Histoire des Sciences nat. —	4
Hall on the Blood —	1
Sketch of Van Dieman's Ld	1
Stevens Catalogue	1
Trans: Soc: of Arts —	3
	75

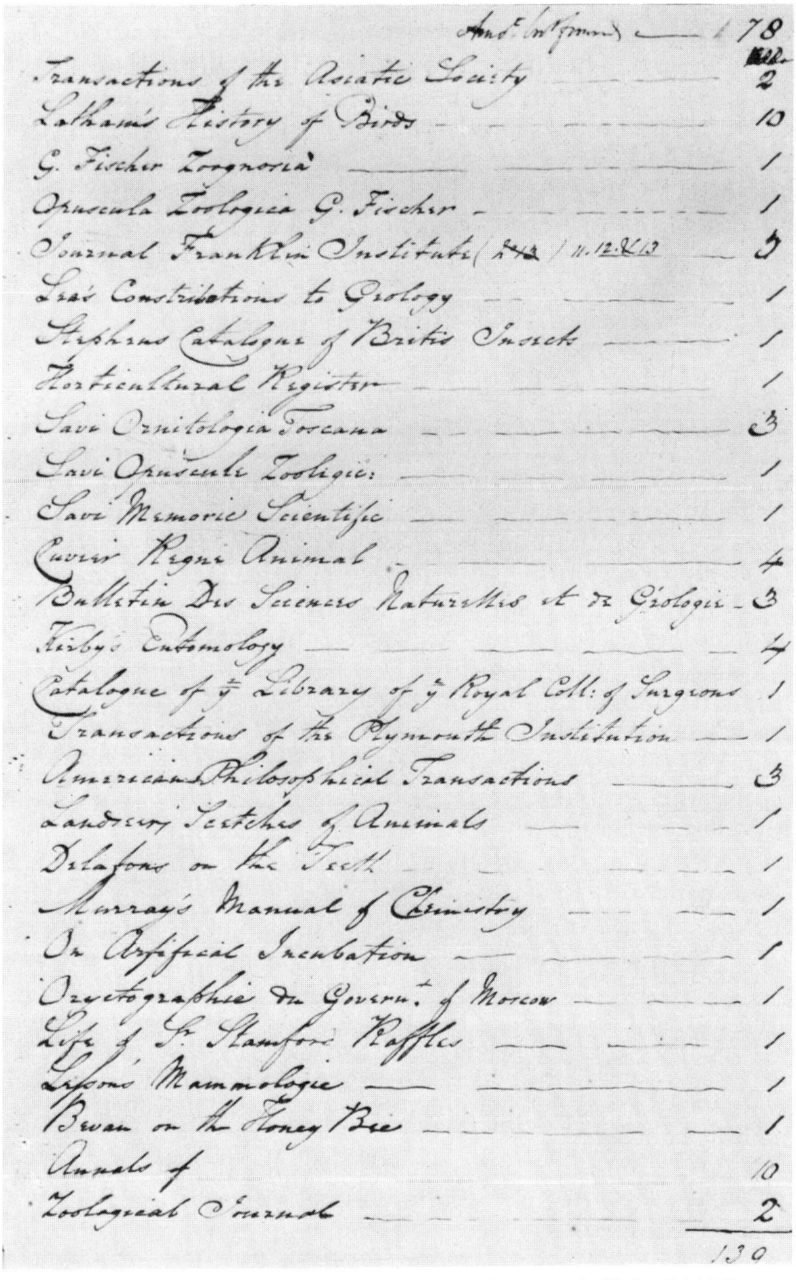

Amt. brt. forward — 178

Transactions of the Asiatic Society — 2

Latham's History of Birds — 10

G. Fischer Zoognosia — 1

Opuscula Zoologica G. Fischer — 1

Journal Franklin Institute (1&2) 11.12.&13 — 3

Lea's Contributions to Geology — 1

Stephens Catalogue of British Insects — 1

Horticultural Register — 1

Savi Ornitologia Toscana — 3

Savi Opuscula Zoologic: — 1

Savi Memorie Scientific — 1

Cuvier Regne Animal — 4

Bulletin Des Sciences Naturelles et de Geologie — 3

Kirby's Entomology — 4

Catalogue of ye Library of ye Royal Coll: of Surgeons — 1

Transactions of the Plymouth Institution — 1

American Philosophical Transactions — 3

Landseer's Sketches of Animals — 1

Delafons on the Teeth — 1

Murray's Manual of Chemistry — 1

On Artificial Incubation — 1

Oryctographie du Govern.t of Moscow — 1

Life of Sr. Stamford Raffles — 1

Lesson's Mammologie — 1

Bevan on the Honey Bee — 1

Annals of — 10

Zoological Journal — 2

139

FIG. 1 a and b. List of the bound volumes in the Library, probably in 1834. There are also four additional sheets listing 313 unbound parts and volumes. The list is signed by William Martin, who was at the time Superintendent of the Museum.

During the first half of the nineteenth century the growth of the Library was slow. The first list of books in the Library dates from 1834 (Fig. 1). By 1836 the stock amounted to only 420 volumes, of which 353 had been presented and only 67 purchased. Among the volumes donated were all the folio volumes so far produced by the Ornithologist of the Society's Museum, John Gould. More than half of the works presented had been given by Edward Turner Bennett, the zealous Secretary of the Society since 1833, in succession to Vigors.

Unhappily Bennett died during 1836, and shamed by his example, the Society decided that it would thenceforth devote to the development of the Library "such sums as may conveniently be spared from the general receipts". Accordingly the following year the sum of £100 was voted for Library purposes, augmented by a donation of 25 guineas from Charles Morris, jun. There is evidence that the intention was to devote the sum of £100 annually to the Library (cf. Annual Report published 1840).

This good resolution was unfortunately never put into effect. In 1837 only £76 19s 1d was actually spent on the Library, and the fact that the figure was as high as this was probably mainly due to the fact that receipts from visitors to the menagerie were particularly high in 1836, because of the interest aroused by the arrival of Thibaut's four giraffes. In only one other year before 1851 was more than £50 spent on the Library.

At first the problem was chiefly accommodation. The office building had to house the Society's offices, meeting room and museum, as well as the Library. The first building at 33, Bruton Street soon proved inadequate and was given up in 1836. The Society moved to a larger house at 28, Leicester Square, where once again problems arose due to lack of space. Until adequate accommodation was available for the books it already possessed

> "it is obviously inexpedient to increase their number, or to risk the loss or injury of valuable and expensive works by exposing them to the insecurity and casualties which necessarily attend them in a crowded room daily used for other purposes". (Annual Report published 1840.)

At last in 1843 the space problem was solved for the time being by the purchase of a more commodious house for the Society's offices at 11, Hanover Square, and by moving the Museum into the converted Carnivore House at Regent's Park. By this time however the Society was suffering from the general economic recession with which the whole country was afflicted. Money was short, and for several years little more was spent on the Library than was necessary to maintain its subscription to serials for which it had standing orders. By 1848 the stock of the Library had reached only 1000 volumes. In spite of Bennett's efforts therefore the rate of growth remained fairly constant from 1826 to 1850.

Even this growth was only sustained by generous donations from the members of the Society. From early in its life and through this lean period the Society had also been the recipient of the publications of societies and institutions which shared its field of interest. The Annual Report for 1836 specifically mentions as gifts the Transactions of the Berlin Academy, of the Royal Academy of Turin, the American Philosophical Society, the Geological Society of London, and institutions in Paris, St. Petersburg, Moscow, Geneva, as well as the Journal of the Royal Geographical Society and the Catalogue of the Royal College of Surgeons. From the date they were first published the *Proceedings* and *Transactions of the Zoological Society* were sent to similar societies throughout the world.

With production costs what they are today it is simply not possible for us to give away our publications with the same liberality, but this practice was the foundation of our present system of exchanges between the libraries of scientific institutions. The Library currently receives over 1000 different journals and of this total more than half come to us in exchange from other institutions in all parts of the world, from New York to Peking, and from Moscow to Madagascar.

The arrival in the menagerie of Obaysch, the first hippopotamus seen in this country, in 1850, which brought large crowds into the gardens, followed by the Great Exhibition in 1851, which brought people to London from all over the world, marked an upward turn in the Society's financial position. The Secretary was David Mitchell, a fine animal artist, who had a strong interest in the Society's Library and publications. For two years expenditure on the Library rose dramatically to £130 and £134, substantial sums in those days. It did not remain at this level, but for the rest of the decade was higher than it had been before 1851. The appointment of P. L. Sclater as Secretary in 1859 marked the start of the strong and sustained growth of the Library, which has continued ever since. The first catalogue of the Library in 1854 listed 460 titles,[a] including titles of scientific journals. A supplement published in 1864 listed 1090 titles added between 1854 and 1864.

[a] These figures bear no relation to total numbers of volumes, which include volumes of journals.

By 1867 it had become necessary to appoint a Librarian. On appointment he was not actually called "Librarian", but bore the chastening title of "Third Clerk". His name was Richard Bowdler Sharpe, he had formerly worked in Quaritch's bookshop, and like Sclater the Secretary he was an ornithologist. He looked after the Library until 1872, when he moved to the British Museum, where he hoped to find more time to devote to "several important works in ornithology" which he had in progress. During his time with the Society he had prepared a second edition of the Library catalogue, which was published in 1872.

His successor, F. H. Waterhouse, was the son of the famous naturalist G. R. Waterhouse, who had been Curator of the Society's Museum. He remained Librarian until 1913. During that period the arrival of the Prince of Wales' Indian collection in 1876 and the sale of Jumbo in 1882 sent the gate receipts soaring, and put the Society into the position where it could once more obtain relief from its perennial problems of accommodation by moving to a larger house, at 3, Hanover Square. The books were arranged on fine new oak shelving (Fig. 2) which remained in use until 1964, surviving the move from Hanover Square to Regent's Park in 1910. By this time the stock of the Library had grown from 460 titles in 1854 to 4200 titles in 1883, in terms of total number of volumes from 1000 in 1848 to 15 000 in 1887.

The move to Regent's Park was brought about not so much by lack of accommodation as the fact that the weight of the books was proving a danger to the building, and even more by the great advantages to be gained from bringing the centre of administration of the menagerie into the gardens. In the early days, Regent's Park was remote from the centre of London, and at night particularly access was difficult and even dangerous. By 1910 all this had changed and there was no reason why evening meetings should not be held there.

Sir Peter Chalmers Mitchell, who was Secretary in 1910, estimates that when the Library was brought to Regent's Park its total stock was 40 000 volumes. The shelving was brought from Hanover Square together with the books. Bookshelves surrounded the Meeting Room to a height of about 15 ft, above was a gallery with another 8 ft of shelving (Fig. 3). The Library also occupied most of the first floor of the building, which contained the Library office and Reading Room (Fig. 4). There was also provision for excavating a basement for increased storage at a later date, and in fact part of this was opened up in 1914 and finally occupied by the

FIG. 2. The Library at No. 3, Hanover Sq. The shelves and furniture, installed in 1883, were moved to Regent's Park in 1910 and were in use until 1964. The small pile of books on the table represents the Society's publications 1831–1840, the larger pile its publications 1891–1900, excluding the *Zoological Record*.

FIG. 3. The old Meeting Room at Regent's Park.

FIG. 4. A corner of the old Reading Room at Regent's Park. The photograph was taken in March 1966 when the books were being moved into the new Reading Room, and rearranged into the new order.

Library in 1921, and another part was excavated and occupied in 1955–56.

In 1910 the system of shelving books was the same as that used at Hanover Square, and was what is known as "fixed location". By this system each shelf is labelled with a number. A new book or journal is placed on a shelf and the number of that shelf is written inside the book, and in the catalogue. To find a book one simply has to look in the catalogue, find the number of the shelf, and then go to the shelf. This system works admirably while there is plenty of space available. However, when space becomes scarce it leads to countless difficulties. If a journal is allocated five shelves and expands beyond these, either whatever is on the sixth shelf has to be moved, or six empty shelves must be found together somewhere else in the Library. By 1964 a situation had been reached where space was so short that whenever a delivery of bound journals was received from the binder an intricate game of musical chairs was necessary to shelve them.

In 1964 it was decided to rebuild the Library. A new Meeting Room had been built so that the old room could be turned over entirely to library purposes. The gallery was to be replaced by a floor built across the old Meeting Room so that the lower half of it became a stack room and the upper half a new Reading Room. The library rooms on the first floor were to become offices.

The stock of the Library was by now approximately 120 000 volumes. The chief problem posed by the reconstruction was that the system of shelving the volumes depended on labelled shelves; now all except those in the basement were to be replaced by shelves which were very different in size. A further problem was that there were some 40 000 volumes in the old Meeting Room which had to be moved somewhere before the building operations could commence. All of these books had to be available for use by readers throughout the operation.

These very problems however provided a wonderful opportunity for tackling the old problems of shortage of space and an archaic system of shelving. As a first step the books in the basement were closed up together, and as the shelves were deep, the books were double-banked, two rows to a shelf. In this way half the basement was cleared of books. The shelving in this half was dismantled and replaced by mobile compact shelving. The same operation was then carried out on the other half of the basement. The compact shelving held three or four times as much as the older shelving, and there was thus space for the books to be moved down from the Reading Room.

The old system of "fixed location" shelf marking was then replaced by the more flexible system whereby each individual volume is marked. Each journal was allocated a number, related to its subject, and that number was lettered onto the spine of every volume.

The books were to be arranged in a classified order according to subject. There was no scheme of classification in existence which was entirely satisfactory for such a large collection on zoology. It was therefore necessary to devise such a scheme, and then to classify the books by that scheme, and letter the new classification numbers onto each book.

When the building work was finished 120 000 volumes had to be completely rearranged into an entirely different order. The possibilities for catastrophe were alarming.

The final stage in the operation is the one on which the library staff is now engaged. This is the production of a new catalogue. As already mentioned, the first edition of the library catalogue was in 1854. The shelf number of each book was simply written into a copy of the printed catalogue and in the early days that was adequate for use in the Library. When the growth of the Library accelerated, interleaved copies of the printed catalogue were kept in the Library, into which additions could be entered by hand. In due time the interleaved sheets became full and a new edition of the printed catalogue was produced. The last edition of the printed catalogue of books was published in 1902, and of journals in 1949. Since those dates it had been necessary to completely retype the catalogues every ten years or so, which in fact meant that retyping was almost continuous.

In 1965 a new serials catalogue was produced on a strip index system, into which additions could be inserted without the need for retyping. The books however were a more difficult problem. Each new edition of the printed catalogue had consisted of the entries from earlier catalogues with additional entries inserted. The old entries were not checked or brought into line with modern ideas on cataloguing. Over the years cataloguing by many different people and no written set of cataloguing rules had introduced many inconsistencies. Moreover, certainly in recent years, there had been no stock check to ensure that the books listed were still in the Library.

As a result of these considerations the only course appeared to be to go back to the books and to recatalogue them on cards by the most recent set of rules—the Anglo-American Cataloguing Rules of 1967. This was done in conjunction with the classification of the books. This project, first embarked on in 1969, is now very near its conclusion.

This brings up to date the story of the books and journals in the Library. There remain two very important collections that have yet to be described—the archive collection and the collection of graphic materials.

The archive collection contains the records of the Society's history, from the Prospectus of 1826 (no copies of the original 1825 prospectus have survived) down to modern times. It includes the Annual Reports of the Society, our most valuable record. In addition, there are the Daily Occurrence Sheets, which were completed daily by the Superintendent of the menagerie and give full details of what occurred—animals added to stock, sick or dead, which members of staff are off duty, what work the men are employed on, the names of special visitors, the temperature and the weather. The sheets are bound in volumes and cover the period from 1828 to 1961. Apart from these there are the minute books of the Council and General meetings and of the various committees, administrative documents of all kinds, and letters to the Society, written over the years.

The idea of keeping a pictorial record of the animals which have been kept in the gardens can be attributed to D. W. Mitchell, the artist Secretary of the Society from 1847 to 1859, and the credit for obtaining the finance to start the collection to the hippopotamus "Obaysch". In the Annual Report for 1851 Mitchell announced that "Council have commenced the formation of a series of Drawings, for which the most interesting subjects will be selected, as the Animals successively arrive at perfection". The artist Joseph Wolf was commissioned to produce the watercolour drawings and he continued to work on them until 1869. At first the watercolours were kept in portfolios, but in 1864 Sclater had them displayed in the old Museum building. They were later hung in the offices at Hanover Square, and having faded, were replaced in the portfolios in 1910.

There were many other additions to the Society's collection of graphic material, and only the most important will be mentioned. In 1875 the Society received two magnificent gifts—the manuscript works and drawings of Brian Houghton Hodgson and those of Samuel Richard Tickell. Hodgson, who was for many years British Resident in Nepal, illustrated and described the mammals and birds of the Himalayas: Tickell, a soldier in the Indian Army, the fauna of India. Both works are of superlative quality.

The quality of their work was however at least equalled by that of Major Henry Jones. He had served in the British Army and retired in 1881. For the next 40 years he worked mainly in the British Museum, producing a treasure-house of over 1000 exquisite watercolours of birds, mainly from India. At his death in 1921 he left this priceless collection to the Society.

The next notable additions to the collection were of photographs. The collection built up over his years of service with the Society from 1903 to 1942 by F. W. Bond, the Society's Accountant, was presented to the Society on his death. In 1948 the Society received the collection formed by Francis Martin Duncan, Librarian of the Society from 1919 to 1939 and in 1959 the collection of David Seth Smith, Curator of Birds from 1909 to 1939. These collections of more than 12 000 photographs between them form an admirable pictorial record of the Society's menagerie during the twentieth century, and of the many rare animals that it has contained.

Before closing this account of the Library there are two other matters which should be mentioned. In 1908 the Assistant Librarian Henry Peavot was also Clerk of Publications. When he succeeded to Waterhouse in 1913 he became Librarian and Clerk of Publications. Though Peavot was tragically killed on the Western Front in 1916 his successors bore that title and were responsible for both the Society's Library and its publications for 50 years, until the retirement of G. B. Stratton in 1964.

Mr Stratton was also Joint Editor of the World List of Scientific Periodicals, which gives the holdings of scientific periodicals of all the libraries in this country up to 1960, and is of immense value to research workers. Mr Stratton and the staff of the Society's Library played a large part in the production of the four successive editions of this work.

When the Society was established 150 years ago its founders intended that the Library should serve the advance of zoological knowledge. Today the main function of the Library is still to serve the zoological research worker. Although the Library staff answers innumerable questions from the public, although interested amateurs and students use the Library as Associates of the Society, and profit from the use, it is the needs of the professional zoologist which the Library is above all designed to satisfy. Since 1965 the Library has also provided a full library service to the Society's research institutes—the Nuffield Institute of Comparative Medicine and the Wellcome Institute of Comparative Physiology.

In its early days zoology was mainly taxonomy and anatomy. With the years the subject has broadened and changed, forming intricate associations with other subjects. One of the ways this sort of development has affected specialized libraries is to increase the need for co-operation between libraries. The Library of the Zoological Society is playing a full part in the library system of the country.

By their efforts and generosity our predecessors have built up a Library which is one of the most important collections of zoological material in the world. The new Reading Room is worthy of such a collection (Fig. 5). The stack rooms with their mobile compact shelving have overcome a space problem which seemed beyond solution. The books are kept clean and in good repair. The efforts of the library staff in classifying, recataloguing and caring for the stock of the Library have been directed towards ensuring that the organization of the Library is appropriate to such a collection in such a setting, and that the reader can make the fullest use of the wealth of material contained in it, and finally that this rich heritage from the past can be preserved and passed on to those that come after.

Notes

1. Letter to Sir R. H. Inglis, Bart, 28th April 1825, published in: *Memoir of the life and public services of Sir Thomas Stamford Raffles, FRS &c. . . .* by his widow. p. 590. London: Murray, 1830.

   "I look more to the scientific part of it, and propose, if it is established on a respectable footing, to transfer to it the collections in natural history which I have brought home with me".
2. This phrase occurs in the Prospectus and also in a letter written to Dr Raffles, his cousin, on the 9th March 1825, published in: *Memoir of the life and public services of Sir Thomas Stamford Raffles, FRS &c. . . .* by his widow. pp. 592–3. London: Murray, 1830.
3. From a letter dated 1st October 1822 reproduced in: *The life of the Rev. William Kirby . . .* by John Freeman. p. 372. London: Longman, Brown, Green & Longmans, 1852.
4. The correspondence between the Zoological Club and the Council of the Linnean Society, and papers regarding the Charter of the Asiatic Society, are preserved in the Linnean Society Library.
5. Letter to Sir R. H. Inglis, Bart, 28th April 1825. (See note 1 above):

   "In the first instance, we look mainly to the country gentlemen for support, in point of numbers; but the character of the institution

FIG. 5. The new Reading Room as it is at present.

must of course depend on the proportion of men of science and sound principles which it contains".

6. *Zoological Journal.* Vol. 3, no. 10, pp. 309–310. April–September 1827.
7. This information is taken from: *A history of the Linnean Society of London*, by A. T. Gage. London: Linnean Soc., 1938.

Symp. zool. Soc. Lond. (1976) No. 40, 253–267.

THE LIBRARY AND SCIENTIFIC PUBLICATIONS OF THE ZOOLOGICAL SOCIETY OF LONDON: PART II

MARCIA A. EDWARDS

The Zoological Society of London, Regent's Park, London, England

SYNOPSIS

The establishment and subsequent development of the Society's scientific publications are reviewed. Possible additional services which could be provided by the computer-assisted system devised for the production of the *Zoological Record* are outlined. The alternatives to conventional publication and the implications these may have for the future of the publications and the library are briefly discussed.

THE SCIENTIFIC PUBLICATIONS

Proceedings and Transactions

The Committee of Science and Correspondence was appointed "for the purpose of suggesting and discussing questions and experiments in animal physiology, of exchanging communications with Corresponding Members of the Society, of promoting the importation of rare and useful Animals, and of receiving and preparing reports upon matters connected with Zoology" (*Proc. zool. Soc. Lond.* 1831). It was also intended that abstracts from the most interesting of the communications would be published and circulated each month. The appearance of the first of these abstracts in January 1831 under the title *Proceedings of the Committee of Science and Correspondence of the Zoological Society of London* and the subsequent opening of the committee meetings to any interested Fellows, established the Scientific Meetings and the *Proceedings*.

By 1832, the Council had decided on some alterations and in the following year, General Meetings for the transaction of Scientific Business replaced the Committee of Science and Correspondence. A further journal, the *Transactions*, was instituted, the first part of which was published on August 14th 1833. As a result of these changes, the name of the Committee was dropped from the title of the *Proceedings* which then became the *Proceedings of the Zoological Society of London*. Familiarly known as the PZS, the title was altered in 1965 to the *Journal of Zoology* although the *Proceedings* was retained as a subtitle.

The present-day Scientific Meetings are still based on papers accepted for publication in the *Proceedings* (now the *Journal of Zoology*) or *Transactions* and although there are but eight meetings each year, these continue to take place on the second Tuesday of the months concerned. In its Report to the General Meeting on 3rd January 1833, the Council expressed the hope "that the Scientific Meetings of the Society will go on increasing in interest and in estimation as a most effectual means of acquiring and imparting Zoological knowledge"; a sentiment which may claim to have been realized in the intervening years.

The early *Proceedings* were modest volumes consisting of abstracts and comprising about 150–200 pages each year, while the *Transactions* were intended to contain rather more important papers, suitably illustrated. This was the usual practice at the time. The Royal Society's *Proceedings* provided abstracts of papers published in full in the *Philosophical Transactions*, and the Linnean Society also issued the *Proceedings* of its meetings, some of the papers communicated appearing subsequently in the *Journal* or *Transactions* of that society. But a gradual change took place in the relative balance of the *Proceedings* and *Transactions*.

The first volume of the *Transactions*, published by the Zoological Society between 1833 and 1835, contained 43 papers and approximately 400 pages but by the time Volume 14 was reached in the 1890s, the number of pages had increased by about 100 and the number of contributions had decreased to 11, much longer, papers. This trend continued and now four or five papers are contained in a volume of some 500 pages. Volume 30, published in 1965, was the last Transaction to appear in the imposing royal quarto and subsequent volumes are crown quarto, matching the *Journal* in size.

Meanwhile the papers in the *Proceedings* had ceased to be just abstracts and by 1848, the volume was illustrated with coloured plates and line figures. Interestingly this change was mirrored in the similar publications of the Royal and of the Linnean Societies.

Currently, about 160–200 manuscripts are submitted annually, all of which must be subjected to critical scrutiny by outside referees, and the Society is most fortunate in the help so generously given. Despite this careful process of selection, the number of accepted papers necessitated the decision to publish four volumes in 1965 and three annual volumes thereafter, so that now about 100 papers are published each year.

With the establishment in the early 1960s of two research institutes in the Society, The Nuffield Institute of Comparative

Medicine and The Wellcome Institute of Comparative Physiology, the amount and variety of scientific work carried out under the Society's supervision greatly increased. The reports of the Pathologist were published in the *Proceedings* but the other research work could be mentioned only briefly in the annual reports of the Society. The specialized nature of this work evidently required an alternative form of publication and it was decided to produce a Scientific Report which would include the reports of the veterinary officers at Regent's Park and Whipsnade, the report of the Pathologist and reasonably full details of work in progress in the Curators' Departments and the Wellcome and Nuffield Institutes. The results of much of this research would subsequently be published in other more specialized journals.

The first two of these reports, covering the periods 1966–1967 and 1967–1969 were produced as separate publications but now they are published in the *Journal of Zoology* and a Scientific Report appears every other year in the April or May issue of the *Journal*. In this way, the work of the Society is brought to the notice of institutions throughout the world, so creating an awareness of the extensive programme of research which it sponsors.

Despite the changes which have taken place over the years, the original intention of the *Proceedings* as a focal point for workers at home and abroad has been maintained and today papers are received in ever increasing numbers from zoologists working in all parts of the world.

After the launching of the *Proceedings* and the *Transactions*, some 30 years elapsed before the Society became involved with the production of another publication, the *Zoological Record*.

Zoological Record

The *Zoological Record* was founded in 1864 on the initiative and under the editorship of Albert Günther and the first volume appeared in 1865 with the title *The Record of Zoological Literature*.

The idea was to produce an annual volume divided into Sections corresponding to the main animal groups. Each Section began with a list of papers and books published in a particular year and the second or special part, arranged systematically, gave details of and comments on, the listed papers. In time this was expanded to include a subject and a geographical index; recorders' comments as to the merits of the included papers were discontinued since these had, not unnaturally, proved a source of friction. With the inclusion where appropriate of a palaeontological/

stratigraphical index, the *Record* has retained this arrangement to the present day. This form of presentation ensures that a reader may use the *Record* in a number of different ways according to his needs:

(i) to trace papers by an author;
(ii) to ascertain what has been published on any subject in which he is interested;
(iii) to see what work has been done on any particular animal.

The origin and subsequent history of the *Record* have been well documented elsewhere (Bridson, 1968; Gunther, 1975) but it is interesting to reflect on two themes which emerged early in the life of the *Record*, have run through its existence ever since and have ultimately been responsible for the recent changes which have had to be undertaken.

Thomas Spencer Cobbold, who was the first recorder for the helminths, was to comment on completion of his contribution to the first volume "I really cannot undertake to re-write my record again. *I have given all the time I could spare*. I can do no more." (Cobbold, 1865.)

Günther, after five years of editing the *Record*, wrote to Newton "... I cannot sacrifice the evenings of six months a year to the trouble and anxiety of this affair alone." (Günther, 1869).

Nearly 100 years later, a recorder remarked on relinquishing his commitment, that he had realized he would have to read and index an average of ten papers every evening of every day in the year in order to complete one issue of the Section for which he was responsible and that did not take into account the time needed to prepare the copy for the printer and to read and correct the proofs.

The *Record* began as a modest volume listing some 3000 titles, compiled by zoologists who were specialists in their fields, for whom the preparation of the *Record* was a spare-time occupation. But the amount of literature published grew steadily and considerably, despite the effect of two world wars, until by the 1950s, the part-time recorders were finding it more and more difficult to cope with the task and many Sections of the *Record* appeared later and later than the year to which they referred.

Eventually it was clear that assistance would have to be provided and gradually the specialist recorders were replaced by full-time staff until at present there are only a very few part-time recorders working on the *Record*. This increase in staff eventually posed accommodation problems. At first the recorders worked in the British Museum (Natural History) but it was not possible to find sufficient space for the growing numbers and now, although the museum continues to provide room for one group, most of the recorders are located in Boston Spa and use the resources of the British Library (Lending Division) which is situated nearby.

The growth of zoological literature in the last two decades has been prolific, and at present, about 6500 different journals and numerous books are scanned each year in order to produce a volume dealing with some 60 000–70 000 titles. Most of these titles require more than one entry and those dealing with the systematic revision of a group of animals may need a considerable number of entries to provide an adequate analysis of the contents.

Clearly this is a major undertaking. Since the *Record* was originally compiled by people working alone and in their own time, the system of production had to be simple and flexible. It consisted of entries written on separate cards and sorted by hand into order for the printer (for details see Bridson, 1968). This method, though tedious, worked well but imposed a great clerical burden on the recorders and created difficulties for the printer who had to work from hand-written copy, which in turn entailed much laborious reading of proofs to correct errors in interpretation. All of this caused delays and increased costs.

A survey of the *Record* was undertaken in 1964 by Dr H. Coblans of Aslib (Association of Special Libraries and Information Bureaux). At that time, most of the *Record* was compiled by the part-time recorders and the introduction of a mechanized system was not recommended as it would have been impractical under the circumstances. Instead, it was suggested that more full-time non-specialist zoologists should be recruited and that efforts should be directed towards relieving the recorders of much of the clerical work involved in compilation.

As more full-time staff were introduced, it became clear that there would have to be a radical change in production methods. In 1970 a further study, sponsored by OSTI (Office for Scientific and Technical Information) was carried out by the Research Department of Aslib. A computer assisted system was suggested as the best solution but it was realized that the specialized system necessary would be extremely costly to provide.

About the time that the *Record* production was under review, consideration was being given by the staff of the American Museum of Natural History in New York to the possibility of recommencing publication of the *Dean Bibliography of Fishes*, the last part of which had been published in 1923. With the support of the Office of Science Information Service of the National Science Foundation, the Research and Development Department of Bio-Sciences Information Services formulated a program suitable to produce the *Dean Bibliography* and to make possible the future retrieval of detailed bibliographic data stored in a computer (Schultz, 1968). The literature covered and the form of indexing closely resembled the Pisces Section of the *Zoological Record* and the staff of the Department of Ichthyology of the American Museum of Natural History were concerned not only that their bibliography might have an adverse effect on the *Record* but also that a great deal of effort would be wasted in producing two very similar services; an opinion shared by many other ichthyologists. Discussions were held between the staff connected with the two publications and it was agreed that as from the volume covering the 1971 literature, the *Dean Bibliography* would continue to be compiled by the Dean Bibliographer, Dr James Atz, but would appear as the Pisces Section of the *Zoological Record.*

To the great disappointment of both institutions it proved impossible for the American Museum to continue this arrangement and so, after the publication of the *Dean Bibliography* as the Pisces Section of Volume 108 of the *Record*, the task of indexing the fish literature once again reverted to the Society. With great generosity, the Directors of the American Museum of Natural History and the Chairman of the Department of Ichthyology, Dr Donn Rosen, donated the system developed for the *Dean Bibliography* to the Society to produce all the Sections of the *Zoological Record.*

The *Dean* system was designed to deal with the relatively small amount of work concerning one animal group but the *Record* must cover the literature published on all animals. Detailed investigations and modifications were therefore necessary to enable the *Dean* system to cope with the much greater volume of data to be processed and also to provide the somewhat different form of presentation required for the *Record.* Much valuable help and advice on the development of the modified system has been given by UKCIS (United Kingdom Chemical Information Service), whose computer facilities are utilized to process the *Record.*

The computer-assisted production methods will not be explained in detail here as a good description has already been given by Dadd (1973). Essentially the information entered on forms by the indexers is keyed onto magnetic tape. These entries are checked by a series of computer programs and any queries are referred to the *Record* staff. When indexing is complete, all entries are computer sorted into the required sequence, and a tape is produced to drive a photo-composing machine which produces film from which the *Record* is printed by off-set lithography.

The introduction of this system has at last relieved the recorders of that great millstone of clerical work which has been a feature of the *Record* from the start and which was no doubt responsible in part at least, for the despairing comments of Cobbold, Günther and many another recorder since.

The other recurrent theme is that of finance. Early difficulties, the Zoological Association, the effect of the *International Catalogue of Scientific Literature*, all these have been admirably described by many authors (Mitchell, 1929; Bridson, 1968; Gunther, 1975). Always it seems, the *Record* has had to cope with a zoological community which reacts with alarm to any suggestion that publication might cease but neglects to give the support necessary to ensure untroubled continuation, a state of affairs obvious from comments in the prefaces to many volumes of the *Record*.

In the preface to Volume 58, the Secretary wrote

> "The increasing cost of producing the Record has thrown a burden on the finances of the Zoological Society which has seriously interfered with sides of its scientific work more closely related to the study and care of living animals. Last year, the Society explained the position in circulars addressed to the leading institutions in the world interested in zoological science, and announced that it would be unable to continue the publication of the Record unless it was assured that a sufficient number of copies would be subscribed for. The response, unfortunately, was insufficient . . . The Zoological Society is prepared to regard an annual loss of £500 or even £600 as a contribution it may reasonably make to the Record in the general interest of Zoological Science and hopes, therefore, that other institutions will amongst them be able to contribute an equal amount in addition to subscribing for copies. Otherwise the Zoological Record must cease." (Mitchell, 1923).

This forthright statement, while it enabled Volume 59 to be published, apparently did little to stir zoological consciences, for again in the preface to that volume, Chalmers Mitchell wrote

> "So far, the only gifts or promises towards the publication of Vol. 60 . . . are £20 from the Society of Tropical Medicine and Hygiene and £5 5s from the Challenger Society. The Imperial Entomological Bureau, which undertook the compilation of Insecta for Vols 59 and 60, on receipt of the usual recorders' fees, has now agreed to

> undertake the same work for Vol. 61 without payment—a most welcome support of moral and material value.
>
> The Zoological Society has received widespread testimony as to the value of the Record to Zoological Science. But it has again to insist that such testimony, although agreeable, will not secure the continuance of the Record unless it assumes the practical form of more subscriptions to the whole volume and the separate divisions, and donations towards the deficit".

The *Record* continued through yet another world war and thanks to the selfless labour of the recorders who devoted many hours at little recompense, and to the degree of subsidy provided by the Society, the price was kept low thus enabling interested zoologists to purchase individual Sections for personal use. But the growth of the literature and consequent employment of a full-time staff, in addition to the rising costs of printing and production, meant that the price of the *Record* had to be increased to a realistic level, if it was to survive.

Grumbles there may always be about the delay in publication, or the cost, or the contents, but this is nothing new for, as the then editor remarked some 75 years ago, "Critics will certainly find it easy to discover deficiencies in the volume, but we may doubt whether they will realize the extent of the work involved in it." (Sharp, 1902).

In a centenary appraisal of the *Record*, Bridson (1968) described it as "one of the great achievements of bibliography in the field of natural history . . . an unparalleled zoological bibliography from 1864 onwards" and he wished it equal success in its second century. A decade of that century has passed and those responsible for the fortunes of the *Record* are determined that, given the support it deserves from the world-wide zoological community, the second century will indeed be the equal of the first.

Nomenclator Zoologicus

In 1934, Dr Sheffield Airey Neave wrote to Sir Peter Chalmers Mitchell, then Secretary of the Zoological Society, setting out proposals for an up-to-date index of generic names of animals, since the lack of this information was seriously impeding the work of systematic zoologists.

A number of previous lists existed and although the *Zoological Record* contained the information, it had not been indexed since 1910 (Waterhouse, 1902, 1912). The Prussian Academy of Science was compiling the *Nomenclator Animalium Generum et Subgenerum* but this only included names published up to 1921 and was, moreover, progressing very slowly.

The original idea was to continue the index of names listed in the *Record* as far as 1935, possibly incorporating the two volumes of the *Index Zoologicus* previously prepared by Waterhouse which covered the period 1880–1910. Eventually it was decided that such an index would not be entirely satisfactory. Instead the Council of the Society approved a scheme for the preparation of as complete a list as possible of all the generic and subgeneric names used in zoology from the 10th edition of Linnaeus' *Systema Naturae* in 1758 up to and including the literature for 1935. The list would be alphabetically arranged under the generic names and would give the author, date and place of description of each name, and also the Class to which the animal was assigned.

The work was undertaken on behalf of the Society by Dr Neave, then Assistant Director of the Imperial Institute of Entomology, and with the help of many members of staff of the British Museum (Natural History) and the Institute.

The *Nomenclator* was published in four volumes, two in 1939 and two in 1940, and contains over 225 000 entries but, in order to provide for the continuation of the work, a fund was created, known as the Neave–Lloyd Nomenclator Zoologicus Fund, which it was hoped would enable further volumes to be published. The first of these supplementary volumes appeared in 1950 and dealt with generic and subgeneric names, some 18 000 or so, published between the years 1936 and 1945. Again the editor was Airey Neave who by this time was Secretary of the Zoological Society. Two more volumes covering the years 1946–1955 and 1956–1965 have been published under different editors and another is in preparation. Together the seven published volumes list about 290 000 entries.

The later volumes are almost entirely based on the *Zoological Record* and the labour of compilation is very great. Since, until quite recently, the *Record* was produced from hand-written copy prepared by part-time recorders, it was impossible to add to their difficulties by asking them to provide an additional set of cards giving details of any new names to be included in the *Record*. Accordingly it was necessary to extract the names and details of new genera, subgenera and new names from the published volumes, each of which lists on average about 2000 such names. Each name is entered on a separate card and all must be checked for accuracy and any queries settled.

The introduction of a computer-assisted system for the production of the *Zoological Record* should considerably ease the clerical task of compiling the *Nomenclator*. A program is being developed which will enable the details of all new names contained in each volume of the *Record* to be printed out in alphabetical sequence by the computer. These details will still have to be checked since queries, particularly those concerning homonyms and synonyms, are often not apparent until the entries extracted from several volumes have been accumulated. When a volume of the *Nomenclator* is complete, it will be printed by the same process used for the *Record*.

Although production methods alter and editors change, it is more than likely that the *Nomenclator* will continue to be referred to simply as "Neave", the name by which it has come to be known and used the world over.

Symposia

In a previous paper in this symposium, Professor Bullough referred to the commencement in 1959 of the Symposia. These meetings on particular aspects of zoology, usually occupying two days, have now become established as an important activity of the Society.

At the time, it was realized that while the rate of advance in scientific knowledge was such that specialization had become inevitable, the chance for people working on different areas of a similar subject to meet to discuss their findings and ideas, would play a valuable rôle in helping to draw together the various threads. Where possible, research workers from overseas are invited to take part and the fact that the meetings are open to any interested scientist ensures that participation and discussion are wide-ranging.

Such has been the variety of subjects selected, that very nearly half the number of Symposia held so far have been arranged in association with some 14 different societies. Quite recently new ground was broken with the publication of a volume representing the proceedings of the Second International Congress of Myriapodology, held at Manchester University and arranged in conjunction with the Centre International de Myriapodologie which is located in Paris. This experiment proved successful and in 1977 the Seventh International Congress on Arachnology will form one of the Symposia for that year.

These Symposia, and the printed volumes which result from them, are undoubtedly of great value in reviewing and placing on record the state of research in various areas at particular times.

International Zoo Yearbook

In November 1830, the Chairman read to a meeting of the Committee of Science and Correspondence, part of an extract from the report of Council which had met a few days previously. This stated that

> "The Council have moreover suggested that letters be sent to the superintendents of the principal Menageries in Europe, *viz* at *Paris, Leyden, Munich, Vienna, Madrid,* etc., proposing mutual communication of all observations . . ." (*Proc. zool. Soc. Lond.* 1831: 2).

Although this suggestion foreshadowed future events, it was not until 130 years later that the present Secretary of the Society, Lord Zuckerman, realizing the need for a special channel for the interchange of information between zoos, wrote to Zoo Directors throughout the world asking for their co-operation in compiling a new publication to be known as the *International Zoo Yearbook*. The response was encouraging. Over 200 replies were received to the detailed questionnaire sent out by the first editors, Dr Desmond Morris and Miss Caroline Jarvis, and the first volume, covering the year 1959, was published in 1960.

The *Yearbook* is intended to further the work of animal conservation, an important aspect of Zoological Gardens and Parks, and it is divided into sections reflecting different features of this intention. The first section deals with a special subject and in Volume 1, this was the keeping of apes in captivity. Over the years various authorities have contributed papers to this section which has now reviewed the treatment in captivity of many of the vertebrate groups, and considered aquatic exhibits, nutrition, and animal trade and transport. The second section is concerned with breeding, husbandry and hand-rearing and includes practical information on handling and caring for animals of all sorts. There are also contributions on buildings and exhibits, on educational schemes in zoos and on conservation. Finally there is a comprehensive reference section. This contains information on vertebrates bred in captivity, including those bred to more than one captive generation, the census of rare animals in captivity and a report on the status of the various rare animal studbooks. There is a biennial directory of zoos and aquaria of the world, with details of senior staff and of the size and type of collection. In the latest volume, there is also a survey of new architectural developments. So popular has the reference section become that it can be obtained as two separate reprints: the directory of zoos and aquaria and the list of species bred in captivity.

It is no small measure of the value of the *Yearbook* that in at least two of the subjects dealt with in the special section, "Small mammals in captivity" and "Nutrition", practical experience and advances in research have been of sufficient importance to review them again in recent volumes.

The maintenance of self-replacing breeding groups of animals is vital and it is hoped to publish more on planned breeding projects, with emphasis on management of family groups and colonies, and connected with this, methods of detecting and avoiding inbreeding problems. The keeping of proper records is an essential part of zoo management, as is the exchange of breeding animals. Articles which explain useful modern techniques in terms the practical animal keeper can appreciate, are of great value, and the education of all—but especially the young—is essential if finance and space are to be found for animals in the modern world. All of these aspects will find a place in future volumes.

CONCLUSIONS: THE FUTURE

The establishment and subsequent history of the library and the scientific publications of the Society have been briefly outlined but what can be said of the future?

We have seen that the computer-assisted production methods devised for the *Zoological Record* can be adapted to produce entries for the *Nomenclator Zoologicus* so saving a great deal of tedious clerical work, and there are other services which can be provided by the new system. It would be possible to issue a monthly bulletin of current titles, details of new taxa, or other subjects, printed on cards for cumulative filing purposes, or even to cut across the present organization of the *Record* by animal groups, to produce information on wider aspects of zoology such as parasite–host relations, conservation, ecology or environmental considerations. The facility is there; how it is used in the future will depend on the style of information zoologists require and whether they are prepared to give sufficient support to new developments.

In the area of conventional journal production, much discussion and experimentation is currently under way. Over the last few years, the cost of producing a journal has substantially increased at the same time that institutions have had to cut subscriptions for economic reasons. Individual workers are unable or unwilling to pay more for a personal copy of a journal and many publishers have therefore to recover mounting costs from decreasing subscriptions. There are the additional points that items of interest to a particular reader may well be dispersed in a considerable number of different journals and that many libraries now provide single photocopies of articles on request.

Various attempts are being made to cope with these difficulties. VINITI (the All-Union Institute for Scientific and Technical Information) implements a scheme in the USSR, whereby manuscripts may be deposited in one of 30 or so centres in the Soviet Union. Titles of these papers are listed in the VINITI catalogue and abstracts appear in the *Referativny Zhurnal*. In the United Kingdom, material considered too extensive for complete publication may be lodged in various institutions but although helpful, such deposition schemes are not a solution.

A more promising approach is the development of the synopsis/microfiche journal. This consists of a conventionally-printed journal containing carefully designed synopses of papers, each of one or two pages in length, fully illustrated and with a list of key references. If a reader wishes to see the full text he may obtain a microfiche which will be a photographic image of the author's manuscript. Libraries will probably take the synopsis journal and the complete set of microfiches whereas individuals can, if they wish, subscribe to the synopsis journal only. This has advantages in that the subscriber can obtain a current awareness journal at moderate cost and for the publisher, there is a saving in production costs, little need of warehousing to accommodate back-stock and great savings in postage costs and distribution time since microfiche can easily be sent by air mail (Line & Williams, 1976; Barr, 1976).

As an optional alternative to microfiche, a miniprint journal can be provided in which the full text is reproduced in miniature print, again, prepared from authors' manuscripts so that no typesetting costs are involved. The Chemical Society has been circulating experimental journals of these types and as a result, it has been decided to commence publication of the synopsis/microfiche journal in 1977, probably as an addition to the society's existing range of research journals.

Another approach is the integrated publication system operated by the International Research Communications System in

Lancaster. Papers from the whole field of biomedicine are submitted to a single editorial office, and, if accepted, they are published in a variety of forms. The articles may appear in one or more specialist journals or, since they are typeset and printed as separates, they may be supplied as individual reprints (Eakins, 1974).

Whether these ideas would be as acceptable to the biological sciences is debatable since previous attempts of a similar nature have met with little enthusiasm (Arnett, 1970; Foote, 1972). Some interesting view points might well emerge from the forthcoming general assembly of the European Association of Editors of Biological Periodicals (ELSE), which is to discuss innovations in journal editing and production.

Possible changes in the way information is presented will obviously affect libraries. Microform journals produced by photographing authors' manuscripts should be considerably cheaper to produce than journals printed by conventional methods. This may be advantageous for a librarian if, as a result, expenditure on journals is reduced and he is able to allocate his budget in other ways. There will be fewer storage problems if journals are produced wholly or in part in microform but more accommodation may be needed for the apparatus necessary to read microfiche or microfilm. If a library is associated with an institution issuing publications, it may have to fulfil the rôle of providing microform copies of articles on demand or of storing original manuscripts from which to produce photocopies.

If the time ever comes when the Society's publications are issued as a combination of synopses and microfiche, then the wheel will, in some ways, have turned full circle to the origin of the *Proceedings* as a journal of collected abstracts.

References

Arnett, R. H. (1970). Data documents: A new publication plan for systematic entomology. *Ent. News* **81**: 1–11.

Bridson, G. D. R. (1968). The Zoological Record—a centenary appraisal. *J. Soc. Biblphy nat. Hist.* **5**: 23–34.

Barr, D. (1976). Access to journal literature: short-term and long-term prospects. *Aslib Proc.* **28**: 116–119.

Cobbold, T. S. (1865). Letter dated 3rd May, quoted in Gunther, A. E. (1975). *A century of zoology at the British Museum* etc.: 290. London: Dawsons.

Dadd, M. N. (1973). Zoological Record. *BLL Review* **1**: 6–13.

Eakins, J. P. (1974). The integrated publication system: a new concept in primary publication. *Aslib Proc.* **26**: 430–434.

Foote, R. H. (1972). Communication in the biological sciences. In *Challenging biological problems—directions toward their solution*: 376–396. Behnke, J. A. (Ed.) New York: Oxford University Press.

Günther, A. C. L. G. (1869). Letter to Alfred Newton dated 15th July, quoted in Gunther, A. E. (1975). *A century of zoology at the British Museum* etc.: 293. London: Dawsons.

Gunther, A. E. (1975). *A century of zoology at the British Museum through the lives of two Keepers 1815–1914.* London: Dawsons.

Line, M. & Williams, B. (1976). Alternatives to conventional publication and their implications for libraries. *Aslib Proc.* **28**: 109–115.

Mitchell, P. C. (1923). Preface. *Zool. Rec.* **58**: iii.

Mitchell, P. C. (1929). *Centenary history of the Zoological Society of London.* London: The Zoological Society of London.

Proceedings of the (Committee of Science and Correspondence of the) Zoological Society of London. Part I, **1830–1831** (1831): 1–2.

Schultz, L. (1968). Re-establishing the direct interface between an abstracting and indexing service and the generator of the literature. *Proc. A. meet. Am. Soc. inf. Sci.* **5**: 101–105.

Sharp, D. (1902). Preface. *Zool. Rec.* **38**: iv.

Waterhouse, C. O. (1902). *Index Zoologicus.* An alphabetical list of names of genera and subgenera proposed for use in zoology as recorded in the "Zoological Record" 1880–1900. London: The Zoological Society of London.

Waterhouse, C. O. (1912). *Index Zoologicus* No. II. London: The Zoological Society of London.

Symp. zool. Soc. Lond. (1976) No. 40, 269–281.

THE CONTRIBUTION OF THE ZOOLOGICAL SOCIETY OF LONDON TO FIELD STUDIES AND PROSPECTS FOR THE FUTURE

P. A. JEWELL

Department of Zoology, Royal Holloway College, University of London, England

SYNOPSIS

Examples are given of the ways in which the Society has supported field studies in the past, and continues to do so at present. Emphasis is placed upon studies in the wild of those larger mammals that have always made an important contribution to the Society's exhibits. The manner in which studies in the field, on captive animals, and in the laboratory, can be complementary is discussed. The suggestion is made that it will be advantageous to the husbandry of captive animals, and to the conservation of wild populations, if the Society is able to develop further its long-standing support of field activities.

INTRODUCTION

I can take up my theme from the lively account that is given by Mr R. Fish and the Hon. Ivor Montagu, on pp. 17–48 of this Symposium, of the famous field naturalists whose work was associated with the Society. My purpose is to emphasize the importance of the link between the Society's activities in London, including the exhibition of animals, and scientific research on animals in their natural environment.

SOME EXAMPLES OF PAST FIELD STUDIES

I am not here concerned with exploration or field expeditions to collect living specimens, although, of course, much new information about the life of animals in the wild has been obtained, incidentally, in this way (see, for example, the account of Ivan Sanderson, 1937). Throughout much of its long history as a scientific society, the research activities of the Zoo have centred round the prosectorium and the pathology laboratories, and have been pursued in collaboration with anatomists and veterinarians in the colleges and universities. Observations on the living collections of animals were not neglected, however. Indeed, our present Secretary's seminal studies of the hamadryas baboons on Monkey Hill, related in *The social life of monkeys and apes* (Zuckerman, 1932), became a pioneering behavioural study in the modern style. Also incorporated in that book are Zuckerman's observations on wild baboons in South Africa, which he made during 1930 on a visit aided by a grant from the Zoological Society.

An example of direct support to fieldwork given by the Society at an early time was a grant to John Budgett, who made an expedition to western Uganda in 1902 to study the fish *Polypterus*. Knowing of Sir Harry Johnston's discovery of the okapi, (see Fish & Montagu, this volume), he informed the Zoological Society that he "might be able to acquire useful information concerning this new animal" as he would be in the Semliki region, near the Congo; the Society awarded him £100 for this purpose. Budgett became so engrossed in his pursuit of *Polypterus*, however, that he redirected his route to the eastern shore of Lake Albert and did not enter the Congo. Budgett, with great conscientiousness, refunded the grant to the Zoological Society, and in an apologetic letter to the Secretary, he explained that he felt he must pursue his main objective and that he had learnt, referring to the okapi, "that there is not much left to be found out about the animal". This was a reference to the news that the Belgians had sent specimens to Europe and that living specimens would probably follow. The Secretary of the Society, P. L. Sclater, was right to reply that:

> "You are quite in error in supposing that the subject of the okapi is exhausted. All that has been ascertained from the specimens sent to Brussels is that the adult male carries horns, but we still want further specimens in England and information of all sorts about the habits and life of this animal".[a]

The promotion of field research was not, in the past, a main commitment of the Society, but it would be remiss not to note the many unobtrusive ways in which fieldwork was fostered.

It was natural that during the period that Dr Harrison Matthews was Scientific Director, several studies in his special field of interest, the reproductive biology of mammals, should have been undertaken. Matthews himself, together with Professors E. C. Amoroso and R. J. Harrison, and Dr I. W. Rowlands, made a study of the grey seal (*Halichoerus grypus*) on the coasts of Britain, and gathered much new information on the physiology of the neonate, and of lactation, in this species (Amoroso *et al.*, 1965). In one particularly audacious episode in the autumn of 1950, Matthews and his colleagues, by the use of a net, caught a female grey seal and her pup in a cave on Ramsey Island, off the coast of Pembrokeshire,

[a]Quotations are from correspondence in the archives of the Zoological Society.

and the female was kept in a large cage at the bottom of the cliff. The research workers used a human breast pump on the female, and they were able to measure the efficiency with which the seal converted blubber to milk, and that with which the pup, in turn, reconverted this into its own fat, as blubber (Matthews, 1950; Amoroso & Matthews, 1951). The Zoological Society gave additional financial assistance to Professor R. J. Harrison in his extension of the work on the common seal (*Phoca vitulina*) that he carried out in the Faeroe Islands, and in the Walsh, East Anglia (Harrison & Young, 1966).

Co-operation between Professor Amoroso's Department at the Royal Veterinary College in Camden Town and the Zoological Society, gave rise to many other field studies, of which the work of Dr Lona Kellas provides an example of an outstanding contribution. She worked in East Africa, financed by a Colonial Office Grant, and produced the first studies of the life cycle and reproductive biology of the dikdik (*Rhyncotragus kirkii*) and also obtained material that revealed new aspects of foetal physiology in the giraffe (Kellas, 1954; Kellas, van Lennep & Amoroso, 1958). During the 1950s there was a growing awareness of the importance of African "game" animals—and an awareness of our lack of knowledge of their biology. It is significant, therefore, that a pioneering study of the population structure and social behaviour of Thomson's gazelle (*Gazella thomsoni*) was carried out at this time by Alan Brooks, it being promoted by the Society with the support of the Colonial Office (Brooks, 1961).

In the late 1950s the Zoological Society fostered another field study of major importance. Under the supervision of Dr Harrison Matthews, the Colonial Office engaged Jasmine Sidney to make a "Fact-finding Survey of the African Fauna". The Society provided accommodation and extensive assistance for the survey and in 1958–1959 Miss Sidney made a comprehensive tour of East, Central and South Africa. The aim of the survey was to compare the density and distribution of game animals at the end of the last century with that prevailing at the present day. This was information much needed by zoologists working in Africa, who readily co-operated in supplying a wealth of data and who, as I discovered in my own visit to Africa in 1962 (also supported by the Society, see Jewell, 1963), eagerly awaited the results. They were rewarded by the monograph published as a volume of the Society's *Transactions* (Sidney, 1965). In this publication, Miss Sidney reviewed the status of the wild asses, zebras and rhinoceroses, and five families of the

Artiodactyla, including, in the Bovidae, all the Alcelaphinae, the Tragelaphinae and the African buffalo.

This work now provides a standard point of reference for the 17 genera of African ungulates with which it deals. The fact that the distribution of so many species has contracted since the survey, so that it is no longer up to date, does not detract from its value because it is most important to have these distributions of the late 1950s placed on record.

In maintaining its collections of animals, especially when these are kept as viable breeding groups, the Society has more than once provided indirect support or stimulus to fieldwork. An example is seen in the work of Vernon Reynolds, who spent three years investigating the social life of the colony of Rhesus monkeys (*Macaca mulatta*) at Whipsnade (Reynolds, 1962). Inspired by this experience he then moved to Uganda to produce one of the earlier studies of the behaviour of chimpanzees in the wild (Reynolds, 1965).

The study of mammals was not the only kind of fieldwork that the Society encouraged, and in quite another field of zoology, that of coral reefs, the Society added its support to Dr Gwynne Vevers' participation in the Royal Society's British Solomon Islands Expedition of 1965. Here Dr Vevers studied the ecology of the reefs and later, in 1969, he went on to do similar work on Rarotonga and Aitutaki in the Cook Islands when he took part in the Royal Society of New Zealand Cook Bicentenary Expedition.

THE SOCIETY'S RESEARCH INSTITUTES

Turning now to the present time, a new era of research activity was initiated with the establishment of the Wellcome Institute of Comparative Physiology (WICP), and the Nuffield Institute of Comparative Medicine (NICM). Their work is fully documented in the Scientific Reports and in Dr Goodwin's contribution to this Symposium. I will, therefore, select only a few examples of the manner in which field studies attained prominence during this period.

Dr I. W. Rowlands and I were early appointed to the WICP, and one of our first activities was to survey the distribution and density of the vole and field mouse on Skomer Island in collaboration with R. M. Lockley of the West Wales Naturalists' Trust and P. J. Fullagar, a research student at the WICP (Fullagar, Jewell, Lockley & Rowlands, 1963). Dr Rowlands continued to collect Skomer voles for a comparative study of their reproductive cycle (Rowlands &

Coutts, 1969) with that of mainland bank voles (both *Clethrionomys glareolus*), Fullagar studied the ecology of the field mouse, and I collected information on environmental factors affecting the breeding seasons of these small mammals (Jewell, 1966).

At this time (in 1960) the Zoological Society joined forces with the Nature Conservancy, and with research organizations in Scotland, to finance a team to study the free-living Soay sheep on the Islands of St. Kilda. This became a major undertaking, in which I was involved, and eventually extended over a period of ten years. As with many field studies, the long time for which it was continued greatly enhanced the value of the data. The work was not unconnected with the Society's activities, since groups of rare breeds of domestic livestock, including Soay sheep, had long been kept at Whipsnade as part of a "gene bank" (Rowlands, 1964). All aspects of the study were eventually published as a monograph (Jewell, Milner & Boyd, 1974).

Amongst its facilities the WICP included a visiting scientists' laboratory, and for some time this was occupied by Ted Smith who was studying grey seals for the Nature Conservancy. His presence continued the interest of the Society in the grey seal, already mentioned, and his major work was on the northern population of the Scottish islands. His presence enabled us to undertake one of the first attempts to capture pinnipeds in the wild using immobilization by drugs (Jewell & Smith, 1965).

The WICP provided a base for many other scientists who spent time there learning techniques or writing-up. Amongst them were Jeremy Grimsdell who worked on the African buffalo, Gerald Clough who studied the hippopotamus, and Clive Spinage who studied the waterbuck. All of them worked in Uganda at the Nuffield Unit of Tropical Animal Ecology. This Unit, which has initiated so many new and important field studies on large mammals was itself established through the initiative of zoologists in this country, and with the active support of the Zoological Society. Sir Landsborough Thomson, then President of the Society, visited the Uganda Parks in 1957 and the Council of the Society strongly backed the subsequent negotiations that led to the Nuffield Unit being set up (Beadle, 1974). Later, the Zoological Society financed the setting up at the Unit of a small meteorological station, which was run by Mr M. Beadle.

I have mentioned only the investigation of certain large herbivores in Uganda, but not all the mammals studied in Africa by research workers at the WICP were large! Hugh Tripp investigated elephant shrews (particularly *Elephantulus myurus*) in South Africa, and established a colony at the Wellcome Institute.

Finally, with reference to this phase of the activities of the WICP, the fieldwork directed to the study of hystricomorph rodents occupies a prominent place. Dr Rowlands, later joined by Dr Barbara Weir, initiated a series of studies on several species of these enigmatic rodents. This necessitated expeditions to South America to capture them, and amongst those investigated were agoutis, acouchis, tuco-tucos, cuis, degu, chozchoris and viscachas. A compilation of much of this work, including accounts of field studies, constitutes Symposium No. 34 of the Zoological Society of London (Rowlands & Weir, 1974).

The Nuffield Institute of Comparative Medicine has sponsored field studies that have covered the widest possible range of organisms, from the elephant to bacteria. Sylvia Sikes studied the ecology of arterial disease in the African elephant, and gathered much other field data that was summarized in her monograph on the species (Sikes, 1971). Dr Michael Crawford has maintained a continuing programme of work on the nutrition and body composition of African game animals, and this has involved not only studies and collaboration with colleagues working in the field in Africa, but also the development of proposals for the much needed new domestication of African large mammals. His work has emphasized the economic importance of these animals and the contribution they can make to human nutrition and health (Crawford & Crawford, 1974); it provides pragmatic and compelling reasons for wildlife conservation.

The tenor of my discussion is concerned particularly with the ecology of mammals, so that I shall not elaborate on the many special studies of their parasites or diseases that form part of the programme of the NICM. These organisms are, of course, a major interest of research workers in comparative medicine, and field investigations of them have been actively promoted by Dr Goodwin. In prosecuting this work Dr Geoffrey Smith and his colleagues have investigated pleuropneumonia in cattle in Nigeria, and the occurrence of the organism causing botulism in aquatic environments in Britain; Alan Young has investigated piroplasmosis in cattle and wild animals in Kenya and Dr A. Voller continues studies on the immunopathology of malaria. Moreover, the long arm of NICM has stretched to Nairobi in Kenya, where Dr Goodwin was, for three years, responsible for the scientific programme of the Wellcome Trust Research Laboratories. These laboratories have made use of baboons as subjects for the study of vitamin requirements. This is not field work in the sense under discussion, but it involves a principle of great significance: the laboratories have been established where the animals are indigenous instead of bringing the animals to an alien environment. This could have great value for the better understanding of the biological processes under investigation.

In concluding these observations on the work of the Research Institutes we may note that the newly appointed Senior Fellow at the WICP, Dr R. D. Martin, has already initiated a new phase of field studies, including his own work on prosimians, and Frances D'Souza's work on tree shrews. He is continuing his studies on lemurs on Madagascar (when political circumstances permit), and John Pollock, studying under him, has just completed a meticulous field study of the indri. In addition, Dr Martin has established a breeding colony of mouse lemurs (*Microcebus murinus*) in the WICP and is observing their behaviour and reproductive biology. This work illustrates an approach to field studies in which the Zoological Society has particular strength, because: (i) field and laboratory studies are integrated and complementary; (ii) the species being studied belong to a rare group that is in special need of conservation; (iii) a breeding colony has been established, which can contribute to conservation; (iv) back-up can be provided to new graduates working in the field and the right conditions created for analysing and writing-up field data.

PRESENT AND FUTURE

Up to the present the Zoological Society has supported a great diversity of field studies, most of which have arisen from the particular interests of individual research workers. The time may have come when these activities should be developed into a more fully integrated programme. Perhaps the most important reason for saying this relates to questions of conservation, particularly of endangered species, and the role of zoological gardens in the preservation of wildlife.

In a sense zoos are indebted to the species of animals that for so long have provided the main attractions in their exhibitions. Now, many of these species are in danger of extinction in the wild, and yet probably for most of these we know next to nothing of their ecology, their social organization, or their habitat requirements. Time is fast running out in which this knowledge may be acquired, yet it could be vital for the successful management of these species, where they survive only in national parks for example, or if they are to be bred in captivity in an attempt to avert their extinction. It is recognized that the proliferation of zoos throughout the world has in itself put adverse pressure on some rare species. The Zoological Society has been in the forefront of those zoos that have adopted strict codes of practice and an ethical approach to the exhibition of animals. The educational role of well run zoos, and their research into animal health, husbandry and breeding, are valuable contributions to conservation. Public opinion on this matter is volatile, however, and could fall out of sympathy with the keeping of wild animals. It is particularly important, therefore, that the Society should have an active presence in conservation endeavours. In one way this is already achieved through its historic association with the Fauna Preservation Society that was founded in 1903, and to which the London Zoo has always acted as host and patron. The Fauna Preservation Society has regularly given small sums of money to assist field research. These sums have often made all the difference between success and failure in an undertaking, or have been catalytic to a more extended study. The Fauna Preservation Society's Oryx 100% Fund, has supported some 39 projects since 1971. Two projects were those of research workers associated with the Zoological Society; in 1971 Dr Michael Crawford was aided in a survey of Ajai's Island white rhinoceros and Toror giant eland; in 1976 Katherine Homewood (attached to the WICP) had support for her study of the Tana River mangabey. Many other extremely valuable field studies could be quoted but there is an ever pressing need for more long-term and sustained support. Perhaps it is now opportune for the Zoological Society to extend in new ways those of its established activities that contribute to the survival of animals in their native environments.

The study of rare species might in the first instance, comprise surveys of their distribution and their surviving numbers. Indeed, in the *Red Data Book* (IUCN, 1972), there is consistent emphasis on the need for more scientific research, and the conservation measures proposed for a large proportion of the most endangered species give high priority to "urgent surveys", "ecological" studies or "field studies". These measures need to be pursued with caution and appropriate selection, however, and, as the late Tom Harrisson so forcefully emphasized shortly before his untimely death, scientific research and survey are not a panacea and by no means automatically command the highest priority of action (Harrisson,

1974). In some few desperately precarious situations an intensive survey or intrusion by scientists might even prove disastrous. Notwithstanding these special situations, I believe that ecological studies should proliferate and a continuing scientific presence can often greatly assist local endeavours in conservation.

There is no lack of enthusiastic young zoologists who are keen to carry out field studies, and even if they are raw graduates, their potential to contribute to new knowledge is great. True, some endeavours would fail, but there have now been enough successful theses produced to suggest that giving support to such studies is an effective way of using money. A concerted effort should be made to assign research workers to study all rare and endangered species, if for no other reason than that for many species, hanging on in degraded habitats, it will soon be too late to understand their ecology.

I do not see this contribution to conservation, through field work, as a one-way process only. I believe that the Society, in facing problems of the management of animals in captivity, will benefit too. Such benefits are already evident and are to be seen in the successful breeding of the cheetah and orang-utan. New knowledge of the highly dispersed social community of the cheetah, and its relatively solitary existence in the wild, was reflected in the successful introduction of male to female at Whipsnade (Manton, 1975), and similarly John Mackinnon's studies of the orang-utan in Borneo revealed new aspects of their strange uncommunicable nature and led to a fresh approach to breeding them in Regent's Park (Brambell, 1975).

We are faced with the most lamentable lack of knowledge of the ecology of so many species of mammals and birds, not just the rare ones, but also the large, conspicuous and common ones, particularly in the tropics. We await, for example, a full ethological study of the two species of elephants, and of the hippopotamus, and the tiger; and we have only just begun to quantify the interaction of most large herbivores with their food resources. Nevertheless, in the past decade there has been a great proliferation of field studies with truly rewarding results. Not all these results have been published, however, and here I come to another aspect of support for field studies that I believe needs attention: that is financial support for the task of writing-up. Research workers in the field usually need to gather data over at least two years in order to properly cover annual cycles and seasons. Often a first year is spent in finding a suitable study site and learning the ways of the animal that is the subject of study. In the outcome a three-year grant may be used up, both profitably and properly, but support for writing-up is lacking. The Zoological Society of London has ideal facilities for writing-up with its unique library and the opportunity to offer bench space for this purpose.

The Zoological Society has always been at the forefront of endeavours to gain new knowledge about wild animals and devise methods for their care. As the need to conduct field studies, to conserve rare species and protect threatened habitats, becomes more urgent, the Society will surely develop its participation in such undertakings further. I would like to think that this effort could be integrated through a co-ordinating centre for field studies, established in the Society's research section. Much work awaits such a centre. It would be useful to catalogue current field studies on the larger mammals and birds in the manner that primate biologists have done in their own field. *The catalogue of current primate field studies* (Chivers, 1976), prepared by the Primate Society of Great Britain, lists 87 field projects in progress, with their locations (Fig. 1) and much information about them. The pattern of effort in relation to type of habitat (from high forest to open savanna) and suborders or superfamilies of primates is of great interest; as is the apparent increase in studies that aim to integrate serological, anatomical, physiological and disease factors with socio-ecology. Similar information about studies on the large herbivores, for example, would be particularly valuable. There are a great many groups for which documentation of their past and present distributions is long overdue, and much material gathered during Miss Sidney's survey remains to be examined and published. The formidable tasks of fund raising and gaining the co-operation of governments need hardly be emphasized. These are matters in which the Zoological Society's standing and established role could play a vital part.

I have emphasized the urgency of studies to support conservation, and the expediency of organizing appropriate research and promoting its publication. This should not detract from the importance of field studies in deepening our knowledge of species and in illuminating the processes of evolution. The knowledge gained from field studies will become an integral part of our assessment of groups of species and their relationships to one another. This knowledge will complete the picture initiated long ago when the first specimens were collected and sent back for exhibition and taxonomic study.

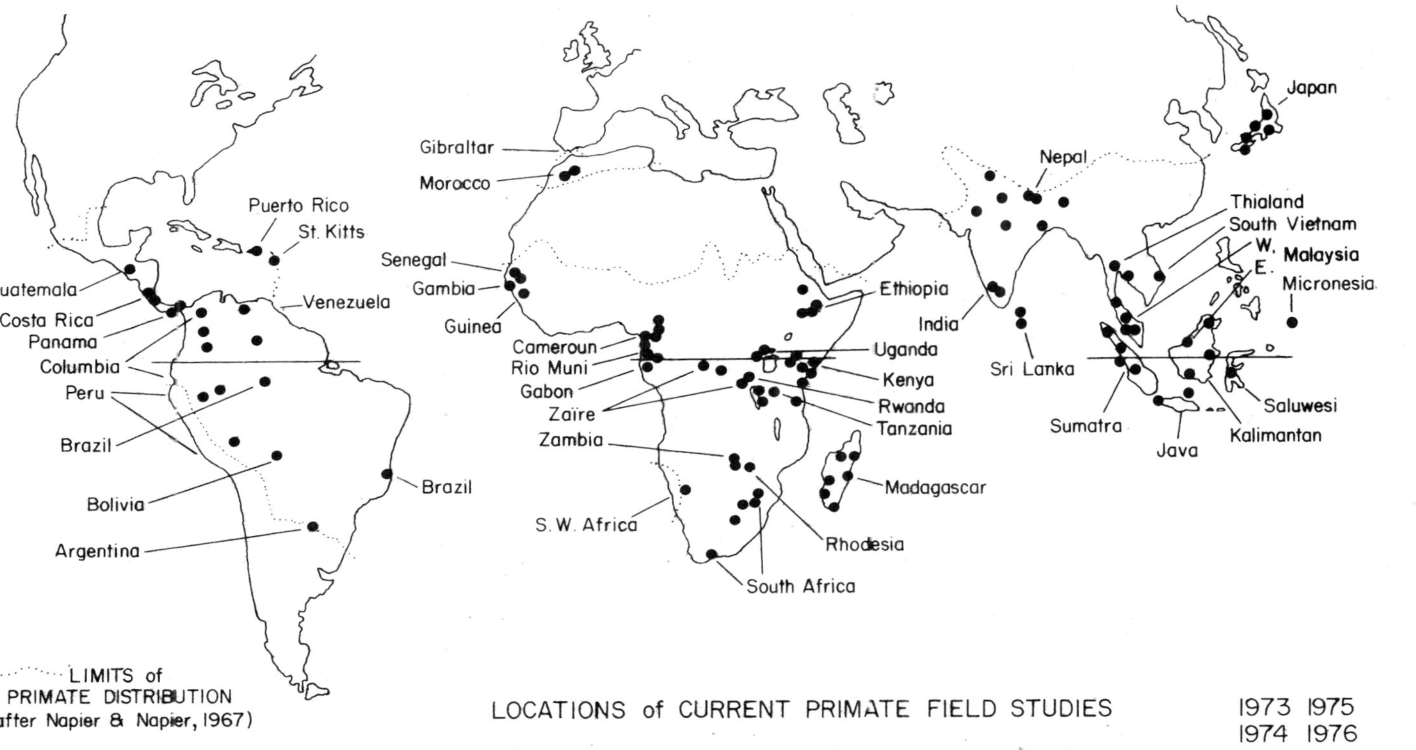

FIG. 1. Location of current primate field studies (from Chivers, 1976).

References

Amoroso, E. C., Bourne, G. H., Harrison, R. J., Matthews, L. H., Rowlands, I. W. & Sloper, J. C. (1955). Reproductive and endocrine organs of foetal newborn and adult seals. *J. Zool., Lond.* **147**: 430–486.

Amoroso, E. C. & Matthews, L. H. (1951). The growth of the Grey seal (*Halichoerus grypus*) from birth to weaning. *J. Anat., Lond.* **85**: 426–428.

Beadle, L. C. (1974). The Nuffield Unit of Tropical Animal Ecology (1961–1971). *J. Zool., Lond.* **173**: 539–548.

Brambell, M. R. (1975). Breeding orang-utans. In *Breeding endangered species in captivity*: 235–243. Martin, R. D. (Ed.) London, New York & San Francisco: Academic Press.

Brooks, A. C. (1961). *A study of Thomson's gazelle in Tanganyika.* (Colonial Research Publication, No. 25). London HMSO.

Chivers, D. J. (1976). Current primate field studies. *Primate Eye* No. 5 (Supp.) Sub-dept. of Veterinary Anatomy, Cambridge.

Crawford, S. M. & Crawford, M. A. (1974). An examination of systems of management of wild and domestic animals based on the African eco systems. In *Animal agriculture*: 218–234. Cole, H. H. & Ronning, M. (Eds.). San Francisco: Freeman.

Fullagar, P. J., Jewell, P. A., Lockley, R. M. & Rowlands, I. W. (1963). The Skomer vole (*Clethrionomys glareolus skomerensis*) and long-tailed field mouse (*Apodemus sylvaticus*) on Skomer Island, Pembrokeshire in 1960. *Proc. zool. Soc. Lond.* **140**: 295–314.

Harrisson, R. J. & Young, B. A. (1966). Functional characteristics of the pinniped placenta. *Symp. zool. Soc. Lond.* No. 15: 47–67.

Harrison, T. (1974). The new Red Data Book. *Oryx* **12**: 321–331.

IUCN (1972). *The Red Data Book: Mammalia.* Morges, Switzerland: International Union for the Conservation of Nature and Natural Resources.

Jewell, P. A. (1963). Wildlife research in East and Central Africa. *Oryx* **7**: 77–87.

Jewell, P. A. (1966). Breeding season and recruitment in some British mammals confined on small islands. *Symp. zool. Soc. Lond.* No. 15: 89–116.

Jewell, P. A., Milner, C. & Boyd, J. Morton (Eds.) (1974). *Island survivors: the ecology of the Soay sheep of St Kilda.* London: Athlone Press.

Jewell, P. A. & Smith, E. A. (1965). Immobilisation of Grey seals. *J. Wildl. Mgmt* **29**: 316–318.

Kellas, L. M. (1954). Observations on the reproductive activities, measurements, and growth rate of the Dikdik (*Rhynchotragus kirkii thomasi* Neumann). *Proc. zool. Soc. Lond.* **124**: 751–784.

Kellas, L. M., van Lennep, E. W. & Amoroso, E. C. (1958). Ovaries of some foetal and prepubertal giraffes (*Giraffa camelopardalis* Linnaeus). *Nature Lond.* **181**: 487–488.

Manton, V. J. A. (1975). Captive breeding of cheetahs. In *Breeding endangered species in captivity*: 337–344. Martin, R. D. (Ed.). London, New York & San Francisco: Academic Press.

Matthews, L. H. (1950). The natural history of the Grey seal, including lactation. *Proc. zool. Soc. Lond.* **120**: 763.

Reynolds, V. (1962). *Social life of a colony of Rhesus Monkeys* (*Macaca mulatta*). PhD Thesis. University of London.

Reynolds, V. (1965). *Budongo: a forest and its chimpanzees.* London: Methuen & Co.

Rowlands, I. W. (1964). Rare breeds of domesticated animals being preserved by the Zoological Society of London. *Nature, Lond.* **202**: 131–132.
Rowlands, I. W. & Coutts, R. R. (1969). The reproductive cycle of the Skomer vole (*Clethrionomys glareolus skomerensis*). *J. Zool., Lond.* **158**: 1–25.
Rowlands, I. W. & Weir, B. J., (Eds) (1974). The biology of hystricomorph rodents. *Symp. zool. Soc. Lond.* No. 34: 1–482.
Sanderson, I. T. (1937). *Animal treasure*. London: Macmillan.
Sidney, J. (1965). The past and present distribution of some African ungulates. *Trans. zool. Soc. Lond.* **30**: 1–396.
Sikes, S. K. (1971). *The natural history of the African elephant*. London: Weidenfeld & Nicholson.
Zuckerman, S. (1932). *The social life of monkeys and apes*. London: Kegan Paul.

Symp. zool. Soc. Lond. (1976) No. 40, 283–319.

A ZOOLOGIST'S VIEW OF RESEARCH ON REPRODUCTION

R. D. MARTIN

Wellcome Institute of Comparative Physiology, The Zoological Society of London, Regent's Park, London, England

SYNOPSIS

The comparative zoological approach is as essential to scientific research today as it was when the Zoological Society of London was founded 150 years ago. This applies to studies of mammalian reproduction, the speciality of the Wellcome Institute of Comparative Physiology, just as much as to other areas of zoology. Research on the reproduction of animals in captivity is of particular practical importance because reliable breeding will in the long term be necessary for maintaining stocks of animals in zoological collections and because such research is required to provide a proper scientific support for conservation measures. However, the future of such research is to some extent in doubt because of relative scarcity of funds, partly resulting from the competing requirements for research on molecular aspects of biology. It is shown that zoology has traditionally attracted, and benefited from, scientists who have combined a broad approach ("divergent thinking") with the ability to exploit specific techniques for the study of individual life-processes ("convergent thinking"). This combination remains as a central requirement for the success of future theoretical aspects of zoological research.

Four examples from the field of mammalian reproduction have been selected to illustrate these main points: (1) monitoring of pregnancy in great apes; (2) comparative aspects of primate placentation; (3) the relationship between stress and reproduction in captivity; (4) some broad ecological aspects of reproduction. In all of these examples, an attempt is made to demonstrate the value of comparative studies, assisted by the application of modern techniques of fundamental research.

With respect to the management of zoological collections, research on mammalian reproduction can yield benefits especially through the monitoring of female cycles and the application of artificial breeding techniques. If associated with some detailed investigations of behaviour, nutrition, genetics and disease, such work can produce results of significance not only for management of animals in captivity, but also for conservation and for theoretical developments in zoology.

INTRODUCTION

The celebration of the 150th anniversary of the foundation of the Zoological Society of London provides not only an opportunity to review past achievements, but also a very appropriate context for assessing future hopes and prospects. Indeed, a review of previous work is of particular value, since it is essential to ensure that in the future one does not lose sight of the guiding principles inherent in the subject of Zoology, which itself owes so much to the activities of the Society.

The aim of this contribution is to discuss possible future developments in the study of mammalian reproduction within the

Zoological Society, drawing upon specific examples which highlight general points. The Wellcome Institute of Comparative Physiology is, in fact, a very recent addition to the research establishment of the Society; it has been in existence for little more than 12 years. Yet, in that short space of time, it has earned an outstanding reputation as a centre for comparative studies of mammalian reproduction, in particular fostering research on little-studied mammal species. In addition to some fine work on classical laboratory rodents and domesticated mammals, new ground has been broken with research on the reproductive biology of more than 50 different species from nine orders of mammals (Insectivora, Rodentia, Artiodactyla, Perissodactyla, Primates, Chiroptera, Carnivora, Pinnipedia, Proboscidea). As has been pointed out by Perry (1971), our knowledge of mammalian reproductive processes is still based on a handful of species—mainly laboratory rodents and other domesticated forms. Any addition to our pool of knowledge is hence of great significance, and the Wellcome Institute's role to date in filling this need for comparative information has been remarkable. I can freely acknowledge both the calibre of the research conducted and the value of the contributions made to zoology, without immodesty, since the credit for these achievements belongs to my predecessor as Senior Research Fellow, Dr. I. W. Rowlands, and to the past and present staff of the Wellcome Institute. As a relative newcomer to the scene, it is my privilege to draw upon the present expertise and past history of the Zoological Society, and of the Wellcome Institute in particular, in formulating future projects for research into mammalian reproduction. In so doing, one of the cardinal principles will be recognition of the fact that comparative zoological studies are still of major importance. Even after 150 years, there is nothing "old-fashioned" about scientific research, properly conducted and assisted by modern techniques, which has as its primary aim the expansion of our detailed knowledge of the animal kingdom. As a zoologist, my fundamental concern is with animals and with studying them in their own right as parts of the natural world. Nowadays, there is a somewhat depressing and unfortunately widespread tendency to dismiss research on "exotic" animal species as a declining vestige of nineteenth century natural history. While it is of course true that techniques of study and theoretical approaches must advance in step with developments in other areas of scientific knowledge, there is nothing "old-fashioned" about the need to study little-known animal species. Indeed, as many of them are acutely threatened with extinction because of man's failure to follow the elementary and long-established rules of ecological balance, the need for such study is more urgent than ever before.

Since the continuing need for broad-based zoological research is not as widely accepted as it should be, the first part of this essay will be concerned with general issues which go far beyond the field of mammalian reproduction. However, those issues have repercussions for future studies in that field, as in many others, and a number of examples from research on the reproductive biology of mammals (mainly primates) will be cited to demonstrate two basic points: (1) the comparative zoological approach offers a number of unique advantages both for expansion of theoretical knowledge and for various practical applications; (2) although rapid development of new techniques has permitted intensive study of particular fine details of the fundamental life-processes, the greatest promise for the future lies in research which combines the broad zoological viewpoint with effective exploitation of these new skills for fundamental investigations.

PROSPECTS FOR RESEARCH ON REPRODUCTION IN ZOOLOGICAL COLLECTIONS

Research on reproduction is an integral part of the contribution that can be made by zoological collections to the advancement of scientific knowledge. Over and above that, there are two pressing practical requirements for such research. Firstly, in order to continue with the educational role of display of animals, zoological collections will increasingly be obliged to breed their own stocks. Supplies of animals from the wild are necessarily decreasing, and there is certainly no justification for importing animals which are to any extent threatened by extinction in their natural habitats, if no attempt is made to breed those animals and to gain scientific knowledge which is vital for competent decision-making by conservation authorities. Secondly, the problem of conservation of many animal species has become so acute that scientific investigations relating to the preservation of endangered species must be undertaken as a matter of priority. On both counts, breeding of rare species in captivity—accompanied by detailed scientific investigation of reproductive processes—is to be regarded as a major function of modern zoological collections. Although eventual reintroduction of stocks bred in captivity cannot be viewed as an alternative to conservation in the wild, it may well prove to be a valuable adjunct to such conservation in certain cases. Reintroduction would, in any case, depend upon preservation or regeneration of natural habitat areas, and for this reason breeding of endangered species in captivity should ideally be closely co-ordinated with conservation programmes in the field (see also Jewell's contribution to this volume). Although those concerned with conservation in the wild may not yet accept the need for breeding in captivity, they will ultimately depend upon studies conducted in captivity for a realistic assessment of the status of endangered species. The encroachment by human populations on wild habitat areas has been so extensive and so rapid (in terms of evolutionary time) that no truly "natural" situations can be said to exist for many species which still occur as substantial free-living populations. It might be said, for example, that the African elephant is not yet "endangered", but the natural range of that species has been so fragmented and circumscribed that the long-term prospects for survival may already be very poor. To assess future prospects, detailed scientific data are sorely needed for many animal species, both from the field and from studies conducted in captivity. In particular, it is necessary to evaluate scientifically the prospects for reintroduction into the wild, since decisions about the requirements and value of breeding in captivity for that purpose are still dependent largely on opinion, rather than fact. Whether or not reintroduction is found to be feasible for certain endangered species, it is certainly true that data on reproduction obtained in captivity are necessary for scientific analysis of the situation in the field. For example, the use of computer programmes for projection of future population trends of wild animal species is dependent not only on the measurement and prediction of natural environmental parameters, but also on the availability of information on basic reproductive parameters (e.g. age at sexual maturity, potential breeding frequency, gestation period, seasonality of reproduction, inhibition of reproduction by population density, etc.). In many instances (e.g. for gestation periods), such parameters have actually been identified with animals maintained and bred in captivity, and it may often be impossible to obtain the necessary data through a field study.

In this connection, it is notable that the Zoological Society is the host for the second international conference on the breeding of endangered species in captivity, to be held during this anniversary year. It is only fitting that the Society should, at this particular time, emphasize its commitment to the cause of conservation. Although there has been some scepticism in various quarters about the contributions made by zoological collections to conservation, it is a fact that zoos and similar institutions are filling a gap which has not been noticeably narrowed by the work of other organizations. At the first conference on the breeding of endangered species (Martin, 1975a), held under the aegis of the Jersey Wildlife Preservation Trust and the Fauna Preservation Society, it was clear that a great deal had already been achieved through the foresight and enthusiasm of numerous people working with the leading zoological collections of the world. If such activities are consolidated through concerted scientific research, with an emphasis on studies of reproduction, genetics, nutrition and disease, then conservation of animal species in the wild will be provided with an invaluable support.

With increasing emphasis in zoological collections generally on the regular breeding of various mammal species, the point has been reached where detailed scientific study can yield benefits in two directions. On the one hand, a scientific input can increase the reliability and long-term effectiveness of breeding programmes; on the other, there is a great deal of basic zoological information which can be extracted in the process. Two particular areas are proving to be of considerable practical value in both respects. The study of female reproductive cycles, using direct sampling of vaginal tract material and the measurement of hormone levels in blood and urine samples, has already yielded a great deal of data. Knowledge of oestrous cycles is of value in determining optimal conditions for mating, and monitoring of pregnancy is useful not only with respect to routine husbandry but also as a means of identifying potential problems. On the male side, one of the most promising areas of research is that of artificial insemination (Rowlands, 1965; Jones, 1971; Rowlands, 1974). This involves not only the development of suitable techniques for semen collection, storage and utilization of insemination, but also the monitoring of oestrous cycles in order to identify optimum times for artificial insemination to be conducted. Other areas of research have also proved to be productive, and observation of the behaviour of captive mammals deserves special mention. Among other things, behavioural studies have the advantage that direct manipulation of the animals is unnecessary for the collection of meaningful data. In many cases, for instance, oestrous cycles can be followed by observing female behaviour, and studies of maternal behaviour should prove to be of increasing significance as so many young mammals born in captivity at present are not properly suckled by their mothers. Individual studies of mammal reproduction in captivity should ideally be viewed in the general framework of the conditions of housing, nutrition and health. Further, consideration must in the long term be given to genetic aspects of breeding in captivity, particularly because of the dangers of inbreeding with small stocks. Overall, investigations of all of these aspects should gradually be integrated into a genuinely scientific set of principles for the breeding of mammals in captivity. This, in turn, would further not only the prospects for maintaining stocks of various species in zoological collections but also the prospects for a major contribution to conservation.

There can therefore be no question but that scientific investigations of the reproduction of animals in captivity, in common with studies of nutrition, veterinary care and genetical aspects, can potentially contribute greatly to zoological knowledge in the future. However, this potential contribution must be viewed in the context of recent developments in zoology generally and in the light of research priorities which have acquired *de facto* currency. It is true to say that the subject of zoology has undergone a virtual "reformation" over the past two decades as a result of developments in molecular biology. These developments have vastly expanded our understanding of life processes, and it is obvious that much zoological research in the future will be conducted at the molecular level. However, these very developments have led to partial neglect of possibilities for studying whole animals. It is, for example, surprising that the opportunities for investigations of rare and endangered animal species in captivity, in collaboration with established zoological collections, have not been grasped more eagerly by university departments of zoology. Ultimately, such investigations can only be fully successful if there is more widespread collaboration between university staff and those concerned with the scientific management of zoological collections. Work on whole animals has tended to become the "poor sister" of zoology in recent years, and it is worthwhile to examine whether this tendency is either desirable or justifiable.

Nobody can deny that there is a widespread tendency to regard classical comparative anatomy as "outdated, old-fashioned and boring", and any work conducted on whole animals now tends to fall under the same cloud. It is obvious that investigations of the fine anatomical details of various animal species are somewhat unrewarding in the absence of any attempt to understand the function of the whole animal. It was, in fact, the failure of nineteenth century anatomists to give sufficient attention to function which eventually led to the decline of this once commanding discipline. Naturally enough, the emergence of the experimental approach, and subsequently of molecular biology, carried with it an excitement which was lacking once classical comparative anatomy had become an established discipline. Yet molecular biology at present carries with it the same seeds of decay, in that the natural functions of whole animals have not been sufficiently considered. As Medawar (1975) pointed out in a recent tribute to the late Sir Julian Huxley, the subject of molecular biology may become by the end of this century "an abuse and an impediment to progress" just like classical comparative anatomy at the end of the previous century. One can safely predict that, by the time classical molecular biology has run its course, exciting new developments will be pioneered by those who are able to synthesize the knowledge that has been gained and relate it to the overall functioning of those animal species which are still extant.

At this point, it is relevant to consider some evidence concerning the psychological background to scientific research. In his studies of the mental aptitudes of schoolboys, Liam Hudson (1966, 1968) drew a crucial distinction between "convergent" and "divergent" thinkers. The former performed well on standard IQ tests requiring specific answers to well-defined questions. The latter performed best on "open-ended" intelligence tests requiring imaginative answers to questions which did not prescribe specific answers. This distinction, which is usually one of degree rather than sharp demarcation in individual cases, is in some ways parallel to a distinction between "specialists" and "generalists". As a rule, Hudson's convergent thinkers were more likely to study science at university, while divergent thinkers were more attracted to arts subjects. It is interesting, however, that *zoology* seemed to attract convergent and divergent thinkers in roughly equal proportions (or people who combined aptitudes in both directions). Different interpretations may be drawn from this observation. It could be said that zoology as a subject attracted a proportion of divergent thinkers because it traditionally lacked the apparently sharp scientific rigour of the physical sciences. At the same time, it is probably true that zoology as a subject has owed many of its major advances to the flair and imagination of those who could both identify and dissect general principles despite the complexities of whole animals in their natural habitats. The question that remains to be answered is: does the future development of zoology depend upon the same aptitudes as the past? For those who view zoology as a domain to be progressively replaced by current approaches to molecular biology, the answer is probably "No". For those who see in zoology a hierarchy of general principles, only some of which may be "explained" in molecular terms, the answer is probably "Yes". Whatever the answer may be, it would seem to be true that the greatest promise for future developments in zoological research lies in a combination of "divergent" and "convergent" thinking. Technical expertise and knowledge of fundamental life-processes should ideally be matched by a comprehensive understanding of whole animals and their interactions in natural environmental systems.

In fact, developments in zoology over the past 20 or 30 years have led to a dichotomy of the subject parallel to that between physical chemistry and organic chemistry. Molecular biology may be regarded as providing the basic "atomic" component of zoology, while the three "E"s of the subject (ecology, ethology and evolutionary studies) are concerned with the natural relationships of intact animals. This distinction has been emphasized by Wilson's new textbook on "sociobiology" (1975). Although a great deal of heated discussion has been generated by a relatively minor aspect of this monumental text (namely, its attempt to provide a basis for interpreting human social behaviour), it has in fact demonstrated that ecology, ethology and evolutionary studies can be fused into a distinct discipline with a rigorous set of explanatory principles and techniques of study. Reference to this discipline as "sociobiology" is somewhat of a red herring; Wilson has simply revived and consolidated whole animal zoology—a subject which has been temporarily overshadowed by the remarkable and stimulating growth of molecular biology. Wilson's discipline is naturally of great value in the interpretation of the biological origins of social behaviour in animals; but it is of far wider significance as a forum for all who are concerned with a full understanding of animals as living organisms. A knowledge of molecular biology is of course essential for such understanding, in the same way that a knowledge of organic chemistry is essential to the molecular biologists; but this does not mean that all zoology is reducible to molecular biology, as has often been implied. The study of whole animals is a discipline in its own right, though dependent on other disciplines.

Hudson's distinction between "convergent" and "divergent" thinking is also revealing with respect to the general advancement of scientific knowledge. In the first place, Kuhn's analysis of the nature of scientific revolutions (1970) indicates that both approaches play a significant part. In the development of any scientific discipline, once a central set of explanatory principles (or "paradigm") has been established, most research tends to be of a "convergent" kind, consolidating the established wisdom with detailed studies. Scientific revolutions, on the other hand, tend to result from a "divergent" approach, through which the assembled body of data is related to a new paradigm of greater explanatory power. When a new paradigm has been accepted (usually after a great deal of resistance from those who helped to consolidate its predecessor), there follows a period of active consolidation of the new wisdom. This interpretation, if it is valid, is important with respect to current re-examinations of governmental support for research, which are taking place in many countries of the world as the prospects for opening up new lines of research exceed the level of financial support which can reasonably be made available. In a very thoughtful article on the planning of scientific research in relation to cancer, Stoker (1975) has suggested that a distinction should be drawn between "strategic research", which expands scientific knowledge in generally useful directions, and "tactical research", which is concerned with the resolution of specific problems. In making this distinction, Stoker quotes H. G. Johnson's concept of "basic research contributing to a general pool of knowledge in which the applied scientist or developer fishes and takes samples as he needs them". If Kuhn is right, this general pool would also be essential to ensure that future scientific revolutions can take place at all. In short, a mixture of "convergent" and "divergent" thinking amounts to the life-blood of science.

In conclusion, to return to the relatively tiny field of research on mammalian reproduction in the context of zoological collections, it may be said that there is a very strong argument for maintaining strategic research which expands the general pool of knowedge. Whilst opportunities for tactical research must be encouraged wherever possible, for example in furthering developments relevant to human medicine, there is little justification for making this the exclusive aim. Indeed, in the long run, the breeding of animals for zoological collections and the development of research relating to conservation will amount to tactical research of great intrinsic value. There is also the consideration that studies of the reproductive functions of whole animals may ultimately contribute in some way to the next scientific revolution in the realm of zoology.

MONITORING OF REPRODUCTION IN THE GREAT APES

Recent work at the Wellcome Institute of Comparative Physiology on the breeding of gorillas (*Gorilla gorilla*) in zoological collections provides one illustration of the potential of comparative investigation of mammalian reproduction as an adjunct to zoo management. The research was, initially, motivated by "divergent thinking", in that the aim was to consolidate comparative work on reproduction of the great apes and man, which are members of the same primate superfamily (Hominoidea). Because of the close zoological relationship between the orang-utan, chimpanzee, gorilla and man, much of the research conducted on human reproduction is relevant to the breeding of the great apes in captivity. Conversely, study of reproduction in great apes may yield results which further our understanding of human reproductive processes.

When it was learned that two female gorillas kept by the Jersey Wildlife Preservation Trust had given birth and that further pregnancies were suspected, arrangements were immediately made for urine samples to be collected from the breeding females as frequently as possible. It was already known that standard human pregnancy test-kits could be used for the diagnosis of pregnancy in great apes. There had also been some prior work showing that patterns of excretion of certain hormones during pregnancy—oestrogens (Hopper, Tullner & Gray, 1968) and chorionic gonadotrophin (Tullner & Gray, 1968)—closely resemble those found in the pregnant human female. At the outset, work was accordingly devoted to detailed evaluation of pregnancy test-kit results and of data from hormone assays, using available urine samples.

As it happened, the preliminary results from these studies were just emerging when arrangements were finalized for London Zoo's female gorilla ("Lomie") to be sent to Bristol Zoo for mating with a proven male (Fig. 1). Although she had been kept together with London Zoo's male "Guy" for almost four years (1971–75), no pregnancy had resulted, and it was hoped that a fertile mating would occur at Bristol Zoo. Collection of regular urine samples from "Lomie", both during her stay at Bristol Zoo (11.6.75–27.2.76) and following her return to London Zoo, was therefore organized. The results already available from analysis of urine samples from the Jersey gorillas permitted early detection of pregnancy in "Lomie" whilst she was at Bristol, and the pregnancy

FIG. 1. London Zoo's female gorilla "Lomie" clinging to the cage wire whilst Bristol Zoo's male displays, during one of the introductions for mating. (Picture reproduced with kind permission from Bristol United Press).

has been continuously monitored since that time. From the zoo management point of view, strategic research has already led on to tactical research.

Collection of body fluid samples from mammals maintained in zoological collections is no easy task. Routine collection of blood samples, as is practised in many current studies of the reproductive endocrinology of various laboratory mammal species, is virtually impossible, and regular gathering of urine samples involves considerable difficulties. Hence, research in zoological collections such as that conducted on gorilla reproduction is necessarily

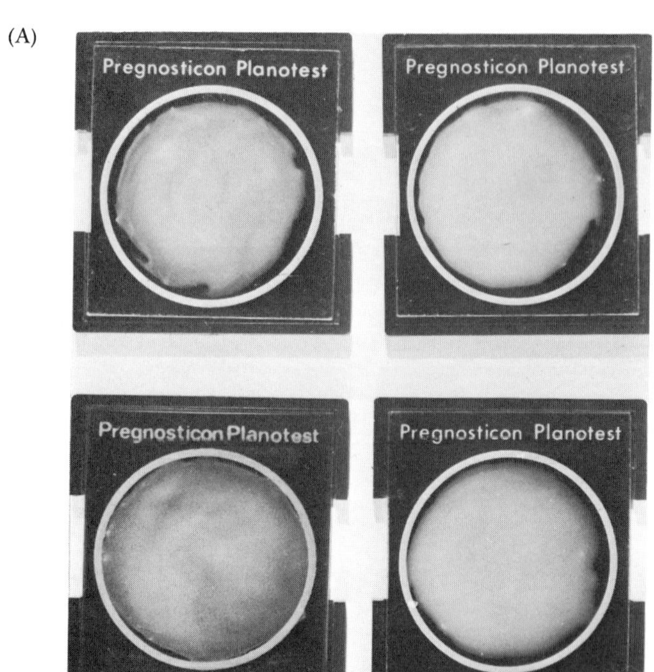
(A)
Pregnosticon Planotest
Pregnosticon Planotest
Pregnosticon Planotest
Pregnosticon Planotest

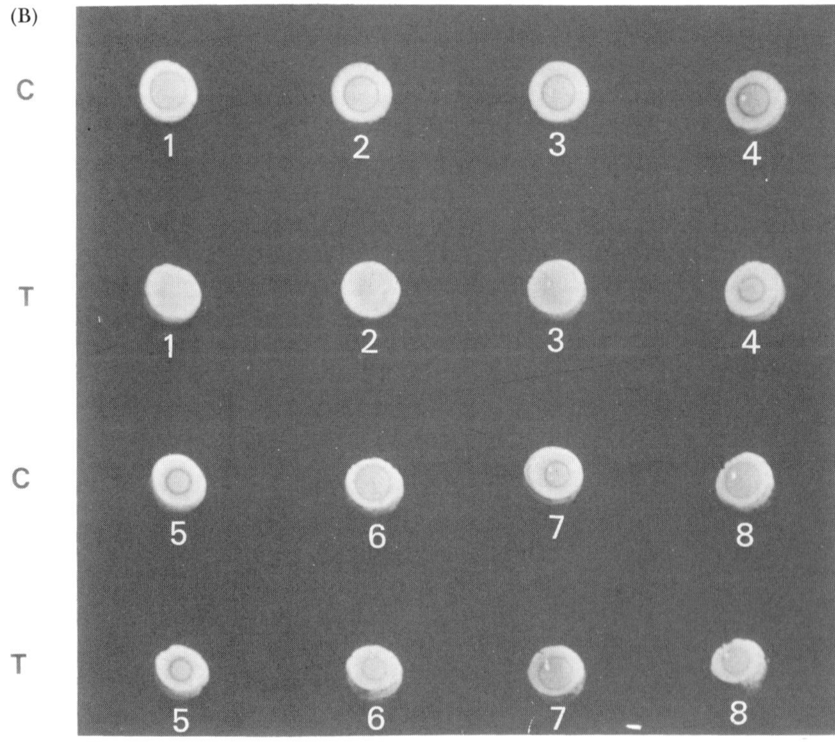
(B)
C
1
2
3
4
T
1
2
3
4
C
5
6
7
8
T
5
6
7
8

opportunistic. However, when one is working with relatively little-studied mammal species, it is quite worthwhile to accept the difficulties and limitations. In fact, techniques developed to cope with these special difficulties may themselves prove to be of value in other contexts.

One major problem with the utilization of urine samples for hormone determinations is that fluctuations in the concentration of the urine may distort hormone level measurements. The usual laboratory way of avoiding this problem is to collect complete 24-hour urine samples, as the total quantity of a hormone excreted in the urine over an entire day is more meaningful than the concentration of that hormone in an isolated urine sample. But it is almost impossible to collect complete 24-hour urine samples from mammals in zoological collections, and some other means of correcting for differential urine concentration is needed. A reliable technique is to measure creatinine levels in the urine samples. Creatinine is a characteristic nitrogenous waste product in the urine of all mammal species, and under uniform conditions of diet and activity the quantity excreted per 24 hours should be approximately constant. Hormone levels can hence be expressed more accurately as a ratio to urinary creatinine content. It is a fairly simple matter to measure creatinine levels in urine samples with a standardized colorimetric technique (Taussky, 1956; Slot, 1965), and this procedure has been used systematically in a number of studies of gorilla reproductive endocrinology (Hopper *et al.*, 1968; Tullner & Gray, 1968; Martin, Seaton & Lusty, 1976). One important practical point which has been underlined by the present study is that although oestrogen and chorionic gonadotrophin levels in gorilla urines do not decline appreciably even after storage for several months in an ordinary refrigerator (+4°C) samples must be deep-frozen (−20°C or below) in order to inhibit decay of creatinine. Hence all urine samples taken from mammals for study at the Wellcome Institute are now deep-frozen as a matter of course.

FIG. 2. A. Specimen results of Pregnosticon Planotest. With a non-pregnant human sample, the latex suspension is precipitated (lower left), whereas with a pregnant human sample the latex remains as a milky suspension (lower right). At 8 weeks after conception, "Lomie" was still giving a negative result (upper left), and the first positive results were obtained after 10 weeks of pregnancy (upper right). B. Subhuman Primate Tube Test results. Control tests (C) give a ring if the urine sample is in good condition. Urine samples taken 10 weeks, 6 weeks and 3 weeks prior to conception (1, 2 and 3) gave negative results on the Test (T), with no ring formation. Samples from 3 weeks, 15 weeks, 17 weeks, 18 weeks and 19 weeks after conception (4, 5, 6, 7 and 8) all gave positive results.

Two pregnancy test-kits have been utilized in the work on gorilla urine samples: the Pregnosticon Planotest (manufactured by Organon Laboratories Ltd), designed for routine use in the diagnosis of human pregnancy, and the Subhuman Primate Tube Test (manufactured by Ortho Diagnostics Inc.), designed for use with macaques and baboons (Fig. 2). Both tests are based on an immunological cross-reaction with chorionic gonadotrophin, but the latter is a far more sensitive test for use with gorilla urines. A series of tests has been conducted on urine samples from the two Jersey females and from "Lomie" (Martin *et al.*, 1976), and the use of the two tests together has proved to provide an extremely useful tool for the diagnosis of pregnancy, even with urine samples stored at +4°C (see also Hobson, 1976). It has been confirmed that the pattern of chorionic gonadotrophin excretion in the female gorilla during pregnancy is very similar to that found in human pregnancy, with a peak in the second to fourth months of pregnancy (Tullner & Gray, 1968; Tullner, 1974; Martin *et al.*, 1976).

An interesting historical note is that one of the earliest investigations on pregnancy tests with urine samples from the great apes was conducted on chimpanzees by the present Secretary of the Zoological Society of London whilst he was working as a research fellow at Yale University (Zuckerman, 1935). In that investigation the technique used was the Aschheim–Zondek test (Aschheim & Zondek, 1928; Zondek, 1931), based on a bioassay of urine samples using laboratory rats or mice. It was shown that this technique could be used for diagnosis of pregnancy in the chimpanzee, and the results were quite comparable to the present results from Planotest pregnancy-testing of gorillas, in that positive reactions were not recorded in the first month after conception and that results in the latter half of pregnancy were equivocal. It would seem that the occurrence of a peak of chorionic gonadotrophin in the second to fourth months of pregnancy is characteristic of the two African great apes and man (see also Tullner, 1974). It remains to be seen whether the same pattern occurs with orang-utans and perhaps with other simian primates such as the gibbons and siamangs.

Tests based on chorionic gonadotrophins suffer from the disadvantage that there is variation in the chemical structure and hence the cross-reactivity of these protein hormones from one mammal species to another (though this variation is, in itself, interesting in an evolutionary perspective and deserves comparative study in its own right). Standard pregnancy tests are further limited by the fact that they are merely qualitative, giving only "positive" or "negative" results. The chemical structure of the three major oestrogens (oestradiol, oestriol and oestrone) does not show this variation between mammal species, and direct measurement of these steroid hormones by radioimmunoassay provides much more detailed data on the progress of pregnancy in man and the great apes. A radioimmunoassay system for oestrogens has now been established on a routine basis at the Wellcome Institute of Comparative Physiology (Seaton, Lusty & Watson, in press; Seaton & Lusty, 1976), and this has permitted intensive study of patterns of excretion in gorilla urines. The assay is based on an anti-serum to oestradiol which also cross-reacts with oestrone (100%) and oestriol (15%) and can therefore be used to obtain some indication of "total oestrogens" in urine samples, though oestriol is incompletely estimated with the assay. Radioimmunoassay of "total oestrogens" has been carried out on urine samples from a total of five gorilla pregnancies to date, confirming the general similarity to the human pattern of excretion (see Ryan & Hopper, 1974). The content of oestrogens in urine rises sharply and continuously during pregnancy in gorillas, chimpanzees and humans (Tullner & Gray, 1968; Clegg & Weaver, 1972; Tullner, 1974).

Unless urinary creatinine levels are measured, the oestrogen content of urine will not reliably indicate pregnancy until about two months after conception in gorillas (Martin *et al.*, 1976). However, if oestrogen levels are expressed as a function of urinary creatinine content, the rise in oestrogens provides a clear indication of pregnancy within two to three weeks of conception (Fig. 3). In the case of "Lomie", radioimmunoassay of oestrogens provided the first clear indication of pregnancy, and this information guided the decision on a suitable date to transfer her back to London Zoo. Further, it is known that, in human pregnancy, complications may be recognized on the basis of a sharp decline in urinary oestrogen:creatinine ratios (Diczfalusy & Mancuso, 1969; Rao, 1976). Oestrogens are produced from maternal precursors by foetal tissue in humans, and doubtless in the great apes as well, so any fall in oestrogen excretion in the maternal urine indicates some problem in foetal development (Ryan & Hopper, 1974).

When the present work was initiated, it was assumed that a great deal was already known about reproduction in gorillas. However, with a few notable exceptions (Hopper *et al.*, 1968; Tullner & Gray, 1968), virtually nothing has been published on the reproductive endocrinology of the gorilla. In fact, the majority of

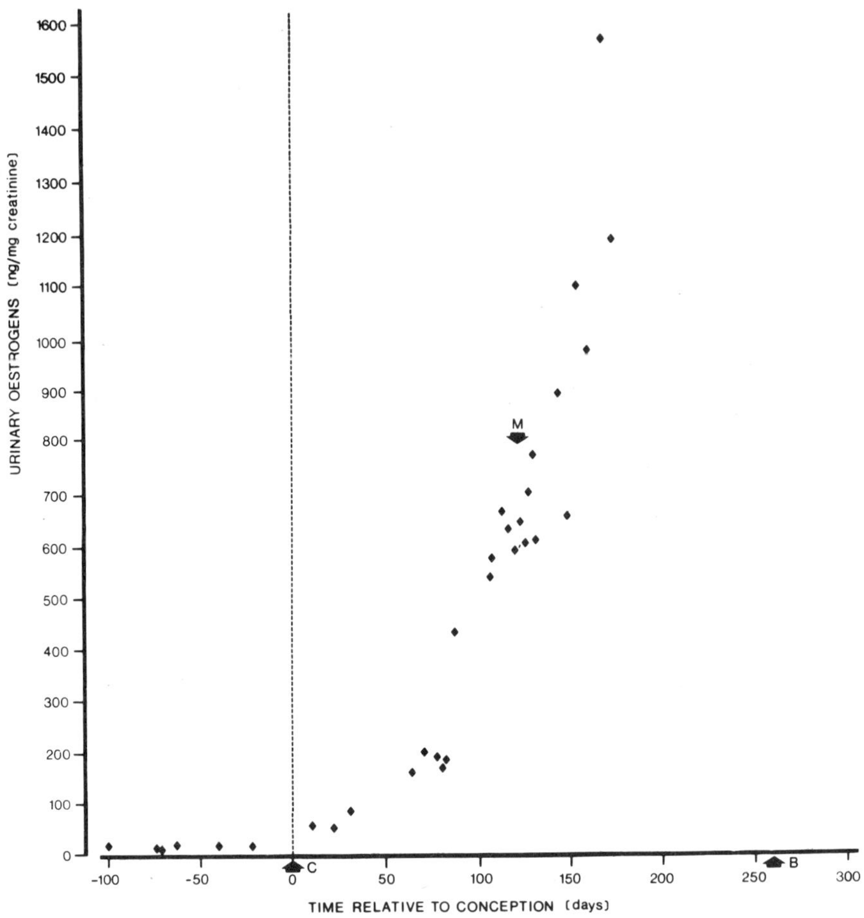

FIG. 3. Graph showing increase in oestrogen:creatinine ratios in urine samples from "Lomie" (July 1975—May 1976). The date of conception (C) is taken as 3.11.75 (matings: 1.11.75–5.11.75). M=Time of transfer from Bristol Zoo to London Zoo. There was no obvious decline of oestrogen levels, indicating that pregnancy was not seriously affected by the move. B=Expected date of birth (±20 days). (Note added in proof: "Lomie" gave birth to a female infant on 16.6.76, after a gestation of 256 ± 2 days.)

papers dealing with gorilla reproduction are concerned mainly with management aspects. Most of them are based on gorillas maintained in zoological collections (Coffey & Pook, 1974; Fisher, 1972; Frueh, 1968; Hardin, Danford & Skeldon, 1969; Kagawa & Kagawa, 1972; Lang, 1959, 1964; Lotshaw, 1971; Mallinson, Coffey & Usher-Smith, 1973; Nadler, 1975; Riddle, Keeling & Roberts, 1973; Rumbaugh, 1965, 1967; Thomas, 1958; Tijskens,

1971). This, in combination with the fact that many of these papers have been published in the *International Zoo Yearbook*, is clear enough proof of the value of scientific research conducted in zoological collections on man's closest zoological relatives.

Such research on gorilla reproduction in captivity is of great importance for at least two reasons. Firstly, gorillas—particularly the mountain subspecies (*Gorilla g. beringei*)—are threatened in their natural habitats. Data on reproduction is essential for sound appraisal of the prospects for conservation in the wild, and it is possible that breeding in captivity may be necessary to assist conservation efforts in the long run. Secondly, the gorilla is one of man's closest zoological relatives, and anything that we may learn about the reproduction of this great ape species is of special relevance to human evolution. Similar arguments would apply to investigation of the reproductive behaviour of the orang-utan (*Pongo pygmaeus*), which is also severely threatened in the wild.

In practical terms, detailed study of gorilla reproduction through urinary hormone determinations has a great deal to offer. Urinary oestrogen levels can be utilized, where necessary, to establish oestrous cycles in order to plan introductions for mating or to decide upon suitable dates for artificial insemination. Semen has already been effectively collected from all three great ape species (Warner, Martin & Keeling, 1974), and artificial insemination has been carried out successfully with the chimpanzee (Hardin, Liebherr & Fairchild, 1975). With the gorilla and orang-utan, oestrous cycles are usually less apparent in simple observation than is the case with the chimpanzees, so artificial insemination in these two species would probably benefit from studies of hormonal variation. In some instances, it may be possible to diagnose abnormalities in female cycles from the results of these studies, as a prelude to suitable treatment. Further, when pregnancies occur they can be monitored in detail. This may permit more accurate prediction of expected birth dates and perhaps recognition of complications at an early stage. In this connection, it is noteworthy that published data (see Martin *et al.*, 1976) on the duration of pregnancy in the gorilla show an unusually wide range of variation (246–288 days). All reported gorilla gestation periods have been calculated from the date of last observed mating, yet it is known that many mammals (including several primate species) will exhibit mating after conception. In the case of "Lomie", the date of conception has been unequivocally established, since she was isolated from the male for 10 weeks prior to the last mating and since the expected rise in oestrogens commenced immediately after that mating. If pregnancy progresses to term without incident, this will permit the first precise determination of the gestation period for the species (see note added to Fig. 3).

Finally, there is the problem that many gorillas which give birth to infants in captivity fail to rear them. At Jersey Wildlife Preservation Trust, for example, all six infants born to date have been hand-reared, and Nadler (1974) has demonstrated that apparent failure of maternal behaviour has occurred with 80% of 94 gorilla births he recorded. It is part of the established mythology of great ape management that early experience within a breeding group is necessary for females to rear their own infants successfully. Although such initial experience is no doubt of value, the fact that some great ape females lacking such experience have reared their infants without problem indicates that other factors are involved. In the future, it may be possible to investigate urinary hormones in late pregnant great apes and to identify signs of hormonal malfunction associated with failure of maternal behaviour. This, in turn, might lead to appropriate changes in management of gorillas and other apes. Babladelis (1975) has also reported that 40% (22 out of 56) infants born to primiparous gorilla females in captivity up to June 1975 died at an early age. It is possible that this high loss rate could be reduced in the future by careful monitoring of pregnancy and appropriate changes in management.

COMPARATIVE ASPECTS OF PRIMATE PLACENTATION

It is widely recognized that the comparative zoological approach can still produce significant information for interpreting mammalian evolution. It is less widely accepted, however, that this "divergent" approach continues to be of value for the planning of "convergent" research directed at the elucidation of specific reproductive mechanisms. Many examples could be cited in support of this view, but one which is particularly apt is the comparative study of primate placentation. Among other things, this can lead to the formulation of reasonable hypotheses which can be put to the test in individual experimental investigations.

All placental mammals show the same basic pattern of foetal membrane relationships (Fig. 4). The outermost membrane, the chorion, is always present and makes intimate contact with the internal surface of the maternal uterus. This contact provides the foundation for efficient passage of nutrients from maternal tissue

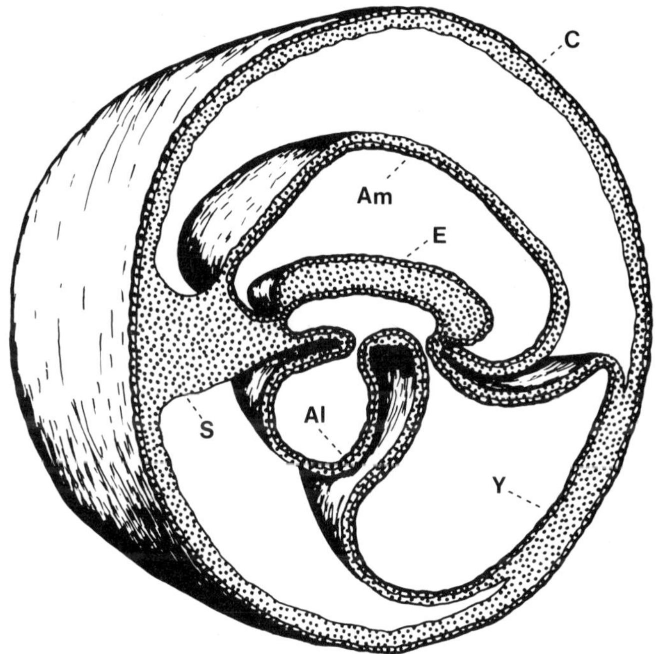

FIG. 4. Diagram (modified from Amoroso, 1952 and Le Gros Clark, 1971), illustrating the foetal membranes in a typical placental mammal: C = chorion; Am = amniotic membrane; Y = yolk-sac (vitelline) membrane; Al = allantoic membrane; E = embryo; S = connecting stalk.

to foetal tissue and for transfer of certain waste products in the opposite direction. As far as nutrients are concerned, there are three basic ways in which the foetus can take up substances of maternal origin: (1) by diffusion of substances from maternal blood vessels to foetal blood vessels; (2) by foetal absorption of the products ("milk") of uterine glands in the lining of the uterus; (3) by actual digestion of maternal cells (cytophagy). The latter source of nutrients is very restricted, and in mammal species generally foetal nutrition depends upon a characteristic balance between maternal blood: foetal blood diffusion and absorption of uterine "milk". In both cases, blood vessels derived from the mesodermal lining of the yolk sac and/or the allantois usually take up the nutrients for transfer to the embryo.

Following Grosser (1909), the definitive placenta formed between the chorion and the internal lining of the uterus is classified according to the degree of breakdown of the maternal

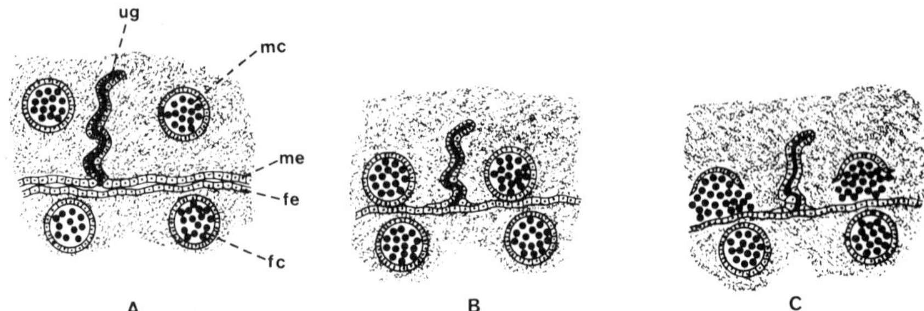

FIG. 5. Diagram illustrating Grosser's classification (1909) of placental structures (simplified from Amoroso, 1952). A: epitheliochorial type; B: endotheliochorial type; C: haemochorial type; mc = maternal capillary; me = maternal epithelium; fc = foetal capillary; fe = foetal epithelium (chorion); ug = uterine gland.

tissue and the resulting approximation between the superficial foetal capillaries and the maternal capillary system (Fig. 5). The simplest form is the *epitheliochorial placenta*, in which the chorion is applied to the epithelium lining the uterus, which persists as a continuous cell layer. If the maternal epithelium is broken down, so that the maternal capillaries come to lie very close to the chorion, an *endotheliochorial placenta* is formed. Finally, if the cells lining the maternal capillaries (i.e. the endothelial cells) are broken down in places, so that maternal blood has direct contact with the chorion, a *haemochorial placenta* is formed.

Primates are most unusual as a group, in that the lemurs and lorises ("strepsirhine primates") have the epitheliochorial type of placenta, while the tarsiers, monkeys, apes and man("haplorhine primates") have the haemochorial type. This has provided one of the major arguments for postulating an early dichotomy in primate evolution (Hill, 1953; Luckett, 1975; Martin, 1975b). Recent biochemical evidence has confirmed this evolutionary interpretation (e.g. Dene *et al.*, 1976). Until recently, however, no functional explanation had been advanced for this divergence in placental structure between strepsirhine and haplorhine primates. A possible basis for functional understanding has now been provided by two comparative zoological studies using the "allometric" approach of plotting the dimensions of particular reproductive features against maternal body weight for a wide range of species, using logarithmic co-ordinates.

Leutenegger (1973) has shown that, for a given maternal weight, the weight of a neonate produced by a haplorhine primate

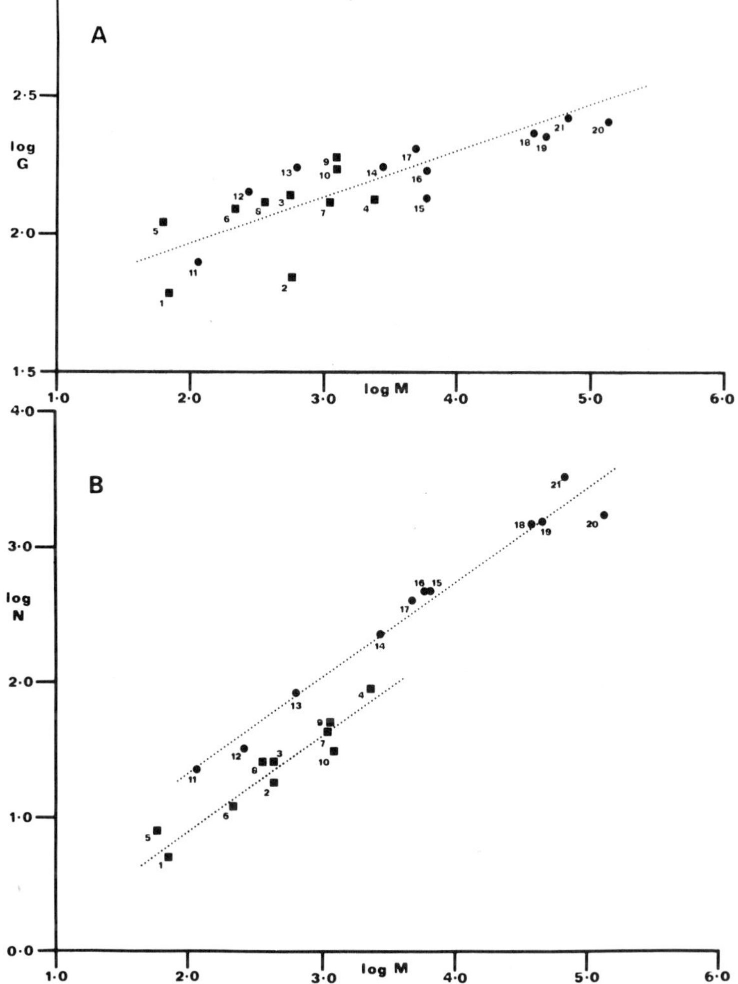

FIG. 6. Logarithmic plots of gestation period (G in graph A) and neonatal weight (N in graph B) against maternal weight (M) for 10 strepsirhine primate species (squares) and 11 haplorhine primate species (circles). Dotted lines are orthogonal regression lines. (From Martin, 1975b, including data from Leutenegger, 1973.)

is approximately three times that of a strepsirhine neonate. Since haplorhines do not have significantly longer gestation periods at any given maternal weight (see Fig. 6), it would seem that this difference in neonate weight between haplorhines and strepsirhines is primarily a function of the rate at which embryonic tissue is developed in the two groups. Sacher & Staffeldt (1974), on the other hand, have produced evidence to show that neonatal brain weight is closely correlated with the gestation period in placental mammals generally. It can therefore be concluded more specifically that, at any given maternal weight, haplorhine and strepsirhine primates will have approximately the same gestation period and produce neonates with approximately the same brain weights, but that the overall body size of the haplorhine neonate will be three times greater. Hence, it may be inferred that the more invasive (haemochorial) placentation of haplorhine primates permits more rapid development of foetal tissues *other than that of the brain.* Obviously, this is a hypothesis which should be checked with other placental mammal groups exhibiting a variety of placental types; but it would certainly explain the data so far examined for the primates.

The information obtained from this "divergent" approach of investigating allometric relationships across a broad range of species is highly relevant to detailed investigations of essential fatty acids currently being conducted at The Nuffield Institute of Comparative Medicine by Dr M. A. Crawford and his colleagues. Polyunsaturated fatty acids with 20 and 22 carbon atoms (arachadonic acid and docosahexaenoic acid, respectively) are accumulated in the foetal brain of mammals during cell division. Mammals are apparently unable to synthesize these fatty acids *de novo,* and they must be either obtained directly in the food or derived from linoleic and linolenic acids, respectively, by *very slow* metabolic processes which increase the number of carbon atoms from 18 (Crawford, Hassam & Williams, 1976). It is therefore unlikely that the rate of diffusion of any of these polyunsaturated fatty acids from the maternal to the foetal circulation would be the limiting factor in brain development. In short, the type of placenta would not be expected to affect the rate of incorporation of arachadonic and docosahexaenoic acids into the foetal brain. This accords well with the observation made by Sacher & Staffeldt (1974) that in placental mammals as a group neonatal brain weight is correlated with time taken for development (i.e. gestation period), regardless of the type of placenta involved.

At this point, comparative zoological information is again important as a source of hypotheses. It is to be expected that, since the magnitude of the barrier to direct diffusion between maternal and foetal blood vessels is greater, the strepsirhine primates might depend to a greater extent on the products of the coiled uterine glands. In fact, in common with other mammals which have

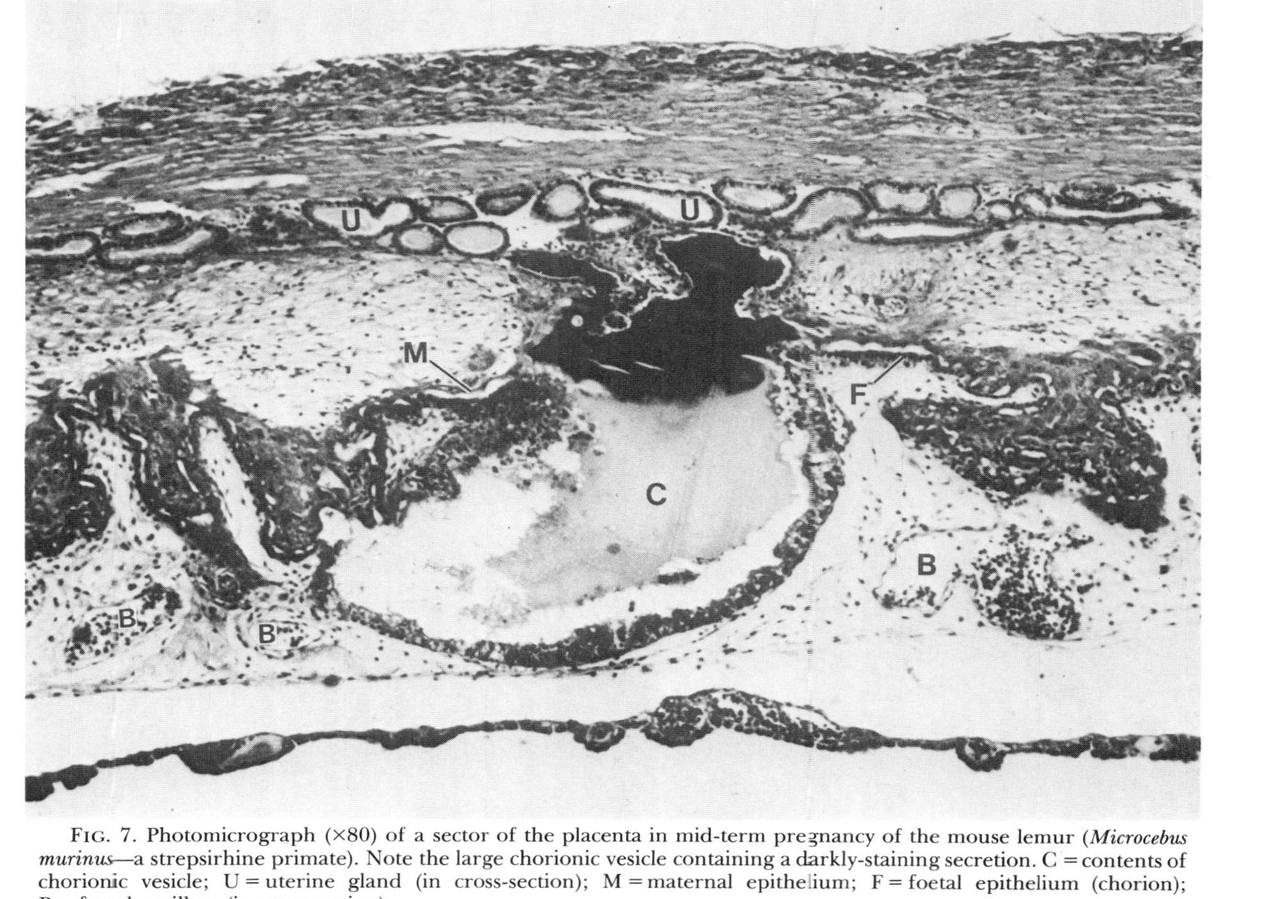

FIG. 7. Photomicrograph (×80) of a sector of the placenta in mid-term pregnancy of the mouse lemur (*Microcebus murinus*—a strepsirhine primate). Note the large chorionic vesicle containing a darkly-staining secretion. C = contents of chorionic vesicle; U = uterine gland (in cross-section); M = maternal epithelium; F = foetal epithelium (chorion); B = foetal capillary (in cross-section).

epitheliochorial placentation, strepsirhine primates have specialized areas of the chorion—in this case referred to as *chorionic vesicles*—which have a particular relationship to the outlets of uterine glands (Fig. 7). The composition of uterine milk has been relatively little studied in the placental mammals, and it remains to be seen whether the chorionic vesicles of the strepsirhine placenta play a special part in the transfer of essential polyunsaturated fatty acids from the mother for foetal brain development. If it is the case, as suggested by Crawford *et al.* (1976), that these essential fatty acids are vital for the normal development of the foetal mammal brain, examination of the placental transfer system by which long-chain polyunsaturated fatty acids reach the foetal circulation could be of great value to medical research. Once again, a combination of "divergent" and "convergent" approaches has a great deal to offer in the process of scientific investigation.

STRESS AND REPRODUCTION IN CAPTIVITY

Laboratory mammals are frequently treated as if they were relatively impermeable containers of certain tissues or molecules of interest in particular investigations. Nowhere in the field of comparative zoology can one find a greater potential source of experimental error and misinterpretation. Animals are complex systems in continuous interaction with the surrounding environment, and this must be taken into account in any work concerned with the elucidation of biological mechanisms. In addition, close attention must be paid to environmental conditions in any attempt to breed animals in captivity. Although this may seem an obvious point, there are many cases where insufficient consideration of environmental factors in captivity has led to acute problems.

A very large proportion of the problems involved in maintaining animals in captivity for scientific study can be summed up in one word—*stress*. This word, which has become so commonplace in recent years, but which is not often clearly understood, owes its scientific value much to the endeavours of one man—Hans Selye. Through his early work on stress in animals, and his eventual establishment of an institute concerned specifically with stress research, a great deal of useful knowledge has been accumulated (e.g. see Selye,1950, 1951, 1959, 1970 and Selye, Collip & Thomson, 1934, 1935). Yet it is important to note that Selye himself, in a semi-autobiographical account of his research on the subject (1956), traces his early recognition of the importance of stress to a particular incident involving a striking example of "divergent thinking". Apparently, as a medical student, he was confronted with a standard examination in which the ailments of a number of patients were to be identified from their specific symptoms. In fact, Selye was impressed more by the symptoms *shared* by all of the patients, because of the fact that they were all ill, rather than by the symptoms specific to each medical condition. His interest was aroused by the fact that all human beings are stressed when affected by illness of any kind.

The concept of stress has been invoked in so many different situations that it is easy to doubt its scientific validity. While it is true that the stress concept must be carefully defined for scientific purposes, as has been pointed out by Selye himself (1956), there can be no doubt of the practical value of recognition of stress as a problem both with respect to the maintenance of animals in captivity and in terms of various human medical conditions. Certainly, in searching for a set of scientific principles for the effective maintenance and breeding of mammals in captivity, it is essential that sources of stress should be acknowledged. Indeed, one of the reasons why the stress concept seems to many people to be so diffuse is that its ramifications are so extensive.

In terms of the transfer of animals from their natural habitats into captivity, it is relatively simple to define what is meant by "stress". The harmful effects of certain aspects of captivity on mammals can be viewed as the result of long-term activation of physiological mechanisms actually adapted for short-term response to emergency conditions in the natural environment. Many of the pathological effects observed with caged mammals, for example, can be attributed to stressful features of the surroundings (stressors), involving both the physical constraints of captivity and interactions with conspecifics. The symptoms themselves are a reflection of physiological processes concerned with the re-establishment of disturbed equilibrium states (hence Selye's concept of the "general adaptation syndrome"—1956). One of the well known side effects of stress is reduced resistance to infection, and for this reason it is somewhat unproductive to consider the health of animals in captivity without taking stress into account (Selye, 1951; Snyder, 1975). The nature of stress, and its relationship to defence mechanisms operating under natural conditions, has been neatly summarized by Snyder (1975) as follows: "The desirable goal of internal stability is not achieved without sacrificing functions less immediately vital to the individual, such as growth, reproduction, and resistance to infectious disease and parasitism". At the physiological level, stress involves increased activity of the autonomic nervous system and of the pituitary-adrenal axis. At the level of the whole animal, stress may be reflected in aberrant body weight, aberrant behaviour, disruption of reproduction and increased susceptibility to disease. In an elegant series of studies on the common tree-shrew (*Tupaia belangeri*), for example, von Holst has demonstrated stress effects at both levels as a function of interactions with conspecifics in captivity. The effects are so far-reaching that in certain cases, such as exposure of a submissive male to threat of attack from a dominant male, death can ensue within two weeks as a result of complete arrest of renal function (von Holst, 1969, 1972a,b, 1974).

Stressors arising from the physical conditions of captivity can have equally far-reaching effects. It has long been established that noise in the laboratory can lead to marked hyperactivity of the physiological mechanisms for "fight, fright and flight", with consequent disruption of reproduction (Arvay, 1967; Pfaff, 1974; Sackler *et al.*, 1959; Zondek & Tamari, 1960, 1967). However, it has not been widely appreciated that even relatively slight exposure to routine laboratory noises can completely upset breeding performance of mammals in captivity. Circumstantial evidence for this was found with a breeding colony of tree shrews (*Tupaia belangeri*) maintained by the author in a standard animal room originally designed for laboratory rats and mice. The reproductive behaviour of these tree shrews is highly unusual in that the female leaves her infants (usually two per litter) in a separate nest and returns to suckle them only once every 48 hours. It was known that stress could lead to disruption of this pattern, in the mildest cases increasing the frequency of suckling while decreasing the amount of milk given every 48 hours, and in extreme cases leading to death of the infants by cannibalism (von Holst, 1969). In the course of routine studies it was noted that a female *T. belangeri* which had been suckling her infants regularly every 48 hours abruptly began to suckle every 24 hours instead. The 48-hour pattern was then re-established, only to break down a second time. On enquiry, it was found that the fire alarm in the corridor outside the animal room had been tested (in each case for only 30 seconds) on the two occasions on which disruption of the suckling pattern had commenced (D'Souza & Martin, 1974). As it happened, the alarm had been tested on days when suckling was not due to occur, and in each case the female had had at least 24 hours to recover before visiting the infants to suckle. Otherwise, the effects of the fire alarm tests could have been more far-reaching.

In fact, the animal technician in charge of the animal rooms was already well aware of the disruption caused by fire alarm tests, since far higher mortality of infants had been recorded with the colonies of rats and mice maintained there. His normal practice was to muffle the fire alarms on the day of the test so that the noise level was greatly reduced. However, with the two tests which affected the tree shrews no advance warning was obtained, and all of the animal rooms were subjected to the full intensity alarm. The effects on the breeding output of the rats and mice can be clearly seen from Fig. 8. The breeding rats and mice were maintained in batches for 35 weeks at a time, and the colony was then totally replaced. The batch affected by the full intensity fire alarm tests (30 seconds duration in each case) showed an immediate sharp rise in infant mortality due to killing and cannibalism of pups. The mortality level dropped only very slowly over the next 26 weeks. Muffling of the fire alarm with a cloth between the clapper and the bell virtually suppressed this effect. With a second batch of rats and mice, it was suggested that the first fire alarm test should be muffled, whilst in the second the bell should be rung at full intensity in two short bursts, like a telephone bell. With this latter type of test (Fig. 8B), a mild increase in mortality was recorded—particularly with the mice—but the effect was by no means as pronounced as with the full 30-second fire alarm test (Fig. 8A). It is therefore clear that routine testing of standard laboratory fire alarms can have a disastrous effect on breeding performance of small mammals in captivity. Still worse, since this fact has not been widely recognized, one can only speculate on the effects which may have been exerted by fire alarm testing in the past during experiments conducted with laboratory rats and mice. Standardization of laboratory experimental procedure must obviously include standardization of environmental stressors which are known to operate, such as sporadic loud noises, introduction of visitors to the premises, alteration of the numbers of animals in the room, and so on. Once again, an understanding of the whole animal and its responses is a prerequisite for reliable experimental investigation of individual physiological mechanisms in the living animal.

As a post-script to this cautionary tale on the maintenance of laboratory mammal colonies, it is worth mentioning that two scientists working at the Medical Research Council Laboratory Animals Centre (Fasham & Clough, 1975; Clough & Fasham,

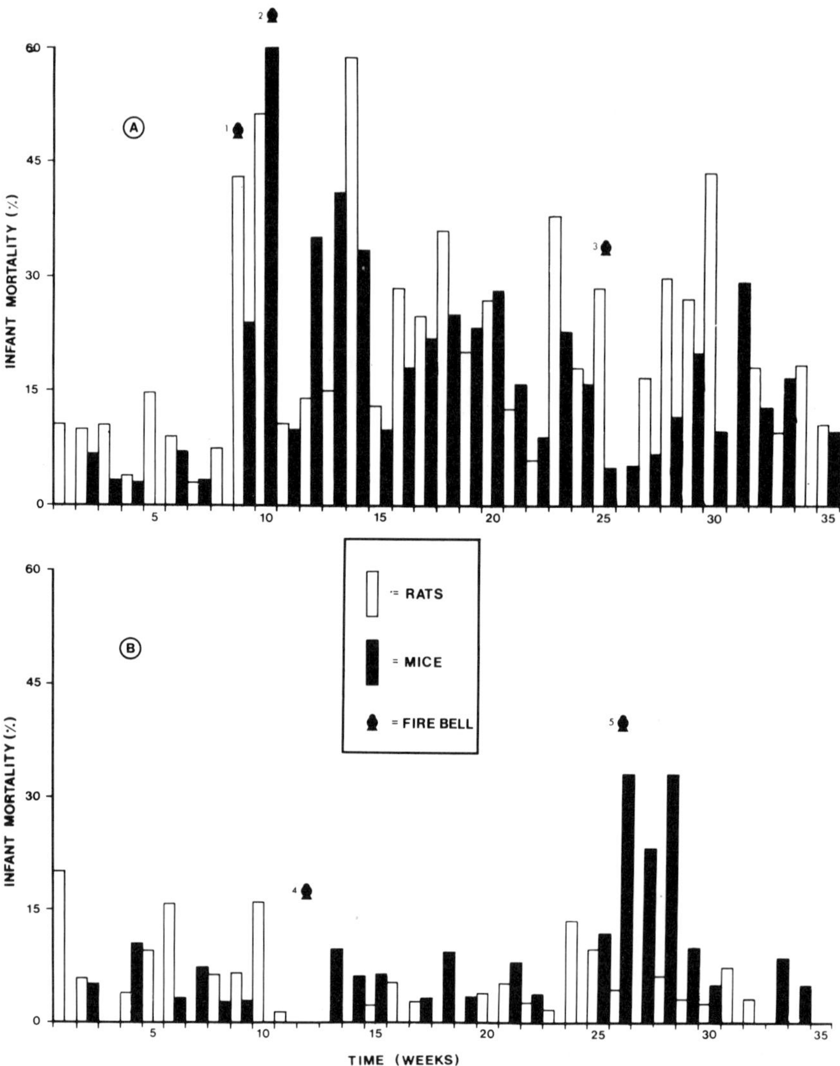

FIG. 8. Histograms showing the effects of fire alarm tests on rat and mouse breeding colonies (36 breeding females of each species). Batch Ⓐ: For 7 weeks prior to the testing of the fire alarm at full intensity (bells 1 and 2) infant mortality was at the usual relatively low level (approx. 10%). When the tests were conducted in weeks 8 and 9, mortality increased to 50–60% and declined only slowly until week 35. A third alarm test (bell 3) with the fire bell muffled did not have a noticeable effect on mortality. Batch Ⓑ: With a different batch of breeding rats and mice kept under identical conditions, mortality was low throughout, even after a muffled fire alarm test (bell 4). When the fire alarm bell was rung at full intensity, but in two short bursts like a telephone (bell 5), there was some increase in mortality, but less than with the full 30-second bursts which affected batch Ⓐ.

1975) have now designed a "silent" fire alarm which produces a loud, low-frequency noise which is easily audible to human beings, but which is outside the most sensitive frequency range of species commonly used as laboratory animals (e.g. mice, rats and rabbits). Preliminary tests indicate that the "silent" fire alarm does not adversely affect the breeding performance of laboratory rats (Gamble, 1976). This is a good example of the way in which "strategic research" can lead on to "tactical research"; the new fire alarm is now commercially available for those who would prefer not to stress their laboratory rodents through periodic testing of alarm bells.

REPRODUCTION IN AN ECOLOGICAL CONTEXT

For the fourth example of the significance of basic research on reproduction, it is particularly apt to consider briefly the ecological context in which the natural breeding of animals takes place. This has already been the subject of a stimulating book written by a former research fellow of the Wellcome Institute of Comparative Physiology (Sadleir, 1969), and it is emphasized in the contribution by Jewell (also a former member of the research staff) in this volume. Ultimately, no reproductive process can be fully understood until its evolutionary background has been properly investigated, and it is in this context that field studies have a major part to play. It is here, too, that the integration of ecological, ethological and evolutionary studies has the most to offer. In practical terms, such studies are essential for long-term planning of conservation, and it is gradually becoming clear that human reproduction itself must eventually be viewed in its original natural setting if we are to ensure that future medical intervention is to be appropriately managed (e.g. see Short, 1974).

There are many aspects of the ecology of reproduction which could be mentioned, but one notable example concerns the overall pattern of reproductive parameters exhibited by each mammalian species. It has long been recognized (see Martin, 1975b, for summary) that mammals generally tend to produce infants which fall into two fairly distinct categories. Some mammal species produce quite large litters of "altricial" infants which are relatively poorly developed at birth. The eyes and ears are closed by membranes, the skull is still at an early stage of development, the pelage has barely commenced its growth, and powers of thermoregulation are still poorly developed. This applies, for example, to most insectivores, most rodents and the smaller carnivores. Other mammals produce small litters, typically only one infant, of a "precocial" type. The eyes and ears are open at birth or soon afterwards, the skull is well developed, the body hair is usually obvious, and thermoregulatory powers are characteristically quite marked. Such infants are typically produced by ungulates, cetaceans and primates, among others. Although this dichotomy in infant types is not absolute, it is clear enough to invite explanation.

Allometric analysis is required to interpret other reproductive parameters. Once the body size of the mother is taken into account, it emerges that altricial infants are produced after a relatively short gestation period and have relatively small brains, while precocial infants are produced after a relatively long gestation period and have relatively large brains. In sum, mothers which produce altricial infants are specialized for quantity rather than quality, while precocial infants are born in very small numbers but gain from increased maternal care and greater elaboration of the central nervous system. In association with this, ovulation in females producing altricial infants is commonly dependent on mating, in combination with a short oestrous cycle (hence increasing reproductive turnover), whereas ovulation tends to be spontaneous in mothers of precocial infants, in combination with a long oestrous cycle (see Martin, 1975b). Finally, the duration of maternal care, the time taken to reach sexual maturity, and longevity are all longer in species with precocial infants.

An explanation of all of these facts is provided by the concepts of *r-selection* and *K-selection* developed by McArthur and Wilson (see Wilson, 1975; Pianka, 1970). The suggestion is that in environments where there is little effective crowding and relatively abundant food, natural selection will favour those animal species which harvest the most food and rear the largest families (r-selection), whereas when there is intense competition selection will favour those animals which can replace themselves with the lowest possible intake of food (K-selection). With r-selection, since food supply is not a limiting factor, small body size and high reproductive turnover is the rule; with K-selection, a population of animals of any given species is consistently close to the carrying capacity of the environment, and low reproductive turnover with greater efficiency in the conversion of food into offspring is encouraged, along with larger body size. It is obvious that unstable, unpredictable environments would tend to be correlated with r-selection, since small-bodied species which can rapidly colonize an area would be at an advantage; local populations would either be drastically reduced by temporary adverse conditions, or they would be in the process of colonizing an area with a temporarily superabundant food supply. Stable, predictable environments (such as primary tropical rainforest and the ocean depths), on the other hand, would be correlated with K-selection. Competition between well established species would be intense, and efficiency of conversion of food sources (enhanced by larger body size) would be at a premium. For example, the primates, as a group, would seem to have been derived from a common ancestor which was subjected to K-selection and hence developed all of the basic characteristics associated with low reproductive turnover (Martin, 1975c). Although there is some variation among modern primates, largely as a reflection of differences in body size, there is a fundamental pattern of reproductive characters common to all members of the order.

Such broad evolutionary interpretation is theoretically interesting in its own right, provided that there are adequate attempts to test the hypotheses involved. However, there are also practical features which are relevant to the central theme of this contribution. Firstly, in planning conservation programmes it is unlikely that we will have the time or the resources to conduct individual field studies on all of the species concerned in order to have a basic framework for scientific interpretation. A general ecological theory which covers various aspects of reproduction and relates them to environmental parameters would be of immense value in predicting future trends with populations of endangered species. Secondly, it is inherent in the concept of stress that the pathological symptoms frequently recorded in captivity are the outcome of natural mechanisms which misfire under unnatural conditions. Although it is unlikely that the adverse effects seen in captivity are common under natural conditions, there is undoubtedly a range of natural response to stressors for any mammal species in the wild. It has already been suggested that stress effects may operate in the natural environment to limit population size in certain mammal species (Christian, 1963), and it would be unusual if there were not some natural system for appropriate response to overpopulation (see also Wynne-Edwards, 1962). There is hence a vast field of research still to be conducted on the relationship between overpopulation and stress in animal populations in the wild, and this must be relevant in some ways to the processes operating in human society at different population densities. To return to the concepts of r-selection and K-selection, it would seem likely that small-bodied species with high reproductive turnover (such as the majority of rodents) would be more prone to respond in a dramatic way to local over-population. This would, in effect, be necessary as a counter-balance to the usual pattern of rapid population expansion with relatively inefficient utilization of food resources. With larger-bodied species having low intrinsic rates of population growth (such as most primates), intense competition for food resources would be present almost continuously, if the K-selection hypothesis is valid. Hence, there would be no requirement for natural spectacular population control mechanisms, and stress effects would be expected to be far more subtle and incorporated into the typical pattern of existence. As a primate, man has been adapted over a long period of time to match reproductive turnover to efficient utilization of food resources in competition with other K-selected species and natural parasites. The present dilemma of human populations is that these other species are no longer effective competitors, and the food resources themselves can be manipulated to attain higher levels of productivity. Out of his natural context, man's reproductive potential is no longer in balance with past evolutionary adaptations for life at a relatively stable population size and density.

CONCLUSIONS

The topics covered in this paper have been deliberately selected to span a broad range, in the attempt to demonstrate that comparative zoology is still as valuable today as it was 150 years ago when the Zoological Society of London was founded. Although the examples have been limited to the field of mammalian reproduction, the special interest of the Wellcome Institute of Comparative Physiology, this observation applies equally to many other areas of research. Zoology has drawn its strength from research which has effectively combined divergent and convergent thinking, laying a basis of strategic knowledge for the design of tactical research programmes. This combination continues to be significant for future developments.

A fusion of research on ecology, ethology and evolutionary biology—which can loosely be termed "whole animal studies"—represents a discrete discipline which deserves considerable support in its own right. This discipline provides a distinct viewpoint from which one can approach specific areas of study, such as that of mammalian reproduction, in terms of adaptations to the natural environment. Research on the effects of stress provides a good example, in that this relates to behavioural and physiological responses to adverse features of the environment. These responses evolved under natural conditions in order to maintain internal equilibria, and in the laboratory the unusual environmental conditions encountered lead to various pathological side effects. Eventually, we must come to understand man in this same light, as a self-domesticated species originally adapted for a given set of environmental conditions and now living in artificial environments of his own making.

Research on mammalian reproduction, in particular, also benefits from a consideration of natural adaptations. As has been shown above, such research conducted from that standpoint can make significant contributions in the future to the breeding of diverse mammal species in captivity, to furthering their conservation in the wild, and to increasing our understanding of human reproductive processes. Finally, it should not be forgotten that a major function of modern zoological collections lies in education, particularly with respect to conservation, and effective breeding and scientific research are vital to this role.

Acknowledgements

The views expressed in this paper have been developed over more than a decade of research on whole animals, during which time I have benefited from instruction and help given by numerous people. Particular acknowledgment is due to the late Professor N. A. Barnicot, whose advice stemmed from a broad approach to zoology combined with remarkable skill in specific techniques of research, and to my former supervisor, Professor N. Tinbergen, who has done so much to establish the point that whole animals can be admired in their natural context and scientifically studied at the same time.

Thanks also go to Dr P. Charles-Dominique, Dr J. Eisenberg, Professor P. Jewell, Dr Devra Kleiman, Mrs Frances D'Souza and Dr D. von Holst for invaluable discussions of the natural background to mammalian reproduction.

The research on gorillas would not have been possible without the enthusiastic co-operation of all the staff directly concerned at Jersey Zoo (Mr J. J. C. Mallinson, Mr P. Coffey, Mr Q. Bloxam and Mr J. Usher-Smith), Bristol Zoo (Mr M. Colbourne, Mr

J. Partridge, Mr M. Chorlton and Mr G. Cawdeary) and London Zoo (Dr M. Brambell, Mr G. F. Callard, Mr R. R. Smith, Mr M. Carman, Mr R. Hutton and Miss B. Radford). Advice and help were also provided at various stages by Dr Beryl Corner, Mr G. Greed, Mr P. McMullin, Miss Margaret Redshaw and Miss A. Glatston.

The radioimmunoassay technique and the creatinine assay were developed by Dr B. Seaton as part of a research project supported by the Medical Research Council. Miss Jane Lusty provided invaluable technical assistance. Sub-Human Primate Pregnancy Test Kits were kindly supplied on a trial basis by Dr G. Bialy of the Contraceptive Development Branch, National Institute of Child Health and Human Development, Bethesda, Maryland, USA.

Information on reproductive performance of rat and mouse breeding colonies in the Dept. of Physiology, University College London, was kindly provided by Mr D. M. Godfrey, from his carefully compiled records.

Finally, special thanks are due to Mr W. V. Holt for histological preparation, to Mr T. B. Dennett for photographic assistance in preparing the illustrations and to Miss Barbara Robertson for patient and skillful typing of the manuscript.

References

Amoroso, E. C. (1952). Placentation, In *Marshall's physiology of reproduction.* Parkes, A. S. (Ed.). London: Longmans, Green & Co.

Árvay, A. (1967). Effects of exteroceptive stimuli on fertility and their role in the genesis of malformations. In *Effects of external stimuli on reproduction.* Wolstenholme, G. E. W. & O'Connor, M. (Eds.). London: J. & A. Churchill Ltd.

Aschheim, S. & Zondek, B. (1928). Die Schwangerschaftsdiagnose aus dem Harn durch Nachweis des Hypophysenvorderlappenhormons. *Klin. Wochenschr.* **7**: 1404–1411.

Babladelis, G. (1975). Gorilla births in captivity. *Int. Zoo-News* **22**: 25–27.

Christian, J. J. (1963). The pathology of overpopulation. *Milit. Med.* **128**: 571–603.

Clegg, M. T. & Weaver, M. (1972). Chorionic gonadotropin secretion during pregnancy in the chimpanzee. *Proc. Soc. exp. Biol. Med.* **139**: 1170–1174.

Clough, G. & Fasham, J. A. L. (1975). A "silent" fire alarm. *Lab. Anim.* **9**: 193–196.

Coffey, P. & Pook, J. (1974). Breeding, hand-rearing and development of the third lowland gorilla, *Gorilla g. gorilla*, at the Jersey Zoological Park. *A. Rep. Jersey Wildl. Pres. Trust* **11**: 45–52.

Crawford, M. A., Hassam, A. G. & Williams, G. (1976). Essential fatty acids and fetal brain growth. *Lancet* **1976** (**i**): 452–453.

Dene, H., Goodman, M., Prychodko, W. & Moore, G. W. (1976). Immunodiffusion systematics of the Primates. III: The Strepsirhini. *Folia primatol.* **25**: 35–61.

Diczfalusy, E. & Mancuso, S. (1969). Oestrogen metabolism in pregnancy. In *Foetus and placenta.* Klopper, A. & Diszfalusy, E. (Eds.). Oxford: Blackwell.

D'Souza, F. & Martin, R. D. (1974). Maternal behaviour and the effects of stress in tree shrews. *Nature, Lond.* **251**, 309–311.

Fasham, J. A. L. & Clough, G. (1975). A fire alarm for laboratories and animal rooms. *Electron. Enginerg* **47**: 27–28.

Fisher, L. E. (1972). The birth of a lowland gorilla, *Gorilla g. gorilla.* at Lincoln Park Zoo, Chicago. *Int. Zoo Yb.* **12**: 106–108.

Frueh, R. J. (1968). A captive-born gorilla, *Gorilla g. gorilla,* at St. Louis Zoo. *Int. Zoo Yb.* **8**: 128–131.

Gamble, M. R. (1976). Fire alarms and fertility in rats. *Lab. Anim.* **10**: 161–163.

Grosser, O. (1909). *Vergleichende Anatomie und Entwicklungsgeschichte der Eihäute und der Placenta.* Vienna: Wilhelm Braumuller.

Hardin, C. J., Danford, D. & Skeldon, P. C. (1969). Notes on the successful breeding by incompatible gorillas at Toledo Zoo. *Int. Zoo Yb.* **9**: 84–88.

Hardin, C. J., Liebherr, G. & Fairchild, O. (1975). Artificial insemination in chimpanzees. *Int. Zoo. Yb.* **15**: 132–134.

Hill, W. C. O. (1953). *Primates. I: Strepsirhini.* London: Edinburgh University Press.

Hobson, B. (1976). The diagnosis of pregnancy in the lowland gorilla, *Gorilla g. gorilla* and the Sumatran orang-utan, *Pongo pygmaeus abelli. A. Rep. Jersey Wildl. Pres. Trust.* **12**: 71–75.

Hopper, B. R., Tullner, W. W. & Gray, C. W. (1968). Urinary estrogen excretion during pregnancy in a gorilla (*Gorilla gorilla*). *Proc. Soc. exp. Biol. Med.* **129**: 213–214.

Hudson, L. (1966). *Contrary imaginations.* London: Methuen & Co.

Hudson, L. (1968). *Frames of mind: Ability, perception and self-perception in the arts and sciences.* London: Methuen.

Jones, R. C. (1971). Uses of artificial insemination. *Nature, Lond.* **229**: 534–537.

Kagawa, M. & Kagawa, K. (1972). Breeding a lowland gorilla, *Gorilla g. gorilla,* at Ritsurin Park Zoo, Takamatsu. *Int. Zoo Yb.* **12**: 105–106.

Kuhn, T. S. (1970). *The structure of scientific revolutions.* Chicago: University of Chicago Press.

Lang, E. M. (1959). The birth of a gorilla at Basle Zoo. *Int. Zoo Yb* **1**: 3–7.

Lang, E. M. (1964). Jambo, the second gorilla born at Basle Zoo. *Int. Zoo. Yb.* **3**: 84–88.

Le Gros Clark, W. E. (1971). *The antecedents of man.* London: Edinburgh University Press.

Leutenegger, W. (1973). Maternal-fetal weight relationships in primates. *Folia primatol* **20**: 280–293.

Lotshaw, R. (1971). Births of two lowland gorillas, *Gorilla g. gorilla,* at Cincinnati Zoo. *Int. Zoo Yb.* **11**: 84–87.

Luckett, W. P. (1975). Ontogeny of the fetal membranes and placenta: their bearing on primate phylogeny. In *Phylogeny of the primates: a multidisciplinary approach.* Luckett, W. P. & Szalay, F. S. (Eds.). New York: Plenum Press.

Mallinson, J. J. C., Coffey, P. & Usher-Smith, J. (1973). Maintenance, breeding and hand-rearing or lowland gorilla, *Gorilla g. gorilla* (Savage & Wyman 1847) at the Jersey Zoological Park. *A. Rep. Jersey Wildl. Pres. Trust* **10**: 5–28.

Martin, R. D. (Ed.) (1975a). *Breeding endangered species in captivity.* London, New York and San Francisco: Academic Press.

Martin, R. D. (1975b). The bearing of reproductive behaviour and ontogeny on strepsirhine phylogeny. In *Phylogeny of the primates: a multidisciplinary approach.* Luckett, W. P. & Szalay, F. S. (Eds.). New York: Plenum Press.

Martin, R. D. (1975c). Strategies of reproduction. *Nat. Hist. (N.Y.)* **84**: 48–57.

Martin, R. D., Seaton, B. & Lusty, J. (1976). Application of urinary hormone determinations in the management of gorillas. *A. Rep. Jersey Wildl. Pres. Trust* **12**: 61–70.

Medawar, P. (1975). Julian Huxley: a tribute. *Nature, Lond.* **254**: 4.

Nadler, R. D. (1974). Determinants of variability in maternal behavior of captive female gorillas. *Symp. Congr. Int. Primate Soc.* **5**: 207–216.

Nadler, R. D. (1975). Second gorilla birth at the Yerkes Regional Primate Research Center. *Int. Zoo Yb.* **15**: 134–137.

Perry, J. S. (1971). *The ovarian cycle of mammals.* Edinburgh: Oliver & Boyd.

Pfaff, J. (1974). Noise as an environmental problem in the animal house. *Lab. Anim.* **8**: 347–354.

Pianka, E. R. (1970). On r- and K-selection. *Am. Nat.* **104**: 592–597.

Rao, L. G. S. (1976). Clinical usefulness of estrogen–creatinine ratios in random samples of urine as a feto-placental function test. *J. Endocr.* **68**: 40–41.

Riddle, K. E., Keeling, M. E. & Roberts, J. (1973). Birth of a lowland gorilla (*Gorilla gorilla gorilla*) at the Yerkes Regional Primate Research Center. *J. Zoo Anim. Med.* **4**: 22–27.

Rowlands, I. W. (1965). Artificial insemination of mammals in captivity. *Int. Zoo Yb.* **5**: 105–106.

Rowlands, I. W. (1974). Artificial insemination of mammals in captivity. *Int. Zoo Yb.* **14**: 230–233.

Rumbaugh, D. M. (1965). The birth of a lowland gorilla at the San Diego Zoo. *Zoonooz* **38**(9): 9, 12–17.

Rumbaugh, D. M. (1967). "Alvila"—San Diego Zoo's captive-born gorilla. *Int. Zoo Yb.* **7**: 98–107.

Ryan, K. J. & Hopper, B. R. (1974). Placental biosynthesis and metabolism of steroid hormones in primates. *Contr. Primat.* **3**: 258–283.

Sacher, G. A. & Staffeldt, E. F. (1974). Relation of gestation time to brain weight for placental mammals: implications for the theory of vertebrate growth. *Am. Nat.* **108**: 593–616.

Sackler, A. M., Weltman, A. S., Bradshaw, M. & Jurtshcuk, P. Jr. (1959). Endocrine changes due to auditory stress. *Acta endocr., Copenh.* **31**: 405–418.

Sadleir, R. M. F. S. (1969). *The ecology of reproduction in wild and domestic animals.* London: Methuen.

Seaton, B. & Lusty, J. (1976). A new approach to radioimmunoassay methodology for steroid hormones. *J. Endocr.* **68**: 36–37.

Seaton, B., Lusty, J. & Watson, J. (in press). A practical model for steroid hormone radioimmunoassays. *J. Steroid Biochem.* **7**.

Selye, H. (1950). *The physiology and pathology of exposure to stress.* Montreal: Acta Inc.

Selye, H. (1951). The influence of STH, ACTH and cortisone upon resistance to infection. *Can. Med. Assoc. J.* **64**: 489–494.

Selye, H. (1956). *The stress of life.* London: Longmans, Green & Co.

Selye, H. (1959). Perspectives in stress research. *Perspect. Biol. Med.* **2**: 403–416.

Selye, H. (1970). *Experimental cardiovascular diseases.* New York: Springer-Verlag.

Selye, H., Collip, J. B. & Thomson, D. L. (1934). Nervous and hormonal factors in lactation. *Endocrinology* **18**: 237.
Selye, H., Collip, J. B. & Thomson, D. L. (1935). Endocrine interrelations during pregnancy. *Endocrinology* **19**: 151.
Short, R. V. (1974). Man, the changing animal. In *Physiology and genetics of reproduction*. Coutinho, E. M. & Fuchs, F. (Eds.). New York: Plenum Pub. Co.
Slot, C. (1965). Plasma creatinine determination: a new and specific Jaffe reaction method. *Scand. J. clin. Lab. Invest.* **17**: 381–387.
Snyder, R. L. (1975). Behavioural stress in captive animals. In *Research in zoos and aquariums*. Washington: National Academy of Sciences.
Stoker, M. (1975). Limits to oncology. *Nature, Lond.* **254**: 547–548.
Taussky, H. H. (1956). A procedure increasing the specificity of the Jaffe reaction for the determination of creatine and creatinine in urine and plasma. *Clin. Chim. Acta* **1**: 210–224.
Thomas, W. D. (1958). Observations on the breeding in captivity of a pair of lowland gorillas. *Zoologica, N.Y.* **43**: 95–103.
Tijskens, J. (1971). The oestrous cycle and gestation period of the mountain gorilla. *Int. Zoo. Yb.* **11**: 181–183.
Tullner, W. W. (1974). Comparative aspects of primate chorionic gonadotropins, *Contr. Primat.* **3**: 235–257.
Tullner, W. W. & Gray, C. W. (1968). Chorionic gonadotrophin excretion during pregnancy in a gorilla. *Proc. Soc. exp. Biol. Med.* **128**: 954–956.
von Holst, D. (1969). Sozialer Stress bei Tupajas (*Tupaia belangeri*). *Z. vergl. Physiol.* **63**: 1–58.
von Holst, D. (1972a). Renal failure as the cause of death in *Tupaia belangeri* exposed to persistent social stress. *Z. vergl. Physiol.* **78**: 236–273.
von Holst, D. (1972b). Die Funktion der Nebennieren männlicher *Tupaia belangeri*. *Z. vergl. Physiol* **78**: 289–306.
von Holst, D. (1974). Social stress in the tree shrew (*Tupaia belangeri*) its causes and physiological and ethological consequences. In *Prosimian biology*. Martin, R. D., Doyle, G. A. & Walker, A. C. (Eds.). London: Duckworth.
Warner, H., Martin, D. W. & Keeling, M. E. (1974). Electroejaculation of the great apes. *Ann. Biomed. Eng.* **2**: 419–432.
Wilson, E. O. (1975). *Sociobiology: the new synthesis*. Cambridge, Mass: Belknap Press.
Wynne-Edwards, V. C. (1962). *Animal dispersion in relation to social behavior*. Edinburgh: Oliver & Boyd.
Zondek, B. (1931). *Die Hormone des Ovariums und des Hypophysenvorderlappens*. Berlin.
Zondek, B. & Tamari, I. (1960). Effect of audiogenic stimulation on genital function and reproduction. *Am. J. Obstet. Gynecol.* **80**: 1041–1048.
Zondek, B. & Tamari, I. (1967). Effects of auditory stimuli on reproduction. In *Effects of external stimuli on reproduction*. Wolstenholme, G. E. W. & O'Connor, M. (Eds.). London: J. & A. Churchill Ltd.
Zuckerman, S. (1935). The Aschheim–Zondek diagnosis of pregnancy in the chimpanzee. *Am. J. Physiol.* **110**: 597–601.

Symp. zool. Soc. Lond. (1976) No. 40, 321–333.

THE INTRODUCTION OF NEW SPECIES OF ANIMALS FOR THE PURPOSE OF DOMESTICATION

R. V. SHORT

MRC Unit of Reproductive Biology, Edinburgh, Scotland

SYNOPSIS

One of the principal objectives of the Zoological Society of London, as set out by Sir Stamford Raffles in the Prospectus, was "the introduction of new varieties, breeds, and races of animals for the purpose of domestication". That aim has never been achieved by any zoo anywhere in the world; the time is now opportune for a new initiative.

Wild animals undoubtedly possess a number of genes that could be exploited to great advantage by our domestic stock. It is unrealistic to attempt to domesticate wild animals, as this would require many generations of selection for "tameness". Results could be achieved far more rapidly by grafting "wild genes" onto stock of proven domestic temperament by interspecific hybridization.

Tropical and polar species may have the most to offer, the former because of their lack of breeding seasons, and the latter because of their large body size, rapid growth rate, and efficient food conversion. Ways are discussed by which an infusion of "wild genes" might improve the performance of existing domestic breeds of geese, ducks and sheep.

INTRODUCTION

It is timely to remind ourselves of the original aims and objectives of the Zoological Society of London, as set out by Sir Stamford Raffles on 1st March 1825, in a document entitled:

> "Prospectus of a Society for introducing and domesticating new Breeds or Varieties of Animals, such as Quadrupeds, Birds, or Fishes, likely to be useful in Common Life; and for forming a General Collection in Zoology."

Here, it is clearly stated that:

> "The great objects should be, the introduction of new varieties, breeds, and races of animals for the purpose of domestication or for stocking our farm-yards, woods, pleasure-grounds, and wastes; with the establishment of a general Zoological Collection, consisting of prepared specimens in the different classes and orders, so as to afford a correct view of the Animal Kingdom at large in as complete a series as may be practicable, and at the same time point out the analogies between the animals already domesticated, and those which are similar in character upon which the first experiments may be made."

Raffles then went on to point out that:

> "When it is considered how few amongst the immense variety of animated beings have been hitherto applied to the uses of Man, and

> that most of those which have been domesticated or subdued belong to the early periods of society, and to the efforts of savage or uncultivated nations, it is impossible not to hope for many new, brilliant and useful results in the same field, by the application of the wealth, ingenuity, and varied resources of a civilized people."
>
> "It would well become Britain to offer another, and a very different series of exhibitions to the population of her metropolis; namely, animals brought from every part of the globe to be applied either to some useful purpose, or as objects of scientific research, not of vulgar admiration. Upon such an institution, a philosophy of Zoology may be founded, pointing out the comparative anatomy, the habits of life, the improvement and the methods of multiplying those races of animals which are most useful to man, and thus fixing a most beautiful and important branch of knowledge on the permanent basis of direct utility."

The 150 years that have elapsed since the foundation of the Zoological Society of London have seen the establishment of a plethora of Zoos and Wildlife Parks all over the world, and yet we have not succeeded in domesticating a single new species of wild animal. It is true that we have a number of new laboratory animals, such as the gerbil and the hamster, and attempts are being made to farm wild animals such as the eland (*Taurotragus oryx*) and the red deer (*Cervus elaphus*), but they are still not truly domesticated, and the enterprises themselves have yet to be proved to be economically viable. In these days of increased Governmental and public demand for more research with a direct relevance to the practical problems that face us, the time has surely come when we should re-examine the whole question of the contribution that the Zoological Society of London could make to the production of new varieties of domestic livestock.

WHAT DETERMINED MAN'S INITIAL CHOICE OF SPECIES TO DOMESTICATE?

Undoubtedly it was expediency. Animals were probably first domesticated in about 7000–8000 B.C. by the early agriculturalists in the Mediterranean basin and western Asia, and it is not surprising that the indigenous wild animals of those regions were man's first choice: the sheep and goat, followed by the cow, horse, pig and donkey as agriculture began to spread further afield (Zeuner, 1963). At a guess, one would imagine that temperament must have been the first character for which selection was unconsciously practised; a wild animal that did not become tame even after hand-rearing could never be domesticated. Although some may think it strange that behaviour should be subject to genetic control, and although admittedly there is little evidence on this point—no doubt because of the difficulty of quantitating behaviour—geneticists are already well acquainted with the marked behavioural differences that exist between various highly inbred strains of mice. The importance of behaviour in domestication is also demonstrated by the fact that male castration is one of the earliest known agricultural practices; it was undoubtedly a very successful way of manipulating an animal's behaviour to man's advantage.

DO WE NEED NEW SPECIES OF DOMESTIC ANIMALS?

If we allow a somewhat liberal interpretation of the word "species", the answer to this question must undoubtedly be "yes".

At the very bottom of the list of factors that influenced early man in his choice of species to domesticate were those economic attributes that we now rate most highly, such as rapid growth rate and high food-conversion ratio. Bequeathed the legacy of early man's purely opportunistic choice of species, the animal breeder of today is faced with the task of maximizing the efficiency of production in these same animals, only to discover that they may be ill-suited to the demands of present-day agriculture.

In the first place, within recent years we have selected so intensely and efficiently for characters of economic importance that much of the natural variability has been eliminated, so the characters themselves now have relatively low heritabilities. An infusion of new genes from the wild might open up exciting new possibilities for selection.

Secondly, since our domesticated stock were derived originally from the temperate regions, it is not surprising that many of them continue to be seasonal breeders. As agriculture spread northwards, the tendency towards seasonal breeding may even have been accentuated. Advances in animal housing, feeding and husbandry have given us such a measure of control over the external environment that seasonal breeding is now becoming a positive disadvantage. For example, it would be highly desirable to develop breeds of sheep that were capable of giving birth to lambs throughout the year, not just in the spring.

WHY HAVE WE FAILED TO DOMESTICATE NEW SPECIES IN RECENT TIMES?

If today's farm livestock have a history of up to 10 000 years selection by man for domesticity, it is naïve to imagine that we could repeat the domestication process in a wild animal after only a few generations of selection. It would be far more realistic to try and graft the desirable genes from the wild species onto a domestic stock background by the process of hybridization. Hitherto, this has posed almost insuperable practical difficulties. Behavioural barriers effectively prevented interspecific matings from taking place, even in captivity. Zoos have been ill-equipped and unwilling to house and rear domestic livestock together with the wild animals in their collections, solely for the purpose of producing interspecific hybrids. Even when hybrids were produced, zoos were not in a position to evaluate their economic attributes. Many interspecific hybrids are either completely sterile, like the mule (*Equus asinus* × *Equus caballus*), or show a severe reduction in fertility, especially amongst the males, as in the catallo (*Bos taurus* × *Bison bison*) or the yakow (*Bos taurus* × *Bos grunniens*) (Gray, 1972), and this makes them unattractive commercially.

A number of recent scientific advances have totally changed the picture, and opened up exciting new opportunities in this field. The development of a range of drugs and projectiles has now made it possible to restrain captive wild animals at will with relative safety. Electroejaculation techniques permit the collection of semen from anaesthetized animals. Avian or mammalian semen can be deep-frozen in liquid nitrogen for long periods of time without losing its fertilizing ability, and the insemination of cattle, sheep, pigs and poultry is now a routine procedure. The stage is therefore set for promoting the flow of "wild" genes into our domestic stock. This is much more likely to be successful, and of direct benefit to agriculture, than *de novo* domestication of any existing wild animal.

THEORETICAL POSSIBILITIES FOR THE EXPLOITATION OF GENES FROM DIFFERENT LATITUDES AND ALTITUDES

It seems likely that we have already exploited to the full the indigenous genes of our domestic livestock which adapted them to their original temperate environment. Whilst having a certain romantic appeal to conservationists, there is probably little point in preserving rare domestic breeds for the sake of their genetic potential. Their very scarcity is an indication that time has passed them by; they are as it were the waste products of the domestication process. But we could now with advantage "borrow" genes from more tropical and polar species for infusion into existing domestic stock.

Tropical species are not usually seasonal breeders, and even when transported to temperate zones their reproduction may continue to be aseasonal. Good examples are the chital or axis deer (*Axis axis*) of India, the Barbary sheep (*Ammotragus lervia*) of N. Africa, and the eland of central and southern Africa, all of which breed throughout the year in their natural habitats, and continue to do so even when introduced to northerly latitudes such as Britain (Zuckerman, 1952). The introduction of "tropical genes" into a domestic species might therefore be expected to extend its mating season.

In contrast to this, animals living in the polar regions, or at high altitudes in the temperate regions, would be expected to have very restricted breeding seasons, an undesirable characteristic as far as domestication is concerned. Nevertheless they would theoretically have a number of highly desirable attributes, such as a large body size, evolved as a way of minimizing heat loss, and a rapid growth rate and high food-conversion ratio, associated with the need to reach maturity in the short summer growing season. So the introduction of "polar genes" or "high altitude" genes into a domestic species might be expected to increase body size and improve growth rates and efficiency of food conversion. The ideal domestic animal could therefore be a man-made blend of desirable genes selected for under environmental extremes, and grafted onto stock of proven domestic temperament.

IMPROVEMENT OF THE DOMESTIC GOOSE

It has recently been pointed out (Kear, 1973; Murton & Kear, 1973; Kear, 1975) that the domestic Embden or Toulouse-type goose (*Anser anser*), which is probably derived from the wild greylag, could probably be improved by an infusion of genes from high-arctic breeding species such as the greater snow goose (*Anser caerulescens*) or the redbreasted goose (*Branta ruficollis*). These wild species have incubation periods of only 23–24 days, as compared with the 33–35 days of the domestic Embden, and have a very rapid growth rate and an excellent efficiency of food conversion. The redbreasted goose, for example, attains 17·7 times its hatching weight by three weeks of age, which is about twice the growth rate of the domestic goose. The bar-head goose (*Anser indicus*), which breeds at high altitudes, has an advantage over the high latitude species in that it has a long breeding season. Even the domesticated Chinese goose (*Anser cygnoides*) has advantages that could be exploited with profit by the larger Embden, since it comes into lay earlier, and will lay again in the autumn on a decreasing daylength, a feature only found in low latitude species (Murton & Kear, 1973). The most spectacular example of this is seen in the ne-ne goose from Hawaii (*Branta sandvicensis*), an endangered species, which actually lays its eggs in winter on short days. Although it is a slow-growing species, it is the only wholly land-dwelling goose; it can copulate effectively on land, and is unlikely to have developed the heavy subcutaneous fat deposits that other geese have evolved as an insulation against the cold water.

There are about 33 wild species and subspecies of geese belonging to the genera *Anser* and *Branta*, all of which produce fertile hybrids when crossed with the domestic goose (Gray, 1958). Unlike chickens, geese have the great advantage that they can be reared on grass protein alone. Surely some genetic engineering with the goose might produce a bird that was far more useful to man, and far less demanding to feed, than the battery hen.

IMPROVEMENT OF THE DOMESTIC DUCK

The domestic duck (*Anas platyrhynchos*) is almost certainly derived from the low temperate zone mallard, whereas the Muscovy duck (*Carinia moschata*) was probably domesticated in central South America. Kear, 1975) has pointed out that the disadvantages of the domestic duck are its dependence on water for successful copulation, an essentially monogamous mating system, the deposition of large amounts of subcutaneous fat below the swimline, a large bone : meat ratio in the carcass, a long incubation period of 28 days, and seasonal breeding confined to the spring. The Muscovy overcomes some of these disadvantages, since it is a larger bird and possibly a better food converter, although its growth rate is slow. The male is twice as large as the female, presumably the result of a polygynous mating system. Although the egg-laying season is long, so is the incubation period, which is 33–35 days. Unfortunately, the Muscovy–domestic duck hybrid only lays small, infertile eggs, so there is no scope for further selection (Gray, 1958).

When we look around the 50 or more wild species of *Anas*, most of which will produce fertile hybrids when crossed with the mallard (Gray, 1958), there appear to be a number of exploitable advantages. Many of the tropical species have a long laying season, and do not go out of lay in the summer if transported to temperate latitudes. The Laysan teal (*Anas laysanensis*) almost never swims, and so probably would not develop so much subcutaneous fat. Many of the arctic species have incubation periods as short as 22 days, fast growth rates, and probably excellent food conversion ratios. Although not as efficient as the goose in being able to subsist entirely on vegetable protein, the mallard needs only 8% of its total 18% protein requirement to be of animal origin, which is a considerable improvement on the hen or turkey (Holm & Scott, 1954). The domestic duck, like the goose, could undoubtedly be improved still further by an infusion of "wild" genes.

IMPROVEMENT OF THE DOMESTIC SHEEP

There are literally hundreds of domesticated breeds of sheep (*Ovis aries*), and they all probably originate from either the European mouflon (*Ovis musimon*) of the Mediterranean basin or the Asiatic mouflon (*Ovis orientalis*) of Turkey and Iran, both of which still exist in the wild state. This supposition is strengthened by the fact that the diploid chromosome number of all domestic breeds of sheep so far examined is 2n = 54, which is identical to that of the European and Asiatic mouflons (see Table I).

There are a number of other wild breeds of sheep in Asia which have different chromosome numbers, and do not appear to have contributed to domestic breeds (see Table I and Fig. 1). These include the urial (*Ovis vignei*; 2n = 58), which is the most southerly; the large argali types (*Ovis ammon*; 2n = 56), such as the Marco Polo sheep, which inhabit the high mountain plateaus of central Asia; and the snow sheep (*Ovis nivicola*; 2n = 52) of eastern Siberia.

The North American continent possesses two distinct species, the dall (*Ovis dalli*; 2n = 54) and the bighorn (*Ovis canadensis*; 2n = 54), both of which show distinct morphological similarities to the snow sheep, whilst having a karyotype that is indistinguishable from the domestic sheep.

All these European, Asiatic and North American wild species of sheep will produce fertile hybrids when crossed with the domestic sheep (Gray, 1972). One interesting exception is the Barbary sheep

TABLE I

Diploid chromosome number and number of chromosome arms (NF) of domestic and wild species of goats and sheep

Species	2n	NF
Domestic goat *Capra hircus*	60	60
Barbary sheep *Ammotragus lervia*	58	60
Urial sheep *Ovis vignei*	58	60
Argali sheep *Ovis ammon*	56	60
Dall sheep *Ovis dalli*	54	60
Bighorn sheep *Ovis canadensis*	54	60
Blue sheep *Pseudois nayaur*	54	60
Asiatic mouflon *Ovis orientalis*	54	60
European mouflon *Ovis musimon*	54	60
Domestic sheep *Ovis aries*	54	60
Snow sheep *Ovis nivicola*	52	60

For references, see Hard, 1969; Nadler, Lay & Hassinger, 1971; Nadler, Korobitsyna, Hoffman & Vorontsov, 1973; Nadler, Hoffman & Woolf, 1974; Korobitsyna, Nadler, Vorontsov & Hoffman, 1974.

of North Africa (*Ammotragus lervia*; 2n = 58) which is taxonomically more closely related to the goats than the sheep; fertile hybrids can occasionally be produced by mating a Barbary ewe with a male domestic goat (*Capra hircus*; 2n = 60), but if domestic sheep are inseminated with Barbary semen, fertilization takes place, but the embryos die at the early blastocyst stage (Gray, 1972). Likewise, if domestic goats are inseminated with semen from a domestic ram, fertilization occurs normally but the embryos invariably die by the second month of gestation (Gray, 1972). Chromosome number is not an invariable guide to successful hybridization, however. For example, the blue sheep (*Pseudois nayaur*) has the same chromosome number as the domestic sheep (see Table I), although they

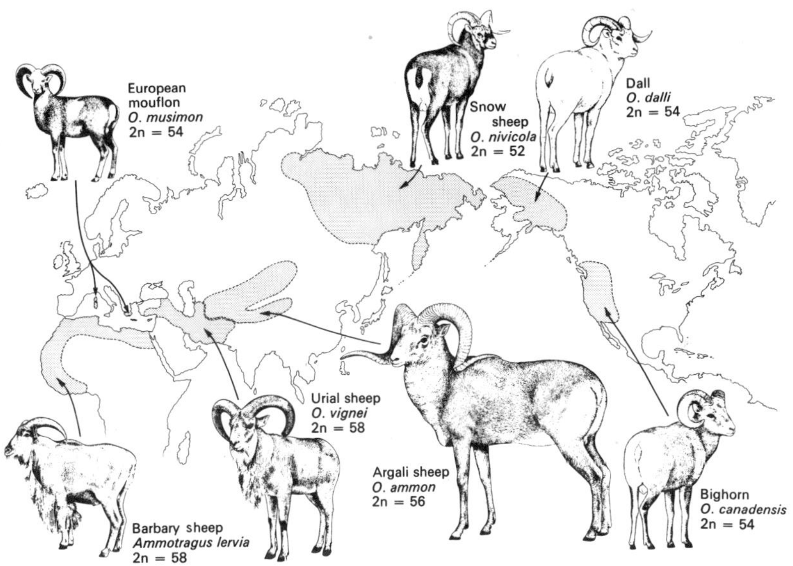

FIG. 1. Geographical distribution of some of the wild species of sheep, drawn to scale, and indicating their respective diploid chromosome numbers.

Reproduced with permission from Short, R. V. (1976). The origin of species. In Austin, C. R. & Short, R. V. (1976). *Reproduction in mammals,* **6**: The evolution of reproduction. Cambridge: Cambridge University Press.

will not hybridize with one another (Gray, 1972). However, full-term offspring have been produced following hybridization of the blue sheep with the domestic goat (Hard, 1969).

Thus it seems that there are a number of possibilities for producing fertile offspring by hybridizing wild species of sheep with our domesticated breeds. Although all these wild breeds, with the exception of the snow sheep, and many of their hybrids have been kept at Regent's Park at one time or another (Flower, 1929), we know extraordinarily little about them.

Although all our domestic breeds may have been derived originally from the mouflon, there could still be some benefit to be gained from back-crossing to this parental stock, since the records show that *O. musimon* has a very extended breeding season in Regent's Park, with births extending from January to November, with a peak in April (Zuckerman, 1952). However, the small body size (620 cm shoulder height) and hairy coat would be disadvantages that would need to be overcome.

The urial is the most southerly of all the wild sheep, and might be expected to show an even more extended breeding season than the mouflon; the information from animals bred at Regent's Park is too scanty to provide conclusive evidence on this point (Zuckerman, 1952). Although the urial is a relatively large animal, with a shoulder height of 850–920 cm (Lydekker, 1898), it might be expected to have a disappointingly slow growth rate, and it has a coat of hair, not wool.

The argali group (*Ovis ammon* sp.) are undoubtedly the most interesting, although they are unfortunately also the most inaccessible, coming from that politically sensitive area where four worlds meet: Afghanistan, Russia, China and Pakistan. They are by far the largest of all the sheep; the rams have massive horns, much prized by sportsmen (see Fig. 1), stand 1·1–1·2 m high at the shoulders, and weigh up to 136 kg. They live on the open grassy slopes of the high Pamirs, often at altitudes in excess of 6000 m, and their large body size is presumably an adaptation to this harsh environment.

Perhaps the best known and most spectacular of the subspecies is Marco Polo's sheep, *O. ammon poli.* It was first described in 1253 by Father William, the Franciscan friar of Rubruck in Flanders. Marco Polo himself saw the skulls of some of these animals in 1273, and the subspecies was named after him in 1758. The animal was "rediscovered" by Burnes in 1834, and Lt. Wood, R.N. collected two heads during the course of his expedition to the source of the River Oxus in 1838, and presented them respectively to the British Museum and the Royal College of Surgeons (Lydekker, 1898; Clark, 1964). Individual animals were occasionally kept in the Zoological Society's collection prior to 1900 (Flower, 1929), but there appear to be none in captivity anywhere in the world today. A recent survey carried out for FAO by R. G. Petocz (1973) suggests that there is a population of about 2900 animals in the Afghan Pamirs, and although sportsmen continue to take a heavy toll, in part of this area hunting is fortunately prohibited.

As one moves further east and north across the central Asian plateau, one encounters other subspecies of *O. ammon,* equally spectacular in size, but about which even less is known: the Tibetan argali, *O. ammon hodgsoni*; the Mongolian argali, *O. ammon darwini*; the species type, the Altai argali, *O. ammon ammon,* and many others. There are a few Mongolian argalis in captivity in Peking Zoo. The only detailed information about hybrids of *O. ammon* with domestic sheep comes from Russia, where the smaller arkhar

(*O. ammon kaselini*) has been crossed with apparent success to produce a hybrid of commercial value (Gray, 1972).

It seems characteristic of British agriculture that we have gone in for miniaturization of our domestic animals, the Dexter cow, the Shetland pony and the Soay sheep being extreme examples. We are now beginning to realize the advantages of larger animals, and have had to import bigger beef and dairy cattle breeds from the continent of Europe at great expense. Perhaps an infusion of "maxi" genes would likewise benefit the sheep industry. If the red deer (which is about the size of the Marco Polo sheep) can survive in the Highlands of Scotland under conditions that are too severe for the small Scottish blackface sheep, it may be that larger sheep would be at an advantage. A reduction in surface area relative to body mass would cut down heat loss, which could be the key to survival on some of those bare, sodden, windswept hillsides. The Marco Polo sheep must almost certainly have developed a rapid growth rate and high efficiency of food conversion if it is to make the most of its transient summer; once again, these would be highly desirable characteristics to incorporate into our domestic sheep breeds. As to its disadvantages, these would be an absence of wool, and almost certainly a very restricted mating season.

The snow sheep of Siberia and the dall and bighorn of North America would appear to be of less potential use. They lack the advantage of a marked increase in body size, although they may well show rapid growth rates and good food conversion ratios. Their northerly latitude would almost certainly mean that they would only come into oestrus for a relatively short period of time late in the autumn.

Another possible advantage to be gained from the wild sheep would be the variation in chromosome number between the various species. Since the number of chromosome arms (nombre fondamentale: NF) of all the sheep and goats is 60 (see Table I) this strongly suggests that the variations in chromosome number are a result of the centric fusion (Robertsonian translocation) of acrocentric chromosomes, or the fission of metacentric ones, a conclusion that is supported by detailed studies of the chromosome banding patterns (Evans, Buckland & Sumner, 1973; Nadler, Hoffman & Woolf, 1974). This variability in chromosome morphology could obviously alter the linkage patterns of some genes and perhaps this could be exploited to advantage in certain situations.

CONCLUSIONS

Zoos, and in particular the Zoological Society of London with its Research Institutes and Whipsnade, could make a most valuable contribution at the present time to the improvement of domestic livestock. By maintaining in the collection pure-bred species of wild animals, and making use of them by natural or artificial means to produce hybrids with domestic breeds, it might be possible to infuse new and highly desirable genes into our domestic stock. As Sir Stamford Raffles so clearly stated over 150 years ago, the great objectives of the Zoological Society of London should be the introduction of new varieties, breeds and races of animals for the purpose of domestication. Animals should be brought from every part of the globe to be applied either to some useful purpose, or as objects of scientific research, not of vulgar admiration. Perhaps we could be forgiven if at the same time a spectacular display of Marco Polo sheep on the Mappin Terraces also happened to draw in the crowds.

REFERENCES

Clark, J. L. (1964). *The great arc of the wild sheep.* Norman, Okla: University of Oklahoma Press.

Evans, H. J., Buckland, R. A. & Sumner, A. T. (1973). Chromosome homology and heterochromatin in goat, sheep and ox studied by banding techniques. *Chromosoma* **42**: 383–402.

Flower, S. S. (1929). *List of the vertebrated animals exhibited in the gardens of the Zoological Society of London, 1828–1927. I, Mammals.* London: Zoological Society of London.

Gray, A. P. (1958). *Bird hybrids.* Slough: Commonwealth Agricultural Bureaux.

Gray, A. P. (1972). *Mammalian hybrids.* Slough: Commonwealth Agricultural Bureaux.

Hard, W. L. (1969). The karyotype of a male Himalayan Blue Sheep, *Pseudois nayaur. Mammalian Chromosome Newsl. 10*: 228.

Holm, E. R. & Scott, M. I. (1954). Studies on the nutrition of wild waterfowl. *N.Y. Fish & Game J.* **1**: 171–187.

Kear, J. (1973). Notes on the nutrition of young waterfowl, with special reference to slipped-wing. *Int. Zoo Yb.* **13**: 97–100.

Kear, J. (1975). How wildfowl could improve our domestic breeds. *Waterfowl Ybk Buyers' Guide* **1975–76**: 37–41.

Korobitsyna, K. V., Nadler, C. F., Vorontsov, N. N. & Hoffman, R. S. (1974). Chromosomes of the Siberian Snow sheep, *Ovis nivicola,* and implications concerning the origin of Amphiberingian wild sheep (Subgenus *Pachyceros*). *Quartern. Res.* **4**: 235–245.

Lydekker, R. (1898). *Wild oxen, sheep and goats of all lands.* London: Rowland Ward.

Murton, R. K. & Kear, J. (1973). The nature and evolution of the photoperiodic control of reproduction in wildfowl of the family Anatidae. *J. Reprod. Fert. Suppl. 19.* 67–84.

Nadler, C. F., Hoffman, R. S. & Woolf, A. (1974). G-band patterns, chromosomal homologies, and evolutionary relationships among wild sheep, goats, and aoudads (*Mammalia, Artiodactyla*). *Experientia* **30**: 744–746.

Nadler, C. F., Korobitsyna, K. V., Hoffman, R. S. & Vorontsov, N. N. (1973). Cytogenetic differentiation, geographic distribution, and domestication in palearctic sheep (*Ovis*). *Z. Säugetierk.* **38**: 109–125.

Nadler, C. F., Lay, D. M. & Hassinger, J. D. (1971). Cytogenetic analysis of wild sheep populations in Northern Iran. *Cytogenetics* **10**: 137–152.

Petocz, R. G. (MS). *Marco Polo sheep* (Ovis ammon poli) *of the Afghan Pamir: a report of biological investigations in 1972–1973.* FAO.

Zeuner, F. E. (1963). *A history of domesticated animals.* London: Hutchinson.

Zuckerman, S. (1952). The breeding seasons of mammals in captivity. *Proc. zool. Soc. Lond.* **122**: 827–950.

Symp. zool. Soc. Lond. (1976) No. 40, 335–336.

AN EARLY LABEL FOUND IN THE ZOOLOGICAL GARDENS, REGENT'S PARK, LONDON

(Half actual size)

This creamware tile label was dug up in the spring of 1970 by the Society's gardeners Mr Gold and Mr Rolls. It was found about 30 yards west of the Main Gate just beside the present electricity substation.

The tile itself cannot be dated more precisely than the late eighteenth or very early nineteenth century, probably being made in Staffordshire. The only mark is a number 5 on the back which is taken to be an indication of its size.

The information on the tile is handpainted (there is some restoration of the bottom lefthand part of the tile). The Society's animal records have been kept since 1828 but do not mention a coati as being presented by S. Cotton. At that time the brown coati was usually called *Nasua fusca* and Storr is the authority who set up the genus *Nasua.* The name *N. narica* seems to have been in general use by the mid-nineteenth century, although it is now included in *N. nasua.*

During the 1820s there were several coatis on exhibition in London, both at Regent's Park and in the Tower Menagerie. According to the first Guide the animals at Regent's Park were labelled. As the label was found at what was then the edge of the Zoo and well away from where coatis would have been kept it is possible that the label came with the animal and was rejected for not comforming to whatever pattern was then in use. On the other hand, it is tempting to speculate that the tile is one of the very earliest labels used at Regent's Park. Whichever of the two explanations is accepted there can be no doubt that it is a very old label.

MICHAEL BRAMBELL

GENERAL INDEX

Numbers in italics refer to pages in the References at the end of each article.

A

Aardwolf, 150
Acclimatization farm of the Zoological Society, 29
Acinonyx jubatus, 173
Acipenser, 117
African buffalo, 273
African elephant, 286
  arterial disease, in, 274
African fauna, survey of, 271
African freshwater fishes, 99, 100
African game animals, 274
Alfred, Professor, 138
Allis, E. P., 101
Allotriognathi
  anatomy and classification of, 100
American bison, 170, 173
Amia, 117
Ammotragus lervia, 325, 328, 329
Amoroso, E. C., 270, 271, 301, *280*
Anas
  laysanensis, 327
  platyrhynchos, 326
Andrade, E. N. da C., 2, *16*
Animal Welfare and Husbandry Committee (1958), 161
Animal and Zoo Magazine, 226
Antennularia antennina, 74
Annual Reports
  (1828), 198
  (1830), 198
  (1836), 241
  (1883), 198
Anser
  anser, 325
  caerulescens, 325
  indicus, 326
Ape Colony, 201
Ape House (1902), 151
Aquarium of the Zoological Society
  building (1924) (Joass), 188, 201
  and circulating water systems, 111, 112
  Crystal Palace, 109

Aquarium—*cont.*
  establishment of (1850), 59
  exhibits in, 117
  and filters, 112
  and filtration, 116, 117
  first attempts at keeping (1852), 107
  first public, 107, 108
  fountain, 110
  freshwater tanks, 113
  history of, 105–107
  and hydrometer readings, 114, 115
  management of, 105–118
  and marine biological stations, 111
  materials suitable for building, 116
  and pH, maintenance of, 117
  and plate glass manufacture, 105
  policy of, 112
  present day, 111–115
  opening of (1924), 93
  sea-water tanks, 113
Aquarium studies, 85
  and ichthyology, 86, 88–94
Aquavivarium, 88
  opening (1853), 88, 89, 93
  first recorded photograph of living fish in, 93
Arabian oryx, 59, 148
Argali sheep, 327, 329, 330
  Altai, 330
  Mongolian, 330
  Tibetan, 330
Argonauta argo L.
  controversy in early discussions of the Society, 77, 78
Arnett, R. H., 256, *266*
Arnold, J. B., 88
Art Studio (1936) (Lubetkin), 190
Arvay, A., 308, *316*
Ascheim, S., 296, *316*
Attenborough, David, 229
Attridge, J., 125, *131*
Atz, J. W., 93, *102*, *104*, 258
Auffenberg, W., 120, *131*
Austin, C. R., 329

Australian lungfish, 90, 97
Avaries, Zoological Society
  assorted birds (1835), 137
  cockatoo, 201
  Eastern (1864) (Salvin), 180, 185, 200
  Great, 185
  new, 140
  new, for Water Birds, 194
  old macaw cage, Zoological Gardens (1831), 136
  owl, 201
  peafowl, 200
  Southern (1905), 185, 201
  small birds (1830), 135
  small parrots, 201
  Snowdon (Northern), 141, 142, 194, 202
Axis axis, 173, 325
Axis deer, 173, 325
Aye aye, 53, 59

B

Babladelis, G., 300, *316*
Bailey, R. M., 100, *104*
Balanced aquarium, concept of, 93
Ball, D. J. (Overseer Reptile House), 120, 121, *131*
Banks, Sir Joseph, 2, 3, 50
Barbary sheep, 327–329
Bar headed goose, 326
Barnett, Burgess (Curator Reptile House), 63, 120, *132*, 215, *222*
Barr, D., 265, *266*
Barrington, E. J. W., 227
Bartlett, A. D. (Superintendent of the Gardens 1859–97), 92, 135, 158, 159, 184
Bass, 89
Bate, C. Spence, 69, 73, 74, *81*
Bayes, Kenneth, 194, 200, 202
Beadle, L. C., 273, *280*
Beadle, M., 273
Beadlet anemone, 105, 106
"*Beagle*", voyage of, 5, 87
Beattie, John 63
Beaver Pond, 201
Beddard, Frank Evers, 63, *66*
Belcher, Sir Edward, 19, 22, 30–34, *48*
Bell, Thomas, 68, 69, 72, 73, 77, *82*, 234
Bellairs, A. d'A. (Honorary Herpetologist), 120, 121, 125, *131*
Bellamy, Thomas, 199
Ben-Tuvia, A., 103
Bennett, E. T. (Secretary 1833–36), 85, 86, 224, 240, 241
Bennett, George, 81
Bennett's wallaby, 167, 168
Bighorn sheep, 327–329, 331
Birds
  breeding, 142, 143, 144, 145
  and conservation of species, 143
  and Darwin's theory of the origin of species, 135
  history of man's association with, 133
  living, studies of, 138
  ornithology, decline of reputation after 1900, 138
  pioneering work after 1920, 138
  policy of keeping
  in the Society's collections (1826–1976), 133–145
  in natural surroundings, 142
  record cards for, 144
Bird House, 200
Birds of Prey House, 201
Bishop, J., *197*
Bison
  bison, 173, 324
  bonasus, 173
Bitis arietans, 129
Black, Sir Misha, 194, 200, 202
Black rhinoceros, 59
Blainville, M. de, 78
Bleeker, Pieter, 94
Blue sheep, 328
Blunt, W., 1, *16*
Blyth, Edward, 19, 34, 35, 36, 40, *48*
Boiler House (Stengelhofen), 191, 201
Bond, F. W. (Society's Accountant 1903–42), 249
Boorer, M. K. (second Education Officer), 226
Bos
  grunniens, 324
  taurus, 324
Botulism, 219, 220

Boulenger, E. G. (first Curator of Reptiles), 92, 119
Boulenger, G. A., 99, 100, 119
Bowdler, Reginald (first Librarian 1867–72), 242
Bowfin, 117
Boyd, J. Moreton, 273, *280*
Bradshaw, M., 308, *318*
Brambell, M. R., 156, 158, *165*, 277, *280*
Branta
  ruficollis, 325
  sandvicensis, 326
Bridson, G. D. R., 256, 257, 259, 260, *266*
Brill, 89
British, the, overseas, 17–48
Brittle stars, 114
Brookes, Joshua, 235, 236
Brooks, Alan, 271, *280*
Bryopsis, 107
Buckland, Frank (Fellow 1862–86), 85, 89, 184, *197*, 198
Buckland, R. A., 331, *332*
Buckley, J. T. C., 216
Budgett, J. S. (Fellow 1898–1904), 98, 99, 270
Building, 150 years of the Zoological Society, 179–202
  architects employed, 179
  costs, 185
  future concepts, 197
  landscape, involvement with, 191, 194
  list of buildings standing, 1976, 200-202
  major development after 1959, 197
  plan of buildings, 196
  proposed extension (Toovey), 195, 197
Burne, R. H., 61
Burton, Decimus (Architect to the Society 1830), 154, 180, 198, 199, 200
Burton Terrace, 154
Busk, George, 60, 61

C

Californian sealion, 148
Canal Bridge, 201
Canis lupis, 173
Capra hircus, 328
Cardwell, D. S. L., 2, *16*
Carinia moschata, 326
Carnivora Terrace (1843) (Elmslie), 180, 182, 185
Carnivores, at Regent's Park, 147–165
  acquisitions, table of, 162
  differences between groups and species, 148, 149
  future of species, 147
  survey of births of, 161, 163, 164
Carp, 88
Casson, Hugh, 190, 191, 194, 197, 199, 201, 202
Catallo, 324
Catering Store, 200
Centenary Address, 198
Central Mammal House (1913), 155
Centre for Life Studies, 2
Ceratotherium simum, 173
Cervus elaphus, 322
Challenger Expedition, 63
Chaplin, Lord (Secretary Zoological Society 1953–55), 9
Charles, Havelock, 61
Cheetahs, 157, 173, 175, 176, 277
  gestation and mating in, 177
Chester Jones, I., 228
Children, J. G., 234
Children's Zoo, 202
Chimaera, 98, 101
Chimpanzee
  rickets and osteomalacia, in, 160
Chinese water deer, 167
Chital deer, 325
Chivers, D. J., 278, *280*
Chlamydoselachus anguineus, 101
Christian, J. J., 313, *316*
Christmas lectures, 229
Cichlidae
  African, 99, 100
  from Lake Malawi, 100
  South American, 100
  Syrian, 99, 100
Clark, J. L., 330, 331
Clean air legislation
  and housing of animals, 155
Clegg, M. T., 297, *316*

Clock Tower (Burton), 180, 200
Clore Pavilion for Small Mammals (1967), 154, 155, 156, 191, 194, 202
Clostridium botulinum, 219
Cloudsley-Thompson, J. L., 125, *132*
Clough, Gerald, 273, 309, *316*, *317*
Clupea harengus, 101
Coati, ancient tile recording, 335
Cobbold, T. Spencer, 97, 256, 259
Coblans, H., 257
Cockatoo Aviary, 201
Coffey, P., 298, *316*
Colebrooke, H. T. (Director of Royal Asiatic Society), 237
Common peafowl, 173
Common seal, 271
Comparative anatomy (nineteenth century), 49–66
  form and function, complementary association of, 50
  fundamental scientific concern of the Society, 51
  role of Prosectorship, 60
Conder, Neville, 190, 191, 194, 197, 199, 201, 202
Conservation, 275, 276, 277, 285, 286, 299
Cooke, A. H., *82*
Corallus caninus, 127
Corresponding members of the Zoological Society, 18, 68, 79, 80
  circular sent to (1827), 20, 21
Cotton Terraces, 189, 191, 202
Couch, Bertha, 74, *82*
Couch, Jonathan, 72, 73, 74
Couch, Richard Quiller, 73, 74, 77, *82*
Coutts, R. R., 273, *281*
Cowan, D. F., 217, *222*
Crab, life history of,
  early research controversy, 77
Crane and Goose Paddocks, 200
Crawford, M. A., 220, 274, 276, *280*, 304, 306, *316*
Crawford, S. M., 274, *280*
Cretaceous shark, 102
Crocodile pools, 122
Crisp, E., 159, *165*
Crocodilians, 126
Crustaceans, 30, 117
Crustaceans—*cont.*
  sessile-eyed, 74
  larvae, J. V. Thompson's developmental series, 76, 176
Crystal Palace aquarium
  design of, 109, 110
  opening of, 109, 110
Cuming, Hugh, 19, 30, 31, 73, 78, 79
Cumingian collection, 30, 78, 79
Curle, R., 71, 73, *82*
Cyprinidae (Indian)
  classification of, 94

D

Dadd, M. N., 258, *266*
Daily Occurrence Sheets (1828–1961), 248
Dall (sheep), 327, 329, 331
Dalyell, Sir John, 105
Dance, S. Peter, 73, *82*
Danford, D., 298, *317*
Darwin, Charles, 5, 6, 7, 80, 87, 135
  presentation of *The Origin of Species*, 135
Darwin, F., 103
Darwinism, advent of, 50
Daubeny, Professor, 107
Davies, Alfred (Fellow Zoological Society), 225
Davy, Sir Humphrey (founder member and Councillor), 2, 3, 6, 26, 27, 28, 29, 88, 103, 236
Dawber, Sir E. Guy (Architect), 120, 188, 189, 199, 201
Day, Francis, 90, 96, 97
  disagreement with A. Günther, 95
Dean, Bashford, 90, 102, *103*
Dean Bibliography of Fishes, 257, 258
Dene, H., 302, *317*
Dexter cow, 331
Diadema antillarum, 114
Diczfalusy, E., 297, *317*
Dikdik, 271
Dolichotis patagonum, 167, 173
Domestic animals
  miniaturization of, 331
Domesticated Chinese goose, 326
Doubleday, E., 80, *82*
D'Souza, Frances, 275, 308, *317*

Duck, domestic
improvement of, 326
Duncan, Frederick Martin (Librarian 1919–39), 249

E

Eakins, J. P., 266
East African tortoise, 120
East Bridge, 200
East Service Gate, 201
East Tunnel (1829) (Burton), 180, 181, 200
Eastern Aviary (1864) (Salvin), 180, 185, 200
Echidna, 59
Echinoderms, 113, 114, 117
Education, role of Zoological Society, 2, 223–231
Christmas Lectures, 229
and conservation lectures, 226
courses for school teachers, 225
first public lectures, 225
future of, 231
of professional Zookeepers, 229
and schoolchildren, needs of, 226
and Sixth Form Symposia, 226, 227
and Studio of Animal Art, 226
and Symposia, 228
Teachers' Centre for Life Studies, 228
and television, 229
and Television and Film Unit, 229
at University level, 228
XYZ club, 229
and "Zoo Quest" expeditions, 229
and Zoological Film Productions Ltd, 226
Education Centre, 155, 202, 227
Education Department, 226, 227
Eels, 88
Eland, 322, 325
Elaphurus davidianus, 173
Elephant shrews, 274
Elephant Houses
(1869) (Salvin), 180, 183, 184, 185
new (1938) (Lubetkin), 188, 190, 194
Elephant and Rhino Pavilion (1965) (Casson and Conder), 190, 194, 202
Elephantulus myurus, 274
Elmslie (second Architect to the Zoological Society), 180, 199
Enteromorpha, 107
European bison, 147
Equus
asinus, 324
caballus, 324
Evans, H. J., 331, 332

F

Fairchild, O., 299, *317*
Fasham, J. A. L., 309, *316*, *317*
Fauna Preservation Society (1903), 276
Feeding of animals, 158–161
expanded pellet feeding, 159
survey of mammal diets, 158
Fellows of the Zoological Society
"FZS Committee", 13, 14, 15
subscriptions, 11, 12, 13
Field Studies
British Solomon Islands Expedition, 272
and conservation, 275, 276, 277
Cook Bicentenary Expedition, 272
grants by the Zoological Society for, 269, 270
and parasitology, 274
primates, current locations, 279
past, examples of, 269–272
present research, 272–275
future possibilities, 275–278
Zoological Society's contribution to, 269–279
Fish
caudal fin skeleton, 101
structure of bone, 101
Fish anatomy
classical works on, 245, 246
Fish collections in Museum of Zoological Society, 86
of Cuming, H. (Chile), 86
of Lowe, R. T. (Madeira), 86
of Poey, Felipe (Cuba), 86
of Raffles, Sir Stamford (Malaysia), 86
sale/gift of to British Museum and confusion arising, 87, 88
of Sibbald (Ceylon), 86

Fish collections—*cont.*
  of Sykes, W. H. (India), 86
  of Telfair, C. and Desjardins, J. (Mauritius), 86
Fish culture, 85
  and ichthyology (1830–32), 86, 88–94
  salmoniculture, 85, 88
Fish House (*see* Aquavivarium)
Fish keeping
  practical aspects of, 88
Fish longevity
  records of, 93, 94
Fisher, L. E., 298, *317*
Fishes
  of Asia Minor, 94
  of Australia, 94
  of Deccan, 94
  outline classification of (T. H. Huxley), 97, 98
  palaeoniscoid, 102
  phallostethine, 100
Flamingo Pond, 202
Flounders, 88
Flower, A., 199, 200
Flower, Professor, 179, 185, 198
Flower, S. S., 93, 94, 159, *165*, 329, 330, *332*
Flower, Sir William (President Zoological Society 1887), 3, 5, 60, 61
Floyd, J. H., 101
Follies (Burton), 180, 181
Forbes, E., 69, *82*
Forbes, William Alexander, 63
Franklin, Sir John, 32
Frueh, R. J., 298, *317*
Frye, F. L., 126, *132*
Fryer, G., 99, 100, *103*

G

Galago, muscles of, 56
Gallus gallus, 173
Gamble, M. R., 311, *317*
Garden Cafe, 201
Gardening Department, 202
Garpike, 6, 13
Garrod, Alfred Henry (second Prosector 1871), 61
Gaur, 59
Gazella thomsoni, 173, 271
Genes
  exploitation of, 324, 325
Gerbils, 322
Giant eland, 276
Giant pandas, 18, 148, 150
Giant tortoise, 122, 123
  enclosure, 123
Gibbons Cage, 201
Ginglymodi, 101
Giraffes
  anatomy of head arteries, 215
  foetal physiology of, 271
  Thibaut's (1836), 240
Giraffe House (1836) (Burton), 180, 200
Globiocephalus, viscera of, 58
Goat, domestic, 328
Goldfish, 88
Goodenough, Edmund, 237
Goodey, T., 101
Goodrich, E. S., 101
Goose, domestic
  improvement of, 325, 326
Gorilla, 54, 55
  first (1897), 158
  breeding in captivity, 292–300
  maternal behaviour, failure of, 300
Gorilla g. beringei, 299
Gorilla House (1933) (Lubetkin), 185, 187, 189
Gosse, Sir Edmund, 109, 118
Gosse, P. H., 69, *82*, 106, 107, 109, 110, *118*
Gould, John (Ornithologist of the Society's Museum), 59, 240
Greenwood, P. H., 101, *103*
Grant, Robert, 80
Gray, A. P., 324, 326, 327, 328, 329, 331, *332*
Gray, C. W., 292, 295, 296, 297, *319*
Gray, John Edward, 60, 68, 70, 73, 74, 77, 78, 79, *82*, 90
Great Exhibition (1851), 241
Greater snow goose, 325
Green, James, 100
Greenwood, P. H., 99, 100, 101, *103*
Grey seal, 270, 271, 273
Grimsdell, Jeremy, 273
Grosser, O., 301, 302, *317*

Guanaco, 173
Günther, Albert C. L. G. (Curator of Fishes, British Museum), 89, 94, 95, 96, 97, 98, 102, 255, 256, 259, *267*
  disagreement with Francis Day, 95
Gunther, A. E., 60, 86, *103*, 256, 259, *267*

H

Hairy fronted muntjac, 59
Halecostomi, 101
Halichoerus grypus, 270, 271
Hamadryas baboons, 269
Hamerton, A. E., 119, 215, *222*
Hammerhead, 59
Hammerhead sharks, 101
Hamster, 322
Hanley, S., 69, *82*
Hard, W. L., 328, 329, 332
Hardin, C. J., 298, *317*
Harmer, S. F., 61
Harrison, R. J., 270, 271, *280*
Harrisson, Tom, 276, 277, *280*
Hartmann's mountain zebra, 178
Harvey, J. B., 81
Hassam, A. G., 304, 306, *316*
Hassinger, J. D., 328, *333*
Hawkey, C. M., 216, 219, *222*
Health of animals, maintenance of, 203–214
  appointment first Medical Attendant, 203
  commercially made diets, 207
  clinical medicine, 211
  and design of animal houses, 204, 205, 206
  emergency service, 208
  keeper, role of, 207, 208
  and modern drug techniques, 211
  and nutritional requirements, 206
  pathology, 210
  planned programme, establishment of, 204
  post-mortem facilities, 208, 209, 210
  preventive aspects, 204–208
  quarantine regulations, 211, 212
  records of, 209
  and research, 209
  veterinary services, 208, 209
Helmeted guineafowl, 173
Hill, J. P., 61
Hill, William Charles Osman (last Prosector, 1950–62), 64, 302, *317*
Hincks, T., 69, *82*
Hippopotamus, 59, 168, 273
Histories, 70, 71, 73
Hobson, B., 296, *317*
Hodgson, Brian Houghton, 19, 34, 35, 37, 38, 248
Hoffman, R. S., 328, 331, *333*
Holdsworth, E. W. H. (Fellow of the Society), 93, *104*
Holm, E. R., 327, *332*
von Holst, D., 308, *319*
Homewood, Katharine, 276
Hopley, Catherine C., 119, *132*.
Hopper, B. R., 292, 295, 297, *317*
Hospital, 201
Housing of animals (*see also individual Houses*)
  future, 155–156
  past, 150–159
  present, 155–156
Hudson, Liam, 289, *317*
Hudson, William Henry, 7, 19, 44, 45, *48*
Hume, Allan, 36, 37, 40
Hummingbird House, 200
Humphreys, Noel, 109
Hunter, John (1728–1793), 49, 50
Hunter, Sir W. W., 38, *48*
Huxley, Julian (Secretary Zoological Society 1935–43), 7, 9, 226, 289
  dividing the *Proceedings* into two parts, 9
  his modernization of Society's image, 9
  research on birds, 138
  resignation, 9
Huxley, L., 81, *82*
Huxley, T. H., 6, 7, 60, 81, 97, 98, 140
Hystricomorph rodents, 274
Hystropotes inermis, 167

I

Ichthyology
  the Society's contribution to, 85–104

Ichthyology—*cont.*
and the Society's Museum (1826–55), 86–88
(*see also* Museum of the Zoological Society)
Iles, T. D., 99, 100, *103*
Indian catfishes, 97
Insect House, 201
establishment of (1881), 59
Interzoo cooperation, 148
International Zoo Yearbook, 224, 263, 264, 299
first volume of, 263
Invertebrate zoology, early, 67–83
in Australia, 80, 81
classification of new material, 78, 79
sources for, 78
controversies in, 75–78
first reference to at scientific meeting of the Society, 75
and natural history studies, 80
in the provinces, 71–75
and zoologists, 78–81

J

Jarvis, Caroline, 263
Javan rhinoceros, 59
Jeffreys, J. G., 69, *82*
Jewell, P. A., 271, 272, 273, *280*
Joass, J. J., 185, 188, 199, 201
Johnston, Alex, 46, *48*
Johnston, George, 69, 72, 74, *82*
Johnston, Sir Harry, 44, 46, 47, 48, 270
Johnstone, James, 107
Jones, David, 216
Jones, Henry, 249
Jones, R. C., 287, *317*
Journal of Zoology, 224, 253–255
Jubilee (first great ape reared in the zoo), 151
Jubilee Address (1887), 179, 198
Jumbo, sale of (1882), 242
Jungle fowl, 173
Jurassic coelacanth, 102
Jurtshcuk, P., Jr., 308, *318*

K

Kagawa, K., 298, *317*
Kagawa, M., 298, *317*
Kagu, 61
Kear, J., 325, 326, *332*
Keeling, M. E., 298, 299, *318*, *319*
Keepers' Lodge, 200
Kellas, Lona, 271, *280*
Kew Gardens
establishment of as a national institution, 3
Kiang, 59
Kingsley, Charles, 109
Kingsley, Mary, 47
Kirby, Rev. W., 233–235
Knowles, R. J., 13, 14, 15
Komodo dragons, 120
Korobitsyna, K. V., 328, *333*
Kuhn, T. S., 291, *317*

L

Lama
glama, 173
guanicoe, 173
Lambert, D. G. (first Education Officer), 226
Land snails, from Jamaica, 80
Lang, E. M., 298, *317*
Lankester, Edwin, 109
Lankester, Ray, 98
Latimeria chalumnae, 102
Laughton, John Knox, 32, *48*
Lavatory buildings, 201, 202
Lawrence, W. E., 215, *222*
Lay, D. M., 328, *333*
Laysan teal, 327
Leach, W. E., 73
Lee, Henry, 110
Le Gros Clark, W. E., 301, *317*
Leiper, R. T., 63, 216
Lemurs, 275
Lennep, E. W. van, 271, *280*
Lepidosiren, 98, 117
paradoxa, 102
Lepidoptera, 80
Lepisosteus, 101, 117
Lester, Jack (Curator Reptile House), 120
Leutenegger, W., 302, 303, *317*
Liassic shark, 102
Library of the Zoological Society, 8, 59, 223, 224, 225, 233–252, 253–256
accommodation for, 240

Library—*cont.*
appointment of Librarian (Third Clerk) (1867), 242
the archive collection, 248
catalogue of
first (1854), 241, 247
second edition (1872), 242
cataloguing system, 247
committee for formation of, 237
donations to, first recorded list of, 237
graphic materials, collection of, 248
growth of, 224, 240
list of bound volumes in (1834), 238-240
move to Regent's Park (1910), 225, 242, 243
new Reading room, 250, 251
old Meeting Room, Regent's Park, 244
old Reading Room, Regent's Park, 245
photographic collection, 249
shelving system, 246, 247
rebuilding of (1964), 246
recataloguing (1966–76), 247, 248
and Wolfson Foundation, 225
Liebherr, G., 299, *317*
Light, action of on plants, 107
Line, M., 265, *267*
Linnaeus (1707–78), 49
animal classification, introduction of, 49
library and collections, sale of, 236
Linnean Society, 2, 3, 233, 234, 235
publications of, 4, 254
zoological club of, 3, 234–236
Lion cubs, zoo born
malformed palates in, 159
Lion House
new (1876) (Salvin), 154, 159, 183, 185
replacement for, 156
old, 179
Lion Terraces, New, 155, 156, 185, 193, 194, 202
Lizards, carnivorous, 126
Llama, 173
Llewelyn-Davis, Lord Richard, 194, 200, 202
Lloyd, W. Alford (Designer Crystal Palace aquarium), 109, 111
Locust, 122
"Lomie" (female gorilla), 292, 293, 295, 297, 299
Longevity, of fish, 93, 94
Lopes, George de Arroyave, 10
Lopes, Georgina, 48
Loricariid catfishes, 100
Lotshaw, R., 298, *317*
Loughran, William, 74
Lowe, Rev. R. T. (Corresponding member 1833–74), 19, 34, 85, 94
Lubetkin, Berthold, 188, 190, 194, 199, 201
Luckett, W. P., 302, *317*
Lunar Society of Birmingham, 2, 3
Lungfishes, 117
Lusty, J., 295, 297, *318*
Luvarus imperialis, 97
Lydekker, R., 330, *332*

M

Macaca mulatta, 272
MacBride, E. W., 61
MacFarlane, R. G., 120, *132*, 215, *222*
Mackinnon, John, 277
Mackintosh, N. A., 101
McLeay, Alexander, 41, 42
Macleay Family, 19, 42, 43
Macleay, Sir George, 41
Macleay, James Robert, 41
Macleay, William Sharp, 41
Macropus rufogriseus, 167
Main Gage (Dawber), 188, 201
Main Gate (Toovey), 202
Main Gate Kiosk, 202
Main Offices (1924) (Joass), 188, 201
Maintenance of the collections, 216, 217
Mallinson, J. J. C., 298, *317*
Malacochersus tornieri, 120
Malthus, 29
Mammals
oldest in the Society's collection, 147, 148
Management of animals, 156–158
Manatee, 61
Mancuso, S., 297, *317*

Manton, V. J. A., 173, 175, *178*, 277, *280*
Mappin Cafe, 201
Mappin Terraces (1914), 155, 156, 185, 186, 201, 202
Mara, 167
Marco Polo, 330
Marco Polo's sheep, 327, 330–332
Marine biological stations, 111
Martin, R. D., 275, 295, 296, 297, 299, 302, 308, 311, 312, 313, *318*
Martin, William (Superintendent, Museum 1834), 241
Matthews, Harrison, 270, 271, *280*
McFadyean, Sir John, 215
Medawar, P., 289, *318*
Meleagris gallopavo, 173
Menagerie of the Zoological Society
  establishment of (1828), 59
Menageries, 18
Metridium senile, 114
Miles, R. S., 91, *103*
Milner, C., 273, *280*
Miller, Agnes E., 102
Mississippi alligator, 129
Mitchell, David W. (Artist, Secretary Zoological Society 1847–59), 241, 248
Mitchell, Sir Peter Chambers (Secretary to the Zoological Society 1902–35), 1, 8, 9, 16, 27, 44, *48*, 86, 87, 93, *104*, 126, *132*, 151, 165, 167, *178*, 185, *197*, 198, 203, 210, *214*, 224, 225, *231*, 235, 242, 259, 260, *267*
Mivart, St. G., 61, 98
Molluscs, 117
Monkey Houses
  (1927) (Dawber), 188, 189
  (1829), 150
  (1839), 150, 185
  (1864) (Salvin), 180, 182, 185
  Monkey Hill (1920s), 151
  new (1927), 151, 152
  outbreak of tuberculosis (1906), 151
  Sobell Pavilions, 147, 153, 155, 156, 192, 194
  Stengelhofen building (1968), 154, 155
Montague, George, 73
Morris, Desmond, 119, *132*, 229, 263
Morris, R., 119, *132*
Mouflon
  Asiatic, 327, 328
  European, 327–331
Mouse lemurs, 275
Moy-Thomas, J. A., 102, *104*
Muntiacus, 167
Muntjac, 167
Murie, James (first Prosector 1865–70), 61, *66*, 85, 89
Murton, R. K., 325, 326, *332*
Muscovy duck, 326
Museum of the Zoological Society, 29, 86–88
  creation of, 5, 86, 223
  dispersal of collections to British Museum, 79
  growth of, 86
  and fish collections (*see* Fish Collections)
  importance of, 86
  progress by 1850, 87
  as a source for early taxonomic research, 78, 79
Myers, N., 175, *178*
Myriapods, 80

N

Nadler, C. F., 328, 331, *333*
Nadler, R. D., 298, 300, *318*
Ne-ne goose, 326
Neave, Sheffield Airey (Secretary Zoological Society 1943–53), 9, 260, 261
Neave-Lloyd *Nomenclator Zoologicus* Fund, 261
Nelson, George, 216
Neoceratodus, 117
  forsteri, 90, 97, 98
Neopterygii, 101
New species
  failure to domesticate, 324
  introduction for domestication, 321–332
  need for, 323
Newby, Frank, 194, 200, 202
Newport, G., 80, *82*
Newton, Alfred, 61, 140

Newton, Isaac, 3
Nomenclator Zoologicus, 224, 260–262
  computer production system, 262, 264
Norman, J. R., 101
North American turkey, 173
North Gate (1926), 201
North Gate Kiosk (1936) (Lubetkin), 189, 190, 201
North Mammal House, 155
Northern Pheasantry, 201
Nuffield Institute of Comparative Medicine, 218, 249, 254, 255, 272, 273, 274
  and animal health, 209
  building of, 194, 202
Numida meleagris, 173

O

Obaysch, first hippopotamus (1850), 241, 248
Ockenden, William, 61
Octopus, 113, 114
Ogilvie-Grant, W. R., 61
Okapi, 270
Oliver, J. H., 113, 115, 118
Ophiocomina nigra, 114
Orang-utan, 148, 277
  reproductive behaviour, 299
Orectolobid sharks, 100
Oryx
  Arabian, 59, 148
  scimitar horned, 149
Ostrich, 235, 236
Ostrich and Crane House (1896) (Trollope), 185, 200
Otaria, viscera of, 57
Otters, 150, 155
Otter Exhibit, 202
Ovis
  ammon, 327–330
  darwini, 330
  hodgsoni, 330
  kaselini, 331
  poli, 330
  aries, 327, 328
  canadensis, 327–329
  dalli, 327–329
  musimon, 327–329
Ovis—cont.
  nivicola, 327–329
  orientalis, 327, 329
  vignei, 327–329
Owen, Richard, 5, 7, 59, 60, *66*, 68, 71, 73, 78, *82*
Owls' Aviary, 201

P

Pavilion Building (1922) (Joass), 188, 201
Peach, Charles William, 73, 74
Peafowl Aviaries, 200
Peaker, M., 131, *132*
Peavot, Henry (Assistant Librarian/ Clerk of Publications), 249
Peccaries, 173
Penguin Cafe, 201
Penguin Pool (1934) (Lubetkin), 185, 189, 201
Père David's deer, 147, 148, 170, 171, 173
Panthera tigris, 173
Palaeopterygii, 101
Palaeospondylus gunni, 102
Panamian coastal fishes, 94
Pantodon buchholzi, 100
Parker, J., 98
Parker, W. K., 98
Parker, W. N., 98
Parrot House, 200
Patterson, C., 101, *103*
Pavo cristatus, 173
Perry, J. S., *318*
Pfaff, J., 308, *318*
Pfeiffer, L., *80, 82*
Phallostethine fishes, 100
Philippine bats, 30
Phoca vitulina, 271
Photoshop, 201
Pian'ka, E. R., 312, *318*
Pictorial record of animals, 248
Pilot whale, 61
Pinnipedes, 61
Plaice, 89
Plants
  action of light on, 107
  growth in enclosed cases, 107

Playground, 202
Pleurobrachia pileus, 80
Plumose anemones, 114
Pocock, R. I., 61
Poeciliine toothcarps, 100
Pollock, John, 275
Polychaetes, 117
Polypterus, 99, 117, 270
senegalus, 99
Polyunsaturated fats, 220, 221
Pond of the Zoological Society, 29
Pook, J., 298, *316*
Power, Jeannette, 78
Price, Cedric, 200, 202
Primates, at Regent's Park, 147–165
acquisitions, table of, 162
differences between groups and species, 148, 149
field studies of current locations, 279
future of species, 147
menstrual cycle of, 215
placentation, comparative, 300–306
survey of births of, 161, 162, 163
Prince of Wales' Indian Collection (1876), 242
Primrose Bridge, 201
Proceedings of the Zoological Society, 4, 8, 89, 90, 241, 223, 224, 253
centennial year, 85
establishment of, 59, 64, 237, 253
history of, 253, 254
Proctor, Joan B. (second Curator, Reptile House), 120, *132*, 151, 188
"Promenades", 7, 10
Prosectorial Committee (1903–1960), 61, 64
Prosectorium, 59, 60
Prosectorship
appointment of first (1865), 61
appointment of second (1871), 61
appointment of third (1879), 63
appointment of fourth (1884), 63
appointment of last (1950), 63, 64
replacement by Research Fellowship, 63
role of, in comparative anatomy, 60
Prosimians, 275
Protopterus, 98, 117
annactens, 90, 91, 97, 99
Prosectorium
organization of, 216
Prospectus of the Zoological Society (1825), 27, 88, 223, 235, 321
(1826), 248
Przewalski's horses, 147, 173, 174, 178
Pseudois nayaur, 328
Publications of the Zoological Society (*see* Scientific publications *and also individual publications*)
Public Relations Department, Zoological Society, 231
Puff adder, 129
Pythons, 123, 126

R

Raffles, S., 86, *104*
Raffles, Sir Thomas Stamford, 2, 6, 19, 23–29, 44, 47, 48, 50, 86, 88, 224, 233, 235–237, 250, 321, 331
Rand, C. S., 102, *104*
Rao, L. G. S., 297, *318*
Ravens Cage (Burton), 180, 200
Ray Society, The, 70, 71, 72, 73
Red breasted goose, 325
Red deer, 322, 331
Regan, C. Tate, 100, 101
Regalecus, 98
Regent Building (1929) (Joass), 188, 201
Repository, 150
temporary, 154
Reproduction, research on, 283–316
ecology of reproduction, 311–314
endangered species, 286, 287, 288
evolutionary background to, 311
foetal membrane relationship, 300, 301–306
genetic aspects, 288
in the great apes, 292–300
r-selection, 312, 313, 314
K-selection, 312, 313, 314
use of human pregnancy test kits, 292, 294, 295, 296
molecular biology, contribution of, 288, 289, 290
primate placentation, comparative, 300–306

K-selection—*cont.*
prospects for, in zoological collections, 285–291
and stress, in captivity, 306–311
Reptiles
and artificial heating, 125
and artificial light/ventilation, 126
bites from, 128, 130
breeding of, 129
cages, 123, 124
feeding, 126–128
handling, 128, 130
history of keeping, 119, 131
house (*see* Reptile House)
husbandry today, 121
maintenance of, for public exhibition, 121
optimum living conditions for, 121
Reptile House
first (1849), 59, 119
(1883) (Trollope), 184, 185
transformation into Bird House (1927), 185
present (1927) (Dawber), 120, 188, 200, 201
contents of, 120–131
plan of, 121–123
security of, 122
second (1889), 119
Research, 215–221
applied aspects of, 217
on botulism, 219, 220
into breeding of wild animals, 217
on giraffe, 215
laboratories for, 218–221
into menstrual cycles of primates, 215
into polyunsaturated fatty acids in the diet, 220, 221
on reproduction, 218
on Russell viper venom, 215
on sickle cells, 219
into vitamin B_{12}, 215
Research Unit, 201
Reynolds, Vernon, 272, *280*
Rhesus monkeys, 272
Rhyncotragus kirkii, 271
Richardson, Sir John, 94
Richmond Park farm
closure of, 89
Richmond Park farm—*cont.*
fish breeding, attempts at, 88
opening of (1829), 88
Riddle, K. E., 298, *318*
Ridewood, W. G., 98
Roberts, J., 298, *318*
Roberts, M., 45
Rosen, Donn, 100, *104*, 258
Rowell, R. E., 215, *222*
Rowlands, I. W., 270, 272, 273, 274, *280*, *281*, 284, 287, *318*
Rumbaugh, D. M., 298, *318*

S

Sacher, G. A., 302, 304, *318*
Sackler, A. M., 308, *318*
Sacred ibis, 143
Sadleir, R. M. F. S., 311, *318*
Salmo salar, 89
Murie/Günther controversy (1868), 89, 90
Salmon, 89
Salmonidae, 90
Salter, Edward, 199, 201
Salvelinus spp., 95
Salvin, Anthony, Jr. (Architect to the Zoological Society 1860–80), 180, 199, 200
Salvin, Osbert, 61
Sanatorium (1907), 203
Sanderson, Ivan, 269, *281*
Savage, Brian (Head Keeper, Reptile House), 120
Schaeffer, B., 102, *104*
Scherren, H., 1, 3, 7, *16*, 88, *104*, 108, 128, *132*, 150, 159, *165*, *198*
Schultz, L. P., 94, *104*, 258, *267*
Scientific publications of the Society, 94–103, 223, 224, 233–252, 253–256
(*see also individual publications*)
future of, 264–266
Scientific Societies Act (1843), 10
Scimitar horned oryx, 149
Sclater, Philip Lutley (Secretary Zoological Society 1859–1902), 8, 44, 60, 99, 135, 138, 241, 242, 248, 270
Scott, M. I., 327, *332*

Sea anemones, 114, 117
Sealion Pond, 185, 201
Sealion Stand, 202
Seaton, B., 295, 297, *318*
Sea-urchins, 113, 114
Seaweeds, 107
Sebates, 101
Selachians, classification of, 100
Selye, Hans, 306, 307, *318*, *319*
Serrasalmine fishes, 101
Service Buildings (Stengelhofen), 190
Seth-Smith, David (Curator of Birds), 140, 249
Sharks, 98
  Cretaceous, 102
  Liassic, 102
Sharp, D., 260, *267*
Shaw's monkey, 59
Sheep
  Argali, 327, 329, 331
  Barbary, 147, 148, 327–329
  Bighorn, 327–329, 331
  Blue, 329
  Dall, 327, 329, 331
  domestic, 328
    improvement of, 327–331
  snow, 327–329, 331
  Soay, 273, 331
Shepheard, Peter, 188, 191, 199, 202
Sheppard, E. M., 101
Sherrington, Sir Charles, 8
Shoebill, 59
Short, R. V., 311, *319*, 329
Siamang, 59
Sickle cells
  research into, 219
Sidney, Jasmine, 271, *281*
Sikes, Sylvia K., 274, *281*
Skates, 98
Skeldon, P. C., 298, *317*
Slabber, M., 75, *82*
Slot, C., *319*
Smelt, 89
Smith, C. E. Gordon, 221
Smith, C. L., 102, *104*
Smith, David Seth (Curator of Birds 1909–39), 140, 249
Smith, E. A., 273, *280*
Smith, Geoffrey, 274
Smith, G. Elliot, 61
Smith, G. R., 220
Smith, James Edward (President Linnean Society), 50, 236
Smith, M. A., 120, *122*
Snake, 119, 126
  bite, death from, 128
  poisonous, 122, 127
Snake venom
  therapeutic uses of, 120
Snowdon, Lord, 200, 202
Snowdon (Northern) Aviary, 141, 142, 194, 202
Snow sheep, 327–329, 331
Soay sheep, 273, 331
Sobell Pavilions for Apes and Monkeys, 147, 153, 155, 156, 192, 194, 202
Social Club, 202
Sole, 89
Sonntag, Charles F. (Anatomist to the Society 1919–25), 63
South Canal Bank, 202
South Gate, 202
South Gate Kiosk, 202
South Mammal House, 155
Southern Aviary, 201
Southern Pheasantry, 201
Sowerby, G. B., 109, 234
Sowerby, J. de C., 234
Spinage, Clive, 273
Spooner, Charles (first Medical Attendant 1829), 203
Squaloraja polyspondyla, 102
Staffeldt, E. F., 302, 304, *318*
Staff flats, 202
Stanhope, Hester, 47
Starfishes, 113
Stengelhofen, Franz, 188, 191, 194, 200, 201, 202
Stoker, M., 291, *319*
Stratton, G. B., 249
Stress, in animals in captivity, 306–311
  physiology of, 308
  relationship to defence mechanisms, 307
  in tree shrews, 308
Stores/Garage, 201
Studio of Animal Art, 226
Sturgeon, 117
Sumatran tiger, 173, 178
Sumner, A. T., 331, *332*

Sutton, John Bland, 61
Swinton, Captain, 185, 188
Symposia of the Zoological Society, 224, 228

T

Talwar, P. K., 96, *104*
Tamari, I., 308, *319*
Tana River mangabey, 276
Tasmanian wolf (*see* Thylacine)
Taurotragus oryx, 322
Taussky, H. H., 295, *319*
Tavern Bar, 202
Tayassu tajacu, 173
Teachers' Centre for Life Studies, 228
Television Building, 201
Television and Films Unit, 229
Thomas, O., 59, *66*
Thomas, Oldfield, 61
Thomas, W. D., 298, *318*
Thomson, K. S., 100, *103*
Thomson, Landsborough (President of the Zoological Society), 273
Thomson's gazelle, 169, 173
Thompson, D'Arcy W., 105, *118*
Thompson, J. V., 73, 75, 76, 77, *82*
Thompson, W., 73
Three Island Pond, 200
Thylacine (or Tasmanian wolf), 230
Tickell, Samuel Richard, 19, 34, 35, 38, 39, 40, 41, 248
Tijskens, J., 298, *319*
Tomes, C. S., 61
Toovey, J., 156, *165*, 192, 193, *198*, 200, 202
Trachinus spp., 98
Transactions of the Zoological Society, 4, 8, 85, 224, 241, 253
  history of, 253, 254
Traquair, R. H., 102
Tree kangaroo, 59
Tree shrews, 275
Trembley, Abraham, 105
Tripp, Hugh, 273
Trollope, Charles Brown (Architect to the Zoological Society), 184, 185, 199, 200
Trout, 89
Tullner, W. W., 292, 295, 296, 297, *319*
Turbot, 89

U

Ulva, 107
Undina, 102
Urial sheep, 327, 328, 329, 330
Uromastyx, 127
Usher-Smith, J., 298, *317*

V

Varanus komodoensis, 120
Veterinary Hospital (1955) (Stengelhofen), 194
Vevers, Geoffrey, 203
Vevers, H. Gwynne (Assistant Director of Science and Curator of the Aquarium), *16*, 93, 112, 113, 118, 120, 272
Vigors, N. A. (first Secretary of the Zoological Society), 233, 234, 235, 236, 237
Voller, A., 274
Volf, J., 173, *178*
van Voorst, John (Publisher), 68, 71
Vorontsov, N. N., 328, *333*

W

Wallace, A. R., 7, 135
Walrus, 52
War Memorial, 201
Warner, H., 299, *319*
Warrington, Robert, 107
Waterbuck, 273
Waterhouse, C. O., 260, 261, *267*
Waterhouse, F. H. (second Librarian 1872–1913), 242
Waterhouse, G. R. (Curator Society's Museum), 242
Watson, D. M. S., 102, 297, *318*
Weaver, M., 297, *316*
Weir, Barbara, 274, *281*
Wellcome Institute of Comparative Physiology, 218, 239, 255, 272, 273, 274, 275, 284
  and animal health, 209
  building (1963) (Stengelhofen), 194, 202
  and comparative mammalian reproduction, 284

Wells, F. R., 101
Weltman, A. S., 308, *318*
Wenyon, C. M., 216, *222*
West African lungfish, 90, 91, 92
West Tunnel, 201
Westwood, J. O., 69, 73, 74, 77, *81*
WHO programme for recording animal disease data, 209
Whitehead, P. J. P., 96, *104*
Whitehouse, R. H., 101
White rhinoceros, 172, 175, 276
Whipsnade, mammals at, 167–178
  export of, 173
  births at Whipsnade, 167, 173
Williams, B., 265, *267*
Williams, G., 304, 306, *316*
Wilson, E. O., 290, 312, *319*
Wilson, D. P., 106, *118*
Woolf, A., 328, 331, *333*
Wolf, Joseph, 248
Wolf Wood (Stengelhofen), 202
Wolfson Foundation, 225, 227, 228
Wolves, 155, 173
Wombwell's Menagerie, 159
Wood, Lt. R. N., 330
Wood Jones, F., 61
Woodward, A. Smith, 102
Workshops, 201
Wray, John (first Warden, Teachers' Centre), 228
Wurtzburg, C. E., *48*
Wynne-Edwards, V. C., 313, *319*

X

XYZ Club, 229

Y

Yakow, 324
Yarrell, William, 68, 69, 72, 73, 75, *83*
Yealland, John (Curator of Birds 1951–69), 142
Young, Allan, 274
Young, B. A., 271, *280*
Youatt, Mr (first Medical Superintendent), 203
Young, G. M., 67, *83*

Z

Zabinski, J., 173, *178*
Zeuner, F. E., 322, *333*
Zondeck, B., 296, 308, *316, 319*
Zooea taurus, 75
Zoo Magazine, 226, 229
"Zoo Quest", 229
Zoological Club of the Linnean Society, 234–236
Zoological Film Productions Ltd, 226
Zoological Journal
  first issue (1824), 234, 237
Zoological Record, 1, 59, 102, 224, 255–260
  form of presentation, 255, 256
  founding of (1864), 255
  modern computer production methods, 258, 259, 264
Zoological Society
  aims of, 223
  charitable status, challenge to, 10, 11
  Charter (1829), 6, 7, 27, 51
  Committee of Science and Correspondence of, 67, 68, 236, 253
  constitutional status of, 1-16
  Council of
    payment to Members, 7, 8
    Report (1855), 5
  establishment of (1826), 51
  farm (Kingston), 5, 88, 89
  foundation of, 26, 27, 29
  future of, 221, 161
  history of, 1–16
  influence of on nineteenth century zoology, 65, 75
  and infringement of the law, 10
  library of (*see* Library of the Zoological Society)
  museum (*see* Museum of the Zoological Society)
  New Charter (1963), 8
  objectives of (1825), 88, 223
  Proceedings (*see Proceedings Zoological Society*)
  and promotion of zoology as a science, 3
  proposed extension to Gardens of, (Toovey), 195, 197
  prosectorium, 59, 60
  (*see also* Prosectorium etc.)

Zoological Society—*cont.*
public lectures, first, 225
rates, exemption from 10, 11, 13
role in education (*see* Education etc.)
Royal Charter of, 4, 10
scientific meetings of, 4, 59, 65, 68, 74
scientific publications of (*see individual publications and* Scientific publications)
shop, 201
Sunday admission, 11, 12, 13
Supplemental Charter (1948), 8
and Zoological Club of the Linnean Society, 234–236
Zoological Studies Centre (1975), 197
Zootomical Committee (1865–1903)
establishment of, 60
Zuckerman, Lord (present Secretary of the Zoological Society), 190
Zuckerman, S., 151, 165, 215, *222*, 226, 231, 269, 270, *281*, 296, *319*, 325, 329, 330, *333*

1828 Clock Tower
1828 Raven's Cage
1829 East Tunnel
1836 Giraffe House
1883 Bird House
1898 Parrot House
1912 Insect House
1914 Mappin Terraces
1924 Aquarium
1927 Reptile House
1934 Penguin Pool

LONDON ZOO
1827 Sketch Plan

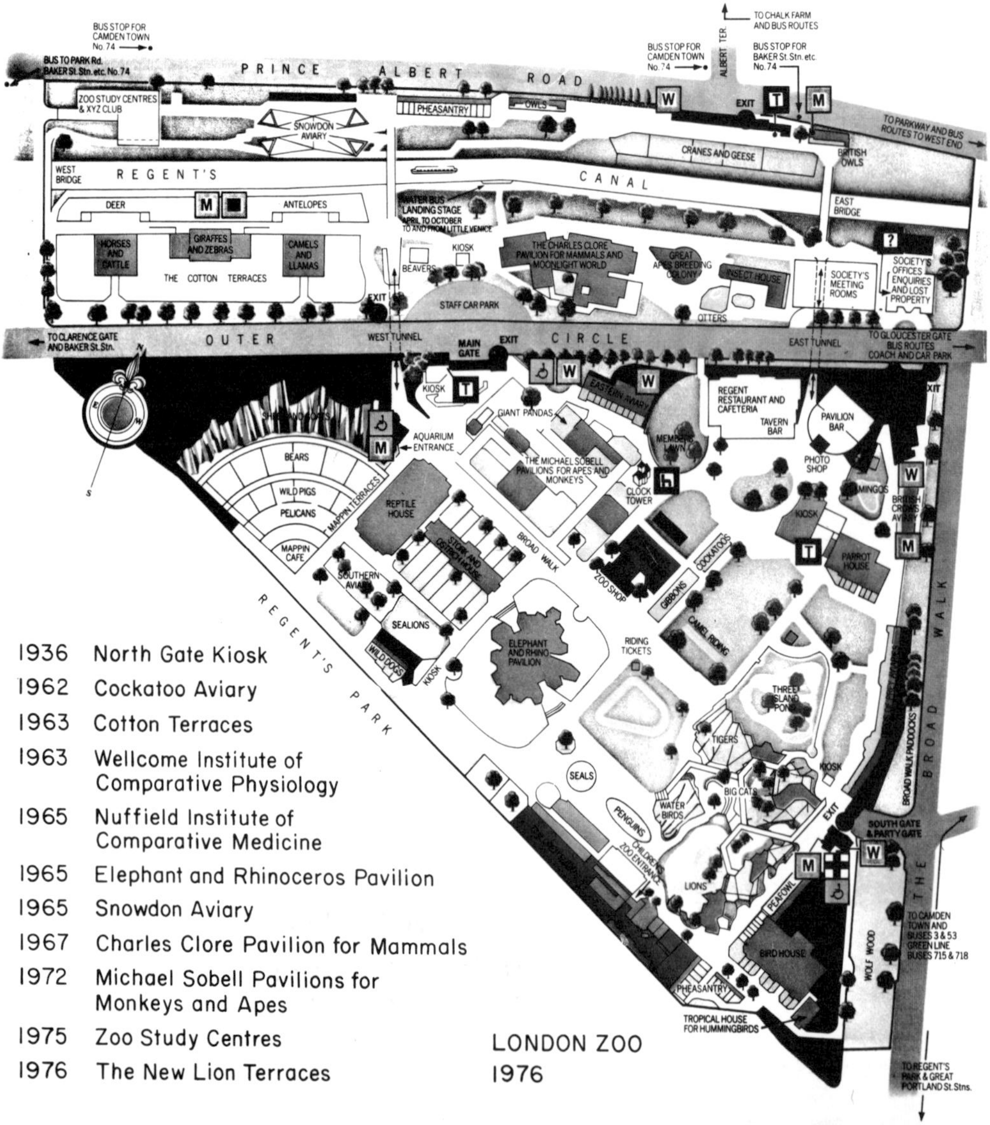

1936 North Gate Kiosk
1962 Cockatoo Aviary
1963 Cotton Terraces
1963 Wellcome Institute of Comparative Physiology
1965 Nuffield Institute of Comparative Medicine
1965 Elephant and Rhinoceros Pavilion
1965 Snowdon Aviary
1967 Charles Clore Pavilion for Mammals
1972 Michael Sobell Pavilions for Monkeys and Apes
1975 Zoo Study Centres
1976 The New Lion Terraces

LONDON ZOO
1976